Lecture Notes in Civil Engineering

Volume 30

Lecture Notes in Civil Engineering (LNCE) publishes the latest developments in Civil Engineering—quickly, informally and in top quality. Though original research reported in proceedings and post-proceedings represents the core of LNCE, edited volumes of exceptionally high quality and interest may also be considered for publication. Volumes published in LNCE embrace all aspects and subfields of, as well as new challenges in, Civil Engineering. Topics in the series include:

Indexed by Scopus

Lecture Notes in Civil Engineering

- Construction and Structural Mechanics
- Building Materials
- Concrete, Steel and Timber Structures
- Geotechnical Engineering
- Earthquake Engineering
- Coastal Engineering
- Hydraulics, Hydrology and Water Resources Engineering
- Environmental Engineering and Sustainability
- Structural Health and Monitoring
- Surveying and Geographical Information Systems
- Heating, Ventilation and Air Conditioning (HVAC)
- Transportation and Traffic
- Risk Analysis
- Safety and Security

To submit a proposal or request further information, please contact the appropriate Springer Editor:

- Mr. Pierpaolo Riva at pierpaolo.riva@springer.com (Europe and Americas);
- Ms. Swati Meherishi at swati.meherishi@springer.com (India);
- Ms. Li Shen at li.shen@springer.com (China);
- Dr. Loyola D'Silva at loyola.dsilva@springer.com (Southeast Asia and Australia/NZ).

More information about this series at http://www.springer.com/series/15087

Arvind Kumar Agnihotri ·
Krishna Reddy · Ajay Bansal
Editors

Sustainable Engineering

Proceedings of EGRWSE 2018

Springer

Editors
Arvind Kumar Agnihotri
Department of Civil Engineering
Dr. B. R. Ambedkar National
Institute of Technology
Jalandhar, Punjab, India

Krishna Reddy
Department of Civil and Materials
Engineering
University of Illinois at Chicago
Chicago, IL, USA

Ajay Bansal
Department of Chemical Engineering
Dr. B. R. Ambedkar National
Institute of Technology
Jalandhar, Punjab, India

ISSN 2366-2557 ISSN 2366-2565 (electronic)
Lecture Notes in Civil Engineering
ISBN 978-981-13-6716-8 ISBN 978-981-13-6717-5 (eBook)
https://doi.org/10.1007/978-981-13-6717-5

Library of Congress Control Number: 2019931865

This Springer imprint is published by the registered company Springer Nature Singapore Pte Ltd.
The registered company address is: 152 Beach Road, #21-01/04 Gateway East, Singapore 189721, Singapore

Preface

Sustainable engineering is the process of using resources in a way that does not compromise the environment or deplete the materials for future generations. Sustainable engineering requires an interdisciplinary approach in all aspects of engineering. All engineering fields should incorporate sustainability into their practice in order to improve the quality of life. Furthermore, with the creation of the Sustainable Development Goals, engineers will continue to play a decisive role in their success. The necessity for environment-friendly technologies in future will require the expertise of engineers. Therefore, the UNESCO Engineering Initiative (UEI) is working with partners to develop engineering curricula that incorporate sustainability as an overarching theme.

The term "Smart City" has now become a buzzword and is extensively used and talked about in the recent past more so than ever before. However, the definition or the concept of a smart city may vary based on how one perceives it. As exemplified by the Smart Cities Mission of the Government of India (GoI), the objective of this initiative is "*to promote cities that provide core infrastructure and give a decent quality of life to its citizens, a clean and sustainable environment and application of 'Smart' Solutions*". The focus is on meeting the minimal needs of a city first and then looking at the opportunities to improve the lifestyle of the people by incorporating innovative technologies and best urban planning practices, by developing public–private partnerships and by making changes to policies to realize sustainable development. The essence of this mission is also aimed at creating a replicable model which can set an example to other aspiring cities to transform themselves into a smart city, thereby spearheading the creation of similar smart cities across the country.

The core elements in a city that would aid in recognizing and further developing it as a smart city are well laid out by GoI, and these include: (i) adequate water supply, (ii) assured electricity supply, (iii) sanitation, including solid waste management, (iv) efficient urban mobility and public transport, (v) affordable housing, (vi) robust IT connectivity and digitalization, (vii) good governance, (viii) sustainable environment, (ix) safety and security of citizens and (x) health and education. In addition, the cities are encouraged to improve as a smart city by incorporating technologies that lead to smart outcomes for the city. However, the above-mentioned features of smart cities could readily be achieved, and it is possible only if there is a due consideration to holistic sustainable development. Each of the propositions to a smart city should be evaluated through the lens of triple-bottom-line sustainability, which aims at protecting the human health and the environment and at the same time providing economic affordability and social viability to the citizens/stakeholders.

There is a pressing need for "Smart" solutions through the initiatives as such in order to cope with the grand challenges of the world including global climate change and increasing population. Application of smart solutions will enable the cities, through states and through nations to bring about a reform in the use of technology, information and data to improve the infrastructure and services and at the same time improve the quality of life and also preserve the planet Earth with its rich resources for the coming generations. This book covers a variety of such multidisciplinary chapters which will be very useful for students, working professionals, practitioners and researchers.

Jalandhar, India — Arvind Kumar Agnihotri
Chicago, USA — Krishna Reddy
Jalandhar, India — Ajay Bansal

Contents

About the Editors

Prof. Arvind Kumar Agnihotri is Professor of Civil Engineering at NIT Jalandhar. He completed his Ph.D. from University of Roorkee (1998), M.Tech. from NIT Kurukshetra (1989) and B.E. from Panjab University Chandigarh (1987). He possesses a work experience of around 29 years in research, teaching and academic administration, with several years spent holding key leadership positions. His areas of interest are Geotechnical and Geo-Environmental Engineering, Reinforced Earth (Geo-Synthetics and Geofibers), Ground Improvement and Soil-Structure-interaction. He has supervised 9 Ph.D. Thesis and 7 more are in progress. He has guided 45 M.Tech. dissertations. He has published more than 100 papers in international and National journals and conferences. He served as Head of Civil Engineering, Dean Academic, Dean (Planning and Development) at prestigious Dr. B. R. Ambedkar National Institute of Technology, Jalandhar. He is fellow/ member of many professional organizations like ASCE, IGS, ISTE, Institution of Engineers, International Society of Soil Mechanics and Geotechnical Engineering and Indian Roads Congress. He is reviewer of many international journals of repute.

Dr. Krishna Reddy is a Professor of Civil and Environmental Engineering, the Director of Sustainable Engineering Research Laboratory and also the Director of the Geotechnical and Geo-Environmental Engineering Laboratory in the Department of Civil and Materials Engineering at the University of Illinois at Chicago. He is the author of 3 books: (1) Geo-Environmental Engineering: Site Remediation, Waste Containment, and Emerging Waste Management Technologies, (2) Electrochemical Remediation Technologies for Polluted Soils, Sediments and Groundwater, and (3) Sustainable Remediation of Contaminated Sites. He is also author of 182 journal papers, 15 edited books, 10 book chapters, and 170 full conference papers. Dr. Reddy has given 160 invited presentations in the USA and 15 other countries. He has served or currently serves as an Associate Editor or Editorial Board Member of over 10 different journals, including the ASCE and ASTM Journals, among others. He has received several awards for excellence in research and teaching, including the ASTM Hogentogler Award, the UIC Distinguished Researcher Award, the University of Illinois Scholar Award, and the University of

Illinois Award for Excellence in Teaching. He is a Fellow of the American Society of Civil Engineers, a Diplomat of Geotechnical Engineering, and a Board Certified Environmental Engineer.

Dr. Ajay Bansal Professor and Head of Chemical Engineering Department at NIT Jalandhar, received his B.Tech. (Gold medallist) from NIT Raipur, M.Tech. from IIT Delhi and Ph.D. from Panjab University Chandigarh. He is working in the area of Environmental Engineering, Multiphase Reactors and Renewable Energy. He has over 23 years of experience in teaching and research. He has supervised 8 Ph.D. theses and 4 are in progress. To his credit, he has 3 books, 5 book chapters and more than 40 international journal publications along with large number of papers in conferences. He is associated with many professional organizations and is Fellow of IIChE Kolkata and Institution of Engineers (India) Kolkata and Life member of Indian Society for Technical Education. He was Vice-President of the Indian Institute of Chemical Engineers Kolkata and is presently Chairman of Doaba Regional Centre of Indian Institute of Chemical Engineers.

Application of Response Surface Methodology to Optimize the Reaction Parameters for Grafting of Cellulosic Fiber

Aanchal Mittal, Sangeeta Garg and Shailendra Bajpai

Abstract In this work, chemical modification of barley husk (BH) was done with palmitic acid to render its hydrophobic property. Optimization of the reaction parameters for grafting of barley husk was performed using response surface methodology (RSM) coupled with central composite design (CCD). Different process parameters were optimized at three levels: reaction time (2.5–5 h), reaction temperature (35–55 °C), solvent ratio (0, 1:3, 1:1, 3:1, 1), and monomer concentration (4.75–11.09 mol/L $\times$ 10^{-3}). Prediction of the optimum reaction parameters was done by producing a quadratic polynomial equation in order to find the maximum value of % graft yield. The adequacy of the regression modeling was tested by using analysis of variance (ANOVA). It was observed that the most effective parameter for chemical modification of barley husk was reaction temperature. The maximum value of % graft yield was 41.53% at reaction time 4.37 h; reaction temperature 40.02 °C; solvent ratio 3:1; and monomer concentration 9.48 mol/L $\times$ 10^{-3}. The characterization of the modified barley husk was done by using Fourier transform infrared spectroscopy (FT-IR) and scanning electron microscopy (SEM). Surface micrographs of grafted barley husk showed that surface of barley husk became rough after grafting of palmitic acid over the cellulosic backbone.

Keywords Grafting · Optimization · Response surface methodology · Regression modeling · Infrared spectroscopy

1 Introduction

Recently, many research efforts have been done to replace the synthetic polymers based materials with natural-based materials (Bharti et al. 2013; Thakur et al. 2013). Polymers produced from natural raw materials are more eco-friendly, cheap, noncor-

A. Mittal (✉) · S. Garg · S. Bajpai
Department of Chemical Engineering, Dr. B. R. Ambedkar National Institute of Technology, Jalandhar 144011, India
e-mail: aanchalm.15.phd@nitj.ac.in

S. Garg
e-mail: gargs@nitj.ac.in

A. K. Agnihotri et al. (eds.), *Sustainable Engineering*, Lecture Notes in Civil Engineering 30, https://doi.org/10.1007/978-981-13-6717-5_1

rosive in nature, and nontoxic relative to the conventional synthetic polymer materials (Jawaid and Khalil 2011; Majeed et al. 2013). In this direction, all the research efforts have been focused on the development of natural based polymer composites with effective utilization of various natural and renewable raw materials.

Cellulose is the most abundant and renewable biopolymer found in nature (Yu et al. 2015; Huang et al. 2017). Cellulose has great significance in the industries due to its utilization in various applications like packaging films, paper, adhesives, textile, and other building materials. Natural cellulosic fibers have been used earlier for the preparation of composites. Recently, they have been used as reinforcing agents for the preparation of polymer composites for packaging applications. Natural fibers have been preferred over synthetic fibers due to their availability, low raw material cost, low density per volume, biodegradability, renewability, and eco-friendly nature (Nguyen et al. 2016; Ryu et al. 2017). However, the main disadvantage associated with natural fiber is their hydrophilic nature, low thermal stability, non-durability, and poor compatibility with polymer matrix (Khoathane et al. 2008). Hence, intense research efforts have been focused on the development of hydrophobic cellulosic material due to their biodegradability and excellent mechanical strength. Various physical and chemical modification methods such as plasma treatment, steam treatment, solvent-vaporized controllable crystallization, alkali treatment, mercerization, acetylation, benzoylation, graft copolymerization, oxidization have been adopted to overcome the disadvantage associated with different bio-based polymers (Morent et al. 2008; Ray et al. 2009; Kaith and Kalia 2008; Garg and Jana 2011; Mittal et al. 2016; Modibbo et al. 2009). Graft copolymerization is the most common method to improve the surface properties of natural fibers due to its simplicity (Teli and Sheikh 2012). Most of the research has been done on graft copolymerization of the different cellulosic backbones with vinyl monomers. However, very limited research is available on the grafting of cellulosic backbone with long-chain fatty acids obtained from the fats and oils of various plants or animals (Kohli et al. 2017; Mittal et al. 2016). The objective of the present study was to perform graft copolymerization of the barley husk (BH) with palmitic acid. Various reaction parameters like reaction temperature, reaction time, solvent ratio, and monomer concentration were optimized through response surface methodology based on central composite design. Graft copolymers produced from the reaction were characterized by Fourier Transform Infrared spectroscopy (FT-IR) and scanning electron microscopy (SEM).

2 Methodology

2.1 Materials

Barley used in the present study was obtained from the local market of Jalandhar. Palmitic acid used for graft copolymerization of barley husk (BH) was obtained from Sisco Research Laboratories Pvt. Ltd. (Bombay, India). Ferrous ammonium sulfate, potassium persulfate and dimethyl sulphoxide (DMSO) were purchased from Loba Chemie Private Limited (Mumbai, India).

2.2 Methods

Barley was mildly ground to separate the husk, washed and dried in hot air oven at 80 °C for 12 h. Extraction with acetone was done for 72 h in Soxhlet extractor to remove the soluble impurities from the husk. The barley husk was grounded and passed through 170 mesh. Graft copolymerization of BH was done by taking a mixture of DMSO/water as the solvent medium. BH was soaked overnight in a solvent medium to activate reactive sites on the surface of the cellulosic backbone. Palmitic acid was added dropwise in the presence of redox initiators potassium persulfate and ferrous ammonium sulfate (KPS-FAS) in the reaction media. The reaction mixture was heated at a specific temperature for a particular time interval with continuous stirring. The grafted copolymer was Soxhlet extracted using ethanol. The sample was dried in a hot air oven at 50 °C for 24 h. Optimization of reaction parameters reaction temperature, reaction time, solvent ratio, and monomer concentration was performed to get maximum graft yield. Percentage graft yield was calculated as

$$Percent\ graft\ Yield = \frac{W_g - W_i}{W_i} \times 100 \quad (1)$$

2.3 Statistical Analysis

The statistical software Design-Expert version 10.0.3.0 (Stat-Ease Inc., Minneapolis, USA) was used for the design of experiments and statistical data analysis. Response surface methodology based on central composite design was used to predict an optimum response by solving multivariate equation. The experimental parameters and their level are given in Table 1. Analysis of variance (ANOVA) was used to analyze the data and to study the effect of individual and combined process parameters on the response. The significance of the process parameters was checked by the i-value and F-value.

2.4 Characterization

Natural BH and grafted BH were characterized by Fourier Transform Infrared spectroscopy (FT-IR) and scanning electron microscopy (SEM). FT-IR spectra of samples were recorded by FT-IR instrument (PerkinElmer, Model RX−1) in the range from 4000 to 400 cm^{-1} at a resolution of 4 cm^{-1} using KBr pellets (Sigma Aldrich). Surface

Table 1 Various factors involved in the optimization of grafting reaction and their levels

Factors	Coded factor	Levels			
		−1	+1	−α	+α
Time (min)	A	187.5	262.5	150	300
Temperature (°C)	B	40	50	35	55
Solvent ratio (DMSO/water)	C	25	75	0	100
Monomer concentration (mol/L × 10^{-3})	D	6.335	9.505	4.75	11.09

morphology of natural BH and grafted BH was observed with a Scanning Electron Microscope (JEOL JSM−6100, Jeol, Tokyo, Japan). The samples were first gold coated and then observed at a magnification of 1000 X.

3 Results and Discussion

3.1 Statistical Study via Response Surface Methodology (RSM)

The interaction and individual effect of different reaction parameters reaction time, temperature, solvent ratio, and monomer concentration on % graft yield were studied. Batch experiments were performed for the graft copolymerization of BH with palmitic acid applying RSM-based CCD. This model consists of 16 factorial points, 8 axial points, and 6 replicates at the center. The 30 rums obtained from experimental design (Eq. 2) obtained using Design-Expert software are presented in Table 2.

$$N = 2^n + 2n + n_c = 2^4 + (2 \times 4) + 6 = 30 \tag{2}$$

A response surface quadratic model was developed by software in terms of process variables and response (Eq. 3).

$$\begin{aligned}\%\ \text{Graft Yield} = {} & 30.93 + (1.30\text{A}) - (0.6138\text{B}) + (2.43\text{C}) \\ & + (3.09\text{D}) - (1.63\text{AB}) + (0.3081\text{AC}) \\ & - (1.03\text{AD}) - (2.08\text{BC}) - (0.4456\text{BD}) \\ & + (1.22\text{CD}) - \left(1.39A^2\right) - \left(2.29B^2\right) + \left(0.0857C^2\right) \\ & + \left(0.2395D^2\right)\end{aligned} \tag{3}$$

The significance of each variable was determined by the corresponding F-value, p-value, and sum of squares (SS). Larger the sum of squares and F-value, the more

Table 2 Experimental design generated from central composite design RSM

Std.	Factor A-time (h)	Factor B-temperature (°C)	Factor C-solvent ratio (DMSO/water)	Factor D-monomer concentration (mol/L × 10^{-3})	Graft yield (%)
1	3.125	40.00	25	6.335	17.3
2	4.375	40.00	25	9.5	41.53
3	4.375	50.00	25	9.5	27.4
4	4.375	50.00	25	6.335	24.6
5	3.125	40.00	75	9.5	25.3
6	3.125	40.00	75	6.335	24
7	3.75	50.00	75	7.92	32.4
8	3.75	50.00	75	7.92	19.8
9	3.75	40.00	25	7.92	35.2
10	3.75	40.00	25	7.92	30.8
11	3.125	50.00	25	9.5	32.1
12	5	50.00	25	7.92	25.7
13	4.375	40.00	75	6.335	26.2
14	4.375	40.00	75	6.335	26.4
15	4.375	50.00	75	9.5	27.6
16	3.75	50.00	75	4.75	25.7
17	3.75	45.00	50	7.92	22.1
18	2.5	45.00	50	7.92	23.4
19	3.125	35.00	50	6.335	23.1
20	4.375	55.00	50	6.335	32.4
21	3.75	45.00	0	11.09	36.4
22	3.125	45.00	100	6.335	23.7
23	4.375	45.00	50	9.5	31.2
24	3.75	45.00	50	7.92	31.9
25	3.125	45.00	50	9.5	35.1
26	3.75	45.00	50	7.92	29.2
27	3.75	45.00	50	7.92	30.1
28	3.125	45.00	50	9.5	30.2
29	3.75	45.00	50	7.92	31.2
30	3.75	45.00	50	7.92	25.67

Table 3 Analysis of variance (ANOVA) and regression coefficients for optimization design

Source	Sum of squares	df	Mean square	F-value	*p*-value	
Model	772.87	14	55.21	29.41	<0.0001	significant
A-time	40.38	1	40.38	21.51	0.0003	
B-temp	9.04	1	9.04	4.82	0.0443	
C-solvent	142.06	1	142.06	75.69	<0.0001	
D-monomer	228.97	1	228.97	122.00	<0.0001	
AB	42.67	1	42.67	22.74	0.0002	
AC	1.52	1	1.52	0.8094	0.3825	
AD	16.95	1	16.95	9.03	0.0089	
BC	69.43	1	69.43	36.99	<0.0001	
BD	3.18	1	3.18	1.69	0.2128	
CD	23.84	1	23.84	12.70	0.0028	
A^2	52.65	1	52.65	28.05	<0.0001	
B^2	143.28	1	143.28	76.34	<0.0001	
C^2	0.2016	1	0.2016	0.1074	0.7476	
D^2	1.57	1	1.57	0.8381	0.3744	
Residual	28.15	15	1.88			
Lack of fit	21.28	10	2.13	1.55	0.3289	Not significant
Pure error	6.87	5	1.37			
Cor. total	801.02	29				
Std. dev.	1.37		R^2	0.9649		
Mean	28.26		Adjusted R^2	0.9321		
C.V. %	4.85		Predicted R^2	0.8346		
			Adeq precision	22.6705		

significant is the corresponding variable. However, a small value of p indicates the rejection of the null hypothesis (Bajpai et al. 2012).

It was observed from Table 3, that model F-value is 29.41 and the *p*-value is very low (<0.0001). The F-value of "Lack of Fit" is 1.55 with a corresponding *p*-value of 0.3289, which implied that lack of fit is not significant. ANOVA study suggested that monomer concentration is the most significant parameter while reaction time, temperature, solvent ratio have a less significant effect on the response (% graft yield). Figure 1 showed the relationship between actual and predicted response for the grafting of BH. The actual data was obtained by experiments, while predicted data was generated by software using Eq. 2. The values of correlation coefficient R^2 and adjusted R^2 were 0.9649 and 0.9321, respectively.

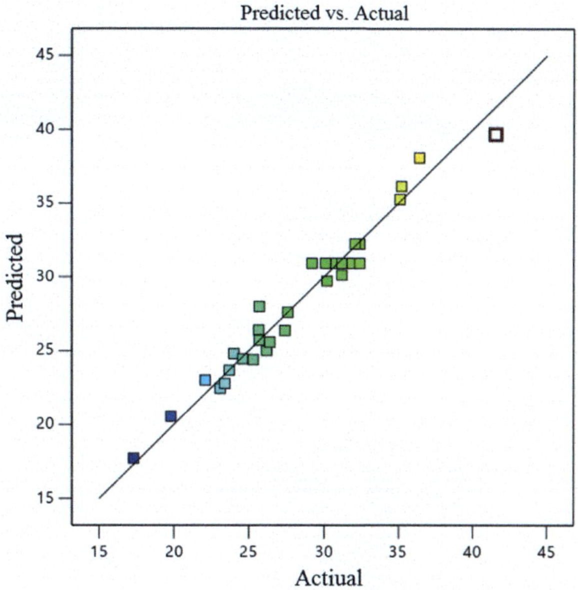

Fig. 1 The actual and predicted response for the graft copolymerization

Three-dimensional (3D) response surface plots were generated to investigate the combined effect of process parameters reaction time, temperature, solvent ratio, and monomer concentration on the response (Fig. 2).

The interactive effect of reaction temperature and time on % graft yield is shown in Fig. 2a. It was observed that % graft yield increased with an increase in reaction time at a lower temperature. Initially, there may be an increase in % graft yield due to the generation of more free radicals and their increased interaction with a cellulosic backbone with an increase in reaction time and temperature. However, at a higher temperature, there was a decrease in % graft yield with an increase in reaction time due to homopolymerization reactions. The combined effect of solvent ratio and time on % graft yield is shown in Fig. 2b. % graft yield increased with increase in solvent ratio and time which may be due to the increase in the movement and interaction of free radicals with cellulosic backbone and resulted in an increase % graft yield.

Figure 2c showed the interactive effect of monomer concentration and time. It was observed that % graft yield increased with the increase in monomer concentration and reaction time. Increase in monomer concentration resulted in the generation of free radicals near the cellulosic backbone. The interaction of these free radicals increased with cellulosic backbone with an increase in reaction time which resulted in high graft yield. The combined effect of solvent ratio and temperature was studied by Fig. 2d. The solvent ratio has a more significant effect on the graft yield at a lower temperature as compared to the higher temperature. At lower temperature, there was an increase in the movement of free radicals near the cellulosic backbone with an increase in the solvent ratio. But at a higher temperature, there was no significant effect of increased solvent ratio on % graft yield due to accelerated homopolymerization reactions.

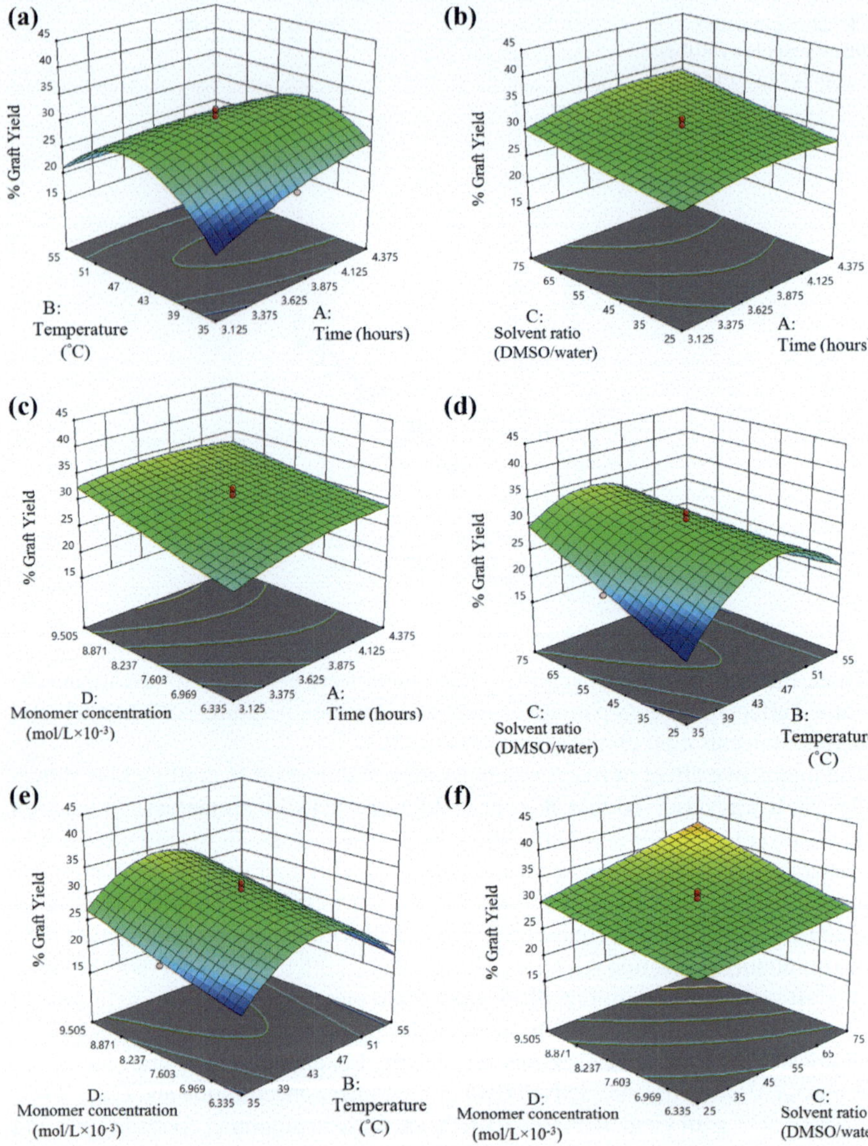

Fig. 2 3D surface graphs of graft copolymerization of BH: **a** effect of temperature and time, **b** effect of solvent ratio and time, **c** effect of monomer concentration and time, **d** effect of solvent ratio and temperature, **e** effect of monomer concentration and temperature, **f** effect of monomer concentration and solvent ratio

Figure 2e depicted the combined effect of monomer concentration and temperature on the % graft yield. Increase in % graft yield with monomer concentration was more significant at a lower temperature relative to the higher temperature. Because at a higher temperature, the free radicals will move forward for homopolymerization reactions in spite of graft copolymerization reactions and result in lower graft yield. The interaction between monomer concentration and solvent ratio was studied by Fig. 2f. It showed that an increase in monomer concentration and solvent ratio led to high graft yield. This may be due to the generation of more active free radicals with an increase in monomer concentration and an increase in the movement on these free radicals near close proximation of cellulosic backbone with an increase in the solvent ratio.

The optimum conditions for the graft copolymerization of BH with palmitic acid were reaction time 4.23 h, temperature 40 °C, solvent ratio 3:1 and monomer concentration 9.50 mol/L × 10^{-3}. At these conditions, 39.7% graft yield was predicted by the response surface model. The graft copolymerization of BH was performed at optimum reaction conditions and resulted in 40.2% graft yield which was close to the predicted value.

3.2 Characterization

3.2.1 FT-IR Spectra

FT-IR spectra of natural BH and grafted BH are presented in Fig. 3.

IR spectra of BH (Fig. 3a) showed strong and broad adsorption bands appeared at 3342.15 cm^{-1} and 1645.15 cm^{-1}, due to the bending of –OH groups present on the surface of cellulosic backbone and stretching of –OH groups derived from intermolecular hydrogen bonds, respectively. The adsorption bands at 2922.15 cm^{-1}, 1428.15 cm^{-1}, and 1047.14 cm^{-1} were accredited to C–H, C–C, and C–O bonds, respectively (Priya et al. 2014; Singha and Rana 2012). The peak intensity at 3343.28 and 1645 cm^{-1} decreased in the IR spectra of grafted BH as compared to the natural BH due to the reduced OH group present on the surface of grafted BH (Fig. 3b). In the IR spectra of grafted BH a new peak was observed at 1728.23 and 1312.28 cm^{-1} which indicated the presence of C=O and C–O bond after grafting of BH with palmitic acid.

3.2.2 Scanning Electron Microscopy (SEM)

SEM micrograph of natural BH and grafted BH are shown in Fig. 4a, b.

SEM images of natural BH showed the cellular and porous structure of the fiber (Fig. 4a). The surface of grafted BH was rough as compared to the natural BH due to the grafting of palmitic acid on the surface of BH (Fig. 4b). Increase in the roughness on the surface of BH increased the surface area for bonding hence it may

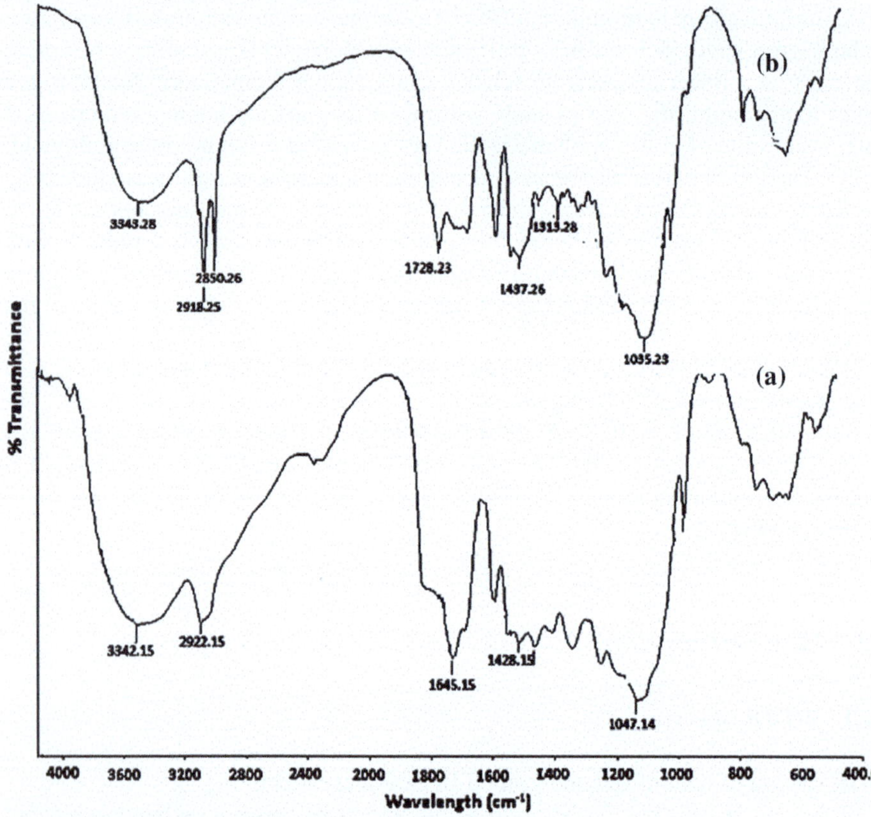

Fig. 3 FT-IR spectra of **a** natural BH, **b** grafted BH

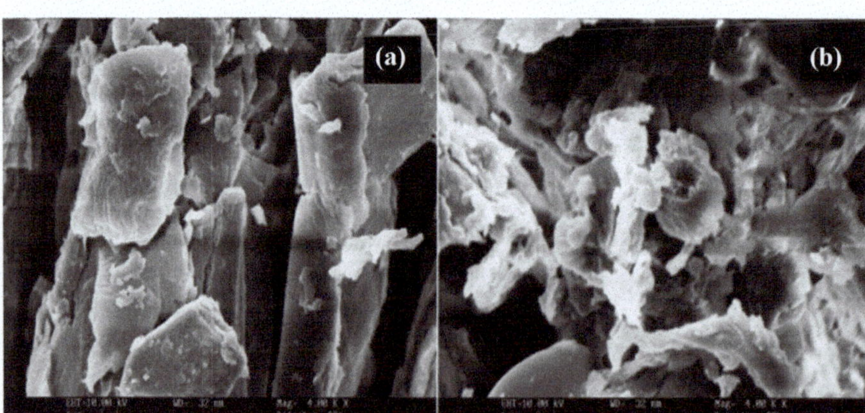

Fig. 4 SEM images of **a** natural BH, **b** grafted BH

lead to improvement in the interfacial adhesion and bonding between grafted BH and polymer matrix.

4 Conclusions

Graft copolymerization of the barley husk (BH) was performed with palmitic acid in the presence of redox initiators. The individual and interactive effect of various reaction parameters reaction time, temperature, solvent ratio, and monomer concentration were studied on the % graft yield. Monomer concentration has the most effect on the graft yield as compared to the other reaction parameters. The maximum graft yield (%) was obtained at reaction time 4.23 h, temperature 40 °C, solvent ratio 3:1 and monomer concentration 9.50 mol/L $\times$ 10^{-3}. The evidence for the graft copolymerization of BH was confirmed by using FT-IR spectroscopy. SEM investigation of natural BH and grafted BH showed that surface of hydrophobically modified BH was rougher as compared to the natural BH.

References

Bajpai S, Gupta SK, Dey A, Jha MK, Bajpai V, Joshi S, Gupta A (2012) Application of central composite design approach for removal of chromium (VI) from aqueous solution using weakly anionic resin: modeling, optimization, and study of interactive variables. J Hazard Mater 227:436–444

Bharti S, Mishra S, Sen G (2013) Ceric ion initiated synthesis of polyacrylamide grafted oatmeal: its application as flocculant for wastewater treatment. Carbohyd Polym 93(2):528–536

Garg S, Jana AK (2011) Characterization and evaluation of acylated starch with different acyl groups and degrees of substitution. Carbohyd Polym 83(4):1623–1630

Huang F, Wu X, Yu Y, Lu Y, Chen Q (2017) Acylation of cellulose nanocrystals with acids/trifluoroacetic anhydride and properties of films from esters of CNCs. Carbohyd Polym 155:525–534

Jawaid MHPS, Khalil HA (2011) Cellulosic/synthetic fibre reinforced polymer hybrid composites: a review. Carbohyd Polym 86(1):1–18

Kaith BS, Kalia S (2008) Graft copolymerization of MMA onto flax under different reaction conditions: a comparative study. Express Polym Lett 2(2):93–100

Khoathane MC, Vorster OC, Sadiku ER (2008) Hemp fiber-reinforced 1-pentene/polypropylene copolymer: the effect of fiber loading on the mechanical and thermal characteristics of the composites. J Reinf Plast Compos 27(14):1533–1544

Kohli D, Garg S, Jana AK, Maiti M (2017) Synthesis of graft copolymers for green composite films and optimization of reaction parameters using Taguchi (L16) orthogonal array. Indian Chem Eng 59(2):136–158

Mittal A, Garg S, Kohli D, Maiti M, Jana AK, Bajpai S (2016) Effect of cross linking of PVA/starch and reinforcement of modified barley husk on the properties of composite films. Carbohyd Polym 151:926–938

Majeed K, Jawaid M, Hassan A, Bakar AA, Khalil HA, Salema AA, Inuwa I (2013) Potential materials for food packaging from nanoclay/natural fibres filled hybrid composites. Mater Des 46:391–410

Modibbo UU, Aliyu BA, Nkafamiya II (2009) The effect of mercerization media on the physical properties of local plant bast fibres. Int J Phys Sci 4(11):698–704

Morent R, De Geyter N, Verschuren J, De Clerck K, Kiekens P, Leys C (2008) Non-thermal plasma treatment of textiles. Surf Coat Technol 202(14):3427–3449

Nguyen HL, Jo YK, Cha M, Cha YJ, Yoon DK, Sanandiya ND, Hwang DS (2016) Mussel-inspired anisotropic nanocellulose and silver nanoparticle composite with improved mechanical properties, electrical conductivity and antibacterial activity. Polymers 8(3):102

Priya B, Gupta VK, Pathania D, Singha AS (2014) Synthesis, characterization and antibacterial activity of biodegradable starch/PVA composite films reinforced with cellulosic fibre. Carbohyd Polym 109:171–179

Ray D, Das M, Mitra D (2009) Influence of alkali treatment on creep properties and crystalinity of jute fibres. BioResources 4(2):730–739

Ryu MH, Park J, Oh DX, Hwang SY, Jeon H, Im SS, Jegal J (2017) Precisely controlled two-step synthesis of cellulose-graft-poly (l-lactide) copolymers: effects of graft chain length on thermal behavior. Polym Degrad Stab 142:226–233

Singha AS, Rana RK (2012) Functionalization of cellulosic fibers by graft copolymerization of acrylonitrile and ethyl acrylate from their binary mixtures. Carbohyd Polym 87(1):500–511

Teli MD, Sheikh J (2012) Antibacterial and acid and cationic dyeable bamboo cellulose (rayon) fabric on grafting. Carbohyd Polym 88(4):1281–1287

Thakur VK, Thakur MK, Gupta RK (2013) Rapid synthesis of graft copolymers from natural cellulose fibers. Carbohyd Polym 98(1):820–828

Yu HY, Chen GY, Wang YB, Yao JM (2015) A facile one-pot route for preparing cellulose nanocrystal/zinc oxide nanohybrids with high antibacterial and photocatalytic activity. Cellulose 22(1):261–273

Bioregeneration of Granular Activated Carbon Adsorbed with Analytical-Grade Hydroquinone Compound Using Mixed Bacterial Culture

Varun Kumar Agarwal, Kaushik Nath, Nikul Maliya, Mufaddal Nathani and Chintan Ghava

Abstract The study was aimed to investigate on "Bioregeneration of granular activated carbon adsorbed with analytical-grade hydroquinone compound in sequentially adsorption and biodegradation process." Initially, the adsorption study of hydroquinone compound was carried out at all optimized conditions until their saturation point, which was achieved within a contact period of 12 h. When the equilibrium level was achieved, a mixed bacterial culture isolated from activated sludge (suited in Rajkot city) has been first acclimatized to the two different concentrations of hydroquinone compound. The experiments are conducted to determine the extent of bioregeneration under specific conditions. It was achieved 37%.

Keywords Bioregeneration · Activated carbon · Hydroquinone · Mixed bacterial culture · Adsorption

1 Introduction

Activated carbon exhibits a high degree of surface reactivity due to the presence of microporous and macroporous structure, which makes it unique and versatile adsorbent, for the adsorption of different organic and inorganic pollutants, but after a prolonged period of contact time between adsorbent and adsorbate, the available adsorption sites on activated carbon get saturated (Aktas and Cecen 2006).

Therefore, there are different conventional ways to regenerate the saturated carbon such as thermal method, chemical method, and many more, but probably 5–10% carbon losses occur due to all such conventional methods (Sodha and Nath 2013). In the view of all such drawbacks, new focus has been established toward bioregeneration. It is defined as the restoration of adsorption capacity of spent activated carbon through the process of metabolism of adsorbed organic compound by microorganisms (Nath and Bhakhar 2011a, b). It is achieved either by providing the exposure

V. K. Agarwal (✉) · K. Nath · N. Maliya · M. Nathani · C. Ghava
Environmental Engineering Laboratory, Department of Environmental Science & Engineering, Marwadi Education Foundation, Rajkot, India
e-mail: varunkumar.agarwal@marwadieducation.edu.in

A. K. Agnihotri et al. (eds.), *Sustainable Engineering*, Lecture Notes in Civil Engineering 30, https://doi.org/10.1007/978-981-13-6717-5_2

of microbes on saturated activated carbon after adsorption or by simultaneous treatment as in biological activated carbon (BAC) (Nath and Bhakhar 2011a, b). Microbes utilized adsorbed organic compound as a substrate for the food and energy and convert into a simple compound. Although, bioregeneration is controlled by various physical and chemical factors such as activation method of carbon, concentration of adsorbate, contact time, type and concentration of microorganism. In this research paper, an attempt has been made to assess the bioregeneration efficiency of activated carbon saturated with hydroquinone compound using mixed bacterial culture. Hydroquinone has been categorized as a persistent organic compound in the list of USEPA and the source of mixed bacteria was the activated sludge of secondary treatment. Additionally, the microbial culture was acclimatized in the known concentration of hydroquinone by providing the nutrient media and the extent of bioregeneration was determined at two different incubation periods.

2 Methodology

2.1 Adsorption Study

Batch adsorption studies were carried out by placing a known amount of granular activated carbon (GAC) in 250 mL conical flask containing 100 mL of hydroquinone with a predetermined concentration at a constant temperature. Thestudy was further optimized at various factors such as adsorbent concentration, adsorbate concentration, pH, and contact time.

The conical flasks were placed on a shaker and rotated at 150 rpm until equilibrium was attained and at the end of saturation, granular were filtered by Whatman Filter Paper (No. 42). The equilibrium concentration of hydroquinone in the supernatant was measured by using UV–Visible spectrophotometer (double beam). The amount of adsorption on GAC was calculated as

$$q = \frac{v(C1 - C2)}{w}$$

q is termed as adsorption amount (mg/g); C_1 and C_2 are initial concentration of hydroquinone and equilibrium concentration, respectively; v is considered as the volume of each batch sample; w is weight of granular activated carbon (g) (Suresh and Srivastava 2011).

2.2 Microorganism and Culture Conditions

For the purpose of bioregeneration, mixed microorganisms from the sludge of secondary treatment plant were cultured in the laboratory and hydroquinone was the sole source of carbon in liquid broth for the acclimation. Bushnell Hass medium was used as the mineral medium, provided to the microbes for their metabolic activities (Sodha and Nath 2013). The composition of the BH medium was as follows.

$MgSO_4.7H_2O$	0.2 g L^{-1}
KH_2PO_4	1.00 g L^{-1}
K_2HPO_4	1.00 g L^{-1}
KNO_3	1.00 g L^{-1}
$CaCl_2$	0.02 g L^{-1}
$FeCl_3$	0.05 g L^{-1}
pH 7.0, Nutrient solution was prepared in distilled water at 25 °C	

The broth was placed in the orbital shaker at 37 °C, 150 rpm, and 5 days incubation period.

2.3 Bioregeneration

Spent granular activated carbons were transferred into a 250 mL conical flask containing 100 mL of sterile BH media and pre-grown mixed culture was inoculated aseptically in laminar air flow. The set was incubated for a specific period of incubation time at 37 °C and 150 rpm. After the decided incubation time, the solution was filtered through Whatman Filter Paper (No. 42). Then, the granular activated carbon was dried at 50 °C in oven for removing the moisture only (Nath and Bhakhar 2011a, b). Again adsorption study of activated carbon was carried out at the same solute concentration and conditions to determine bioregeneration efficiency. The bioregeneration study was optimized at two different incubation periods and volumes (mL) of mix culture broth. The extent of bioregeneration can be determined by using the following equation:

$$RE(\%) = \frac{Ar}{Ao} \times 100 \tag{1}$$

A_o is considered as the original capacity of carbon that quantity of particular solute adsorbed per unit weight of carbon at equilibration and exhaustion. A_r is the capacity of regenerated carbon, quantity of the same solute adsorbed per unit weight of carbon at the same equilibration and exhaustion (Nath and Bhakhar 2011a, b).

3 Results and Discussion

3.1 *Adsorption of Hydroquinone on Granular Activated Carbon*

Batch adsorption study of analytical-grade hydroquinone on granular activated carbon was conceded at the initial adsorbate concentration, adsorbent dose, pH, contact time, where temperature was kept constant. The effect of each parameter on adsorption was studied within a suitable range, while other parameters values were kept constant.

A maximum of 89% adsorption was achieved at the optimized value of pH 7, adsorbent dose of 10 g L^{-1}, and adsorbate concentration of 600 mg L^{-1} in 10 h of contact time (Fig. 1).

The result shows that maximum adsorption was achieved at a natural pH of 7 and with 600 mg L^{-1} initial adsorbate concentration. However, with a further increase of adsorbate concentration, the percentage adsorption drops significantly.

3.2 *Bioregeneration*

Microorganism regenerates the surface of exhausted activated carbon, using the substrate as a sole source of carbon. The bioregeneration study was optimized at two different volumes of mixed microbial volumes and incubation periods. The maximum bioregeneration efficiency of spent activated carbon was 35% using Eq. (1) at 4 mL of microbial culture and 6 days of incubation period.

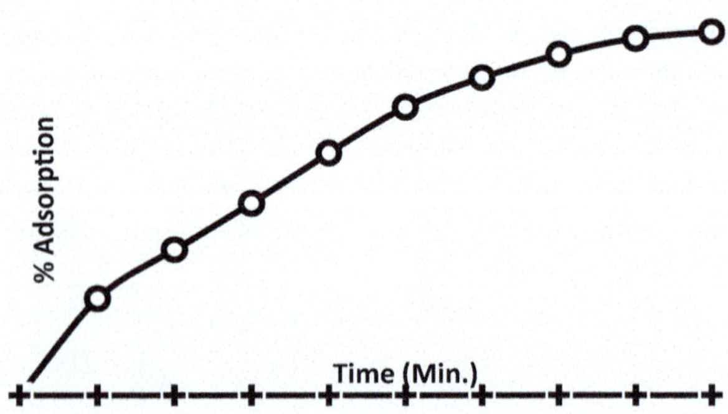

Fig. 1 Maximum % adsorption of hydroquinone by granular activated carbon (GAC) (pH—7, adsorbent dose—10 g L^{-1}, adsorbate concentration—600 mg L^{-1}, and temperature—25 °C)

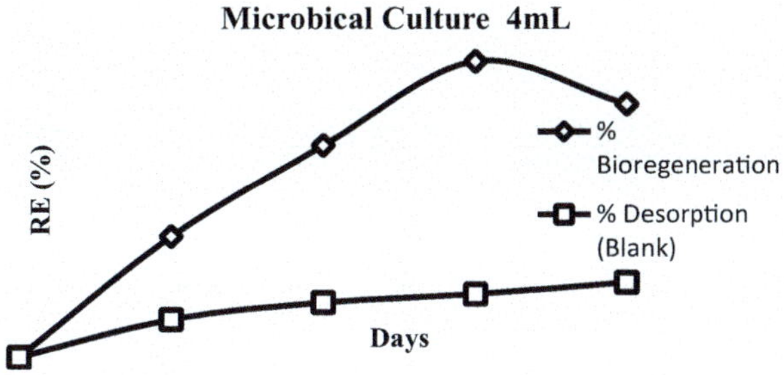

Fig. 2 Maximum microbial regeneration efficiency of spent granular activated carbon (GAC) (Microbial culture—4 mL and incubation period—6 days)

4 Conclusions

The study shows the feasibility of mixed bacterial culture to regenerate the spent activated carbon adsorbed with hydroquinone. The extent of bioregeneration is estimated to be 35%. In essence, bioregneration can compete with the conventional regeneration process due to its regeneration safety and eco-friendly nature (Fig. 2).

References

Aktas O, Cecen F (2006) Effect of type of carbon activation on adsorption and its reversibility. J Chem Technol Biotechnol 81:94–101

Nath K, Bhakhar M (2011a) Bioregeneration of spent activated carbon: effect of physico-chemical parameters. J Sci Ind Res 70:487–492

Nath K, Bhakhar M (2011b) Microbial regeneration of spent activated carbon dispersed with organic contaminants: mechanism, efficiency, and kinetic model. Environ Sci Pollut Res 18:534–546

Sodha K, Nath K (2013) Feasibility study of microbial regeneration of spent activated carbon sorbed with phenol using mixed bacterial culture. Indian J Chem Technol 20:33–39

Suresh S, Srivastava V (2011) Adsorption of hydroquinone in aqueous solution by granulated activated carbon. J Environ Eng 1147–1157

Bearing Capacity Improvement of Sand Bed Reinforced with 3D Geogrids of Rectangular Pattern

Femy M. Makkar, S. Chandrakaran and N. Sankar

Abstract In this paper, the effect of three-dimensional geogrid as a reinforcing material in enhancing the bearing capacity and settlement behaviour of reinforced soil is investigated. Laboratory-scaled plate load tests have been conducted to study the performance of square footing resting on the sand bed reinforced with 3D geogrids of rectangular pattern. The results show that a single layer of 3D geogrid placed at a depth of $0.25B$ from the bottom of footing provides a threefold increase in bearing capacity compared to unreinforced soil. The optimum spacing between two consecutive layers of 3D geogrids was obtained as $0.75B$. Soil reinforced with two layers of 3D geogrid gives higher bearing capacity improvement (6 times) and a further increase in the number of layers does not show any significant improvement in bearing capacity. The results obtained from this study shows that 3D geogrids can be used as an effective reinforcing material in improving the mechanical properties of reinforced soil structures such as embankments, retaining walls, shallow foundations, stabilization of steep slopes, etc.

Keywords 3D geogrids · Bearing capacity · Footing settlement · Plate load test · Shallow foundation

1 Introduction

The provision of geosynthetics for improving the bearing capacity and settlement characteristics of the soil are well documented (Ghosh and Bera 2005; Latha and Somwanshi 2009a; Vinod et al. 2009; Abu-Farsakh et al. 2013). Generally, conventional horizontal or planar reinforcements such as geotextiles and geogrids are used in many areas of geotechnical engineering. Recently, a few studies have been reported on the performance of reinforcements in three-dimensional forms (Zhang et al. 2006; Khedkar and Mandal 2009; Sireesh et al. 2009; Harikumar et al. 2016; Mozellanezhad et al. 2016). The beneficial effect of geosynthetic reinforcement

F. M. Makkar (✉) · S. Chandrakaran · N. Sankar
Department of Civil Engineering, NIT Calicut, Calicut 673601, Kerala, India
e-mail: femy.makkar@gmail.com

A. K. Agnihotri et al. (eds.), *Sustainable Engineering*, Lecture Notes in Civil Engineering 30, https://doi.org/10.1007/978-981-13-6717-5_3

largely depends on the form in which it is used (Latha and Somwanshi 2009b). In the case of geotextiles-reinforced soil beds, the friction between the soil and surface of reinforcement enhances the strength of reinforced soil. In geogrid-reinforced soil, the friction between soil and surface of geogrid ribs, and the interlocking of soil particles within the apertures of geogrid contributes to the improvement in strength. Whereas, the 3D reinforcement forms provide additional confinement along with the advantages of conventional forms.

In this paper, geogrid in three-dimensional (3D) form is used to reinforce the soil. Laboratory-scaled plate load tests were conducted to study the performance of sand bed reinforced with 3D geogrids of the rectangular pattern. The results are expressed in terms of bearing capacity improvement and settlement reduction.

2 Materials

Locally available river sand with an effective particle size (D_{10}) of 0.33 mm was used as the sand bed. According to the unified soil classification system, the sand was classified as poorly graded sand (SP). The grain size distribution curve for sand is shown in Fig. 1. The sand bed was prepared under the medium dense condition with 50% relative density.

The reinforcement used in the study was three-dimensional geogrid. The 3D geogrid was made from conventional biaxial geogrid of 100 kN/m tensile strength. It was made by placing two conventional geogrid sheets of required size, one above the other. The bottom layer was placed horizontally and the top geogrid layer was folded in a rectangular pattern and it was tied to the bottom layer at every junction using nylon thread. Figure 2 shows the 3D geogrid used in the study.

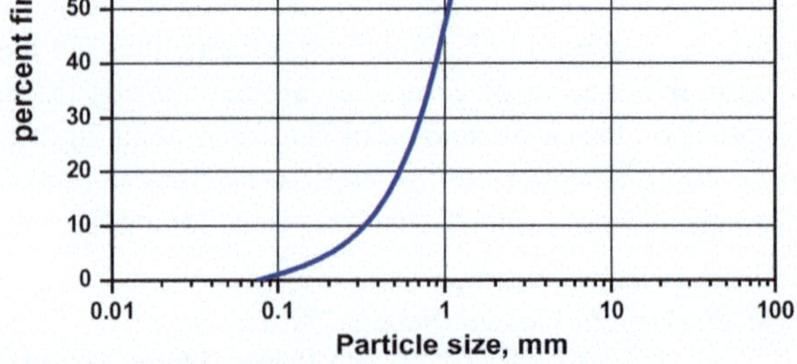

Fig. 1 Grain size distribution curve for sand

Fig. 2 3D geogrid of rectangular pattern (3DGR)

3 Experimental Setup

The plate load tests were conducted on a square steel tank of 750 mm side and 750 mm height. A square steel plate of 150 mm × 150 mm size and 25 mm thickness was used as the model footing. The load was applied to the model footing with a manually operated hydraulic jack. The applied load was measured with a pressure gauge of 200 kg/cm^2 capacity. The settlement of footing was measured through dial gauges provided at two diagonally opposite points on the footing. In the case of the reinforced sand bed, the 3D geogrids were placed at predetermined locations with respect to its mid-height.

The typical layout of 3D geogrid-reinforced sand bed is shown in Fig. 3, where, B—width of footing, b—width of 3D geogrid, H—height of 3D geogrid, u—depth of first layer of reinforcement, h—spacing between two consecutive layers of 3D geogrid and d_r—depth of reinforcement. The study aims to find out the optimum depth of the first layer of reinforcement, spacing between two consecutive layers and optimum number of layers. In the present investigation, the width of reinforcement was 4B (600 mm) and height was 40 mm.

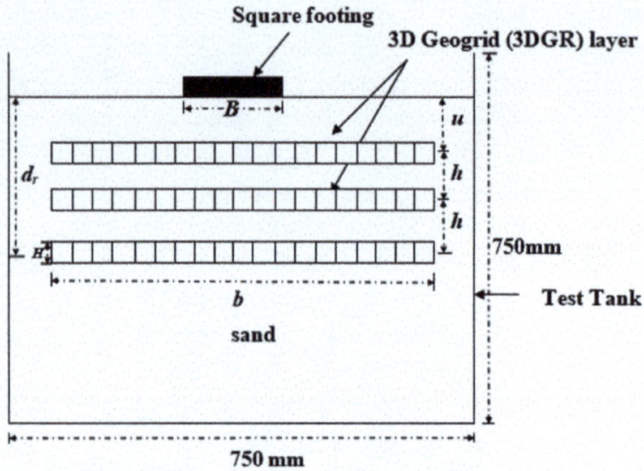

Fig. 3 Typical layout of 3D geogrid-reinforced sand bed

4 Results and Discussion

The improvement in bearing capacity and reduction in settlement due to the provision of 3D geogrid in sand bed was quantified in terms of non-dimensional parameters, improvement factor (I_f) and settlement reduction factor SR_f.

$$I_f = \frac{q_r}{q_u} \tag{1}$$

$$SR_f = \frac{(S_0 - S_r)}{S_0} \tag{2}$$

where q_r is the bearing pressure of reinforced sand at a given settlement and q_u is the bearing pressure of unreinforced sand at the same settlement. S_0 is the settlement of unreinforced sand bed at a given pressure and S_r is the settlement of reinforced sand bed at the same pressure. SR_f was calculated for a normalized settlement of 30% of the unreinforced sand bed.

4.1 Effect of Depth to First Layer of 3D Geogrid

Figure 4 shows the bearing pressure versus normalized settlement curve for unreinforced and 3D geogrid-reinforced sand bed for various depths of the first layer of reinforcement (u). It can be seen that the bearing pressure increases with the provision of 3D geogrid compared to unreinforced sand. When the 3D geogrid was placed at a depth of 0.25B from the base of the footing, the maximum improvement in bearing pressure was observed. The frictional resistance mobilized by the overburden pressure along with the confinement of soil particles within the cells of 3D

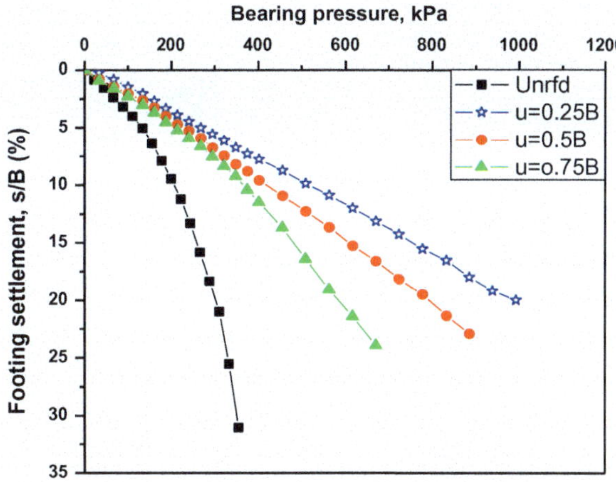

Fig. 4 Variation of settlement with bearing pressure for different u values

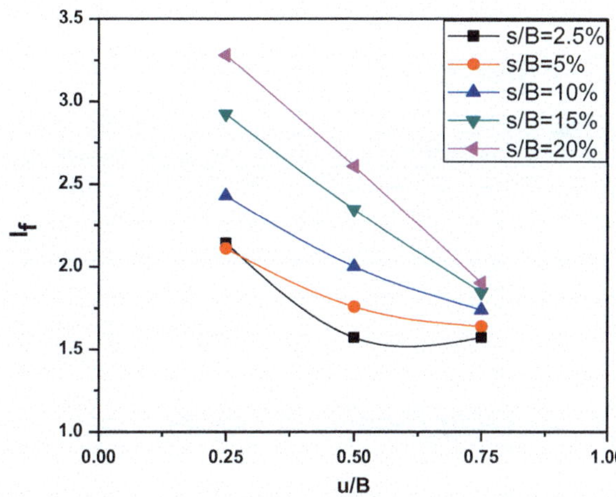

Fig. 5 Variation of improvement factor with *u*/*B* ratios for different normalized settlements

geogrid improves the performance of 3D geogrid-reinforced soil. 79% reduction in the settlement was observed with single-layer 3D geogrid-reinforced sand.

Figure 5 shows the variation of improvement factor (I_f) with *u/B* ratios for different normalized footing settlements. It can be seen that the bearing capacity improvement is maximum, when the 3D geogrid was placed at 0.25*B* at all levels of footing settlement. The ultimate bearing capacity of unreinforced and reinforced case was calculated corresponding to a settlement of 25 mm. About three times improvement in ultimate bearing capacity was observed with a single layer of 3D geogrid at an optimum depth of 0.25*B*.

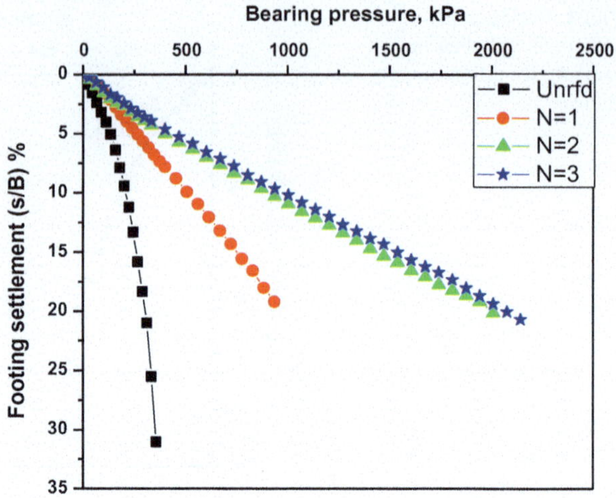

Fig. 6 Bearing pressure versus footing settlement curve for different number of layers

4.2 Effect of Spacing Between Two Layers

In order to study the effect of spacing between two consecutive layers, the first layer was placed at an optimum depth of 0.25*B* and second layer was placed at 0.5*B* and 0.75*B* spacing. At these spacing ratios, the improvement factor was 5.36 and 6, respectively. Therefore, the optimum spacing between two consecutive layers is taken as 0.75*B*.

4.3 Effect of Number of Layers

Figure 6 shows the bearing pressure versus footing settlement curve for different number of layers. The maximum improvement in bearing pressure (6.4 times) was observed with three layers of 3D geogrids. When the number of 3D geogrid layers increases from 1 to 2, the bearing capacity improvement was 97%, and from 2 to 3, the improvement was only 6%.

Further increase in the number of layers will not show any significant improvement, since the maximum depth of influence of vertical load is governed by the depth of the pressure bulb, which is equal to 2B. Also, second and third layers of 3D geogrid give 87% settlement reduction. Considering the percentage improvement in bearing capacity and settlement reduction, the optimum number of 3D geogrid layers was taken as two. Figure 7 shows the variation of improvement factor with the number of layers. Table 1 shows the settlement reduction factors for various parameters.

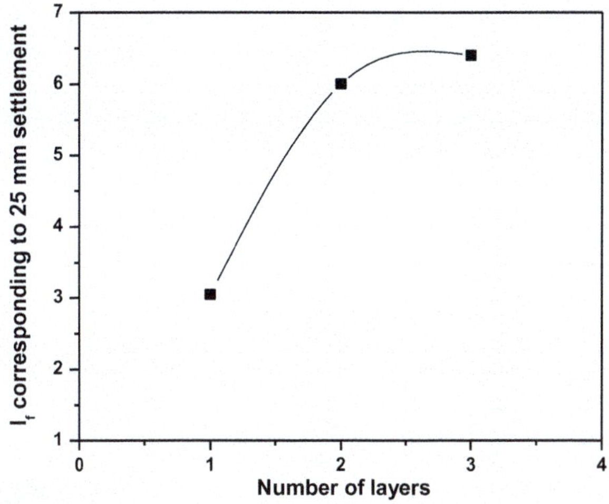

Fig. 7 Variation of improvement factor with number of layers

Table 1 Settlement reduction factor for various parameters

Parameters		SR_f
u/B	0.25	0.79
	0.5	0.73
	0.75	0.72
h/B	0.5	0.86
	0.75	0.87
N	1	0.79
	2	0.87
	3	0.87

5 Conclusions

In this study, a series of plate load tests were conducted to understand the bearing capacity and settlement characteristics of a model square footing resting on 3D geogrid-reinforced sand bed. The major conclusions from the study are the following:

- 3D geogrid-reinforced sand bed performs far better than an unreinforced sand bed in improving the bearing capacity and reducing the settlement.
- The optimum depth of the first layer of reinforcement was obtained as 0.25*B* from the base of the footing, where B is the width of footing.
- A single layer of 3D geogrid at the optimum depth improves the bearing capacity by three times and the settlement of footing was reduced by 79% compared to unreinforced sand.
- The optimum spacing between two consecutive layers was 0.75*B*.

- The optimum number of 3D geogrid layers was taken as 2 with an improvement factor of 6, and a further increase in the number of layers does not show any significant improvement.
- The settlement of the footing was reduced by 87% by the provision of two layers of 3D geogrid.

References

Abu-Farsakh M, Chen Q, Sharma R (2013) An experimental evaluation of the behavior of footings on geosynthetic reinforced sand. Soils Found 53(2):335–348

Ghosh A, Bera AK (2005) Bearing capacity of square footing on pond ash reinforced with jute-geotextile. Geotext Geomembr 23(2):144–173

Harikumar M, Sankar N, Chandrakaran S (2016) Behaviour of model footing resting on sand bed reinforced with multi-directional reinforcing elements. Geotext Geomembr 44:568–578

Khedkar MS, Mandal JN (2009) Pullout behavior of cellular reinforcements. Geotext Geomembr 27:262–271

Latha GM, Somwanshi A (2009a) Bearing capacity of square footings on geosynthetic reinforced sand. Geotext Geomembr 27(4):281–294

Latha GM, Somwanshi A (2009b) Effect of reinforcement form on the bearing capacity of square footings on sand. Geotext Geomembr 27:409–422

Mosallanezhad M, Alfaro MC, Hataf N, Sadat Taghavi SH (2016) Experimental and numerical studies of the performance of the new reinforcing system under pullout conditions. Geotext Geomembr 44:70–80

Sireesh S, Seetharam TG, Dash SK (2009) Bearing capacity of circular footing on geocell-sand mattress overlying clay bed with void. Geotext Geomembr 27:89–98

Vinod P, Ajitha BB, Sreehari S (2009) Behavior of a model square footing on loose sand reinforced with braided coir rope. Geotext Geomembr 27:464–474

Zhang MX, Javadi AA, Min X (2006) Triaxial test of sand reinforced with 3D inclusions. Geotext Geomembr 24:201–209

Rain Garden—A Solution to Urban Flooding: A Review

Osheen and K. K. Singh

Abstract Rain gardens or bioretention systems are the best storm water management practices, which use natural processes of the hydrological cycle such as infiltration and evapotranspiration. Rain gardens were first coined for residential use in 1990 in Prince George's County, Maryland which was an alternative to the conventional system of sidewalks and gutters. However, countries like Japan, China, Australia and U.S.A. are encouraging the use of rain garden these days for the sustainable development of the country. From the last few decades, the world is witnessing harmful results of urbanization. This has led to a rapid increase in impervious land cover and deterioration of the quality of the ecosystem. The impervious surface of concrete and asphalt seal the soil layers and causes excessive surface runoff, which leads to the problem of urban flooding. Also, chemical and oils falling from vehicles on roads get washed away with storm water and enters the natural water bodies leading to their pollution. Rain garden reduces and delays the flood peaks as well as helps in groundwater recharge and enhances the biodiversity. Moreover, its vegetation works as a filter media for storm water treatment. Rain garden is a low impact development (LID) technique having a long-term performance and is aesthetically pleasing. This paper quotes the benefits and chronological order of implementation of bioretention systems in urban cities having the problem of storm water management with an aim to create awareness among scientific communities.

Keywords Rain garden · Flooding · Urbanization · BMPs · Storm water

1 Introduction

Storm water management has been an important issue since many years. Urbanization leads to an increase in the volume of storm water and peak flows as vegetation is removed and soil surfaces are replaced by concrete and asphalt (Swain et al. 2018a, b). For most of history of urban drainage, wastewater and storm water have

---

Osheen (✉) · K. K. Singh
Department of Civil Engineering, NIT, Kurukshetra, Kurukshetra 136119, Haryana, India
e-mail: osheen101094@gmail.com

A. K. Agnihotri et al. (eds.), *Sustainable Engineering*, Lecture Notes in Civil Engineering 30, https://doi.org/10.1007/978-981-13-6717-5_4

been managed together. Storm water and wastewater are considered as a hazard to human health and environment, therefore need to be removed as quickly as possible from urban areas. *Cloaca maxima* is one of the ancient drains in Rome, which was fabricated in around 600 BC, still, exist today. These ancient drains were similar to combined sewers. Combined sewers are designed to overflow in heavy rainfalls, so they constitute a danger to human health and environment (Tang et al. 2016; Verma et al. 2016). In the 1950s, programs started to separate combined sewers. Separate pipe networks for storm water and wastewater have become a standard practice.

A paradigm shift occurred in the last decade, when it was realized that urban water bodies are getting deteriorated day by day due to poor water quality and flooding. This led to a change in storm water management system. Techniques were investigated to replace conventional drainage system with some method that not only attenuates the flow but also treat the storm water contaminants. Some common examples of techniques that were investigated include detention ponds and wetlands, infiltration surfaces, permeable paving, swales (vegetated ditches), media filters, bioretention and green-roofs. These management systems are generally called water sensitive urban design (WSUD), low impact design (LID), sustainable urban drainage systems (SUDS) and best management practices (BMPs).

Storm water BMPs are used to mitigate the impacts of urban development on the quantity and quality of storm water. BMPs are of two types: structural and non-structural BMPs (USEPA 1999). Non-structural BMPs focus on protection of natural systems and preserving open spaces. Non-structural BMPs uses existing stream corridors and wetlands like landscape features to manage and treat storm water at its source. These techniques attempt to minimize the disturbance to nature and minimize the expanse of the impervious area. Structural BMPs are based on natural systems of soil mechanism and vegetation to treat storm water.

The structural BMPs commonly used were storage tanks, detention basins and wet ponds. They were simple to design and provided an effective reduction in peak flow and satisfactorily improved the quality of water. But these methods were not sufficient to meet the storm water quality requirements of National Pollutant Discharge Elimination System (NPDES), so wetlands and other techniques which depend on ecosystem functions to improve the quality of storm water gained popularity (Bratieres et al. 2008).

Constructed wetlands use natural landscape for releasing or storing water in the situation of drought and flood (Swain et al. 2017). Soil and plants of wetland filter the storm water naturally by filtration, sedimentation, microbial decomposition and plant uptake (Champagne 2008). The limitation of wetland for using storm water treatment is that some specific hydrological conditions are required for vegetation adapted to environments of wetland (USEPA 2003). Also, lack of maintenance impacts the effectiveness of wetlands and become a breeding ground for mosquitoes.

Rain garden or bioretention system is one of the BMPs, which has acquired significant attention in the last decade. Rain gardens are shallow excavated landscapes which uses natural processes of infiltration and evapotranspiration to sustainably control the surface runoff. In a bioretention system, soil is excavated and replaced by a material having high permeability and organic matter to increase the rate of

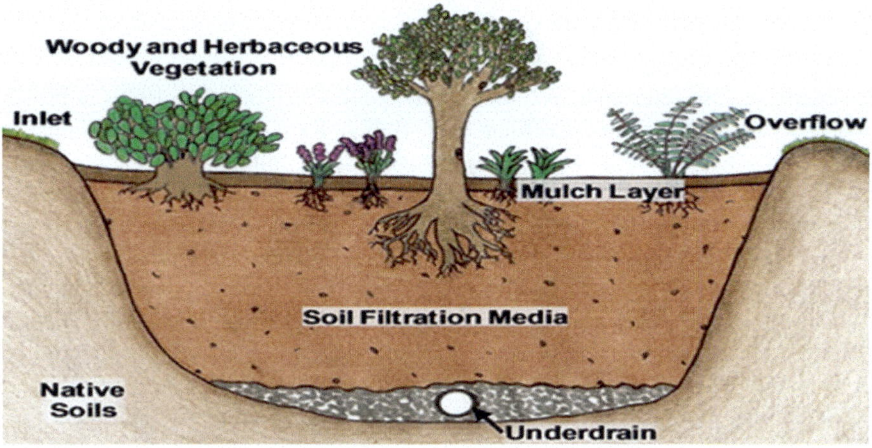

Fig. 1 Schematic of a typical rain garden

infiltration. Native vegetation is generally grown in the rain garden. The vegetation is a very important part of a rain garden because it holds the soil, filters the pollutants and makes the rain garden aesthetically pleasing. Small plants, shrubs and trees can be grown in a rain garden, choice of vegetation also depend on the size of the rain garden. A layer of organic matter and mulch is generally overlaid on soil media. An inlet structure is provided to direct the storm water of surrounding area towards the rain garden and an outlet is also provided to discharge the water excess of the ponding capacity of the rain garden. Outlet is generally connected to the sewerage system. Rain garden in an area having the soil of low permeability can be constructed by providing an underdrain at the bottom to prevent the standing of water for an excess time. Figure 1 shows the cross-section of a typical bioretention system with the labelled design elements.

Rain garden, like wetlands, depends on ecological interactions in a natural way to reduce the quantity of runoff and improves the quality of runoff. But unlike wetlands, rain garden depends on the terrestrial forested ecosystem and can be drained within hours (PGC 2007). The main advantage of the rain garden is that it uses natural processes of infiltration and evapotranspiration to reduce the storm water volumes. Thus, bioretention system can mainly be used in urban areas to counteract the problem of increased storm water volume that arises due to urban development (Tang et al. 2016).

This review paper outlines the evolution of rain garden, from its initial development to future improvements. This paper highlights the reasons that motivated the expansion of rain garden and improvements made in system design.

2 Development of Rain Garden

Rain gardens, also referred to as bioretention systems, originated in 1990s in Prince George's County, Maryland (Coffman et al. 1993a). Rain gardens were designed to treat storm water before it enters the sewerage system. Initially, rain gardens were made by excavating the land and refill it by planting soil, a layer of sand is applied at the bottom to increase infiltration (Prince George's County 1993). Generally, shrubs and native grass were planted in rain garden. Rain gardens were introduced to treat the "first flush" of the storm water from urban areas that carry very heavy pollutant load. The concept of rain garden for treatment of storm water was influenced by natural systems that were used to treat wastewater. Rain garden is an attempt to "maximize all available physical, chemical, and biological pollutant removal processes found in the soil and plant complex of a terrestrial forested community" (Coffman et al. 1993b).

The growth and expeditious adoption of rain garden were persuaded by its beneficial characteristics. Rain gardens not only help in reducing the volume of runoff, but it also improves the aesthetical appearance of the urban areas. Rain gardens are easy to install, inexpensive and require little maintenance after installation (Aravena and Dussaillant 2009; Yuan et al. 2017). The cost of construction of a rain garden to treat the storm water from a parking lot of area 0.3 ha was approximately $6500 (1993 USD). This is approximately one third the cost of an oil and grit separator fabricated to treat the same volume of storm water (Coffman et al. 1993a). According to guidelines established by manual of Prince George' County (1993), a rain garden covering 5% of the impervious area could allow 0.5 in. of storm water from area to infiltrate. Prince George's County manual states that in a commercial and industrial area a total of 6% of total green space can be easily converted into rain garden (Coffman et al. 1993b).

3 Research on Rain Garden's Performance

Considerable research has been conducted in field and laboratory to evaluate the performance of rain garden in retaining pollutants and reducing the volume of runoff. Rain garden is a structural storm water BMPs. Best Management Practice is generally described as efficiency in the reduction of pollutant percentage. This section reviews the main findings of studies related to bioretention system.

3.1 Hydrologic Impact

A series of column experiments were conducted to study the infiltration behaviour of a rain garden for a long time period (Le Coustumer et al. 2007). The permeability of soil decreases considerably over the first 4 weeks of testing, after then it attains a constant value. To depict undersized rain gardens, runoff inflow volumes were doubled in some columns. The declination in permeability is obtained in all columns was expedited in undersized columns due to the high rate of hydraulic loading.

In Xi'an city of China three rain gardens (1, 2 and 4%) of the study area were established to study the influence of the size of rain garden's control effect on urban runoff and the effect of discharge facility on its working. Li et al. (2014) reported that a larger proportion of area for a rain garden gives better results in controlling urban runoff. Also, he states that bioretention systems without discharge facility are better than systems with discharge facility under the identical conditions of rainfall and ratio of area.

Ishimatsa et al. (2016) reported that bioretention systems are more suitable for dealing with low flow volumes of storm water, however, their suitability could be altered by changing the properties of soil or catchment area. They also added that like any other green infrastructure, bioretention systems require proper maintenance. The most important maintenance is of disposal of garbage as garbage blocks the running channel between sewerage system and rain garden.

To identify the effect to evapotranspiration in volume reduction potential of rain garden three lysimeters mimic three rain gardens were fabricated at Villanova University, Pennsylvania. The rain gardens with unconstricted drainage systems: sandy loam and sand with Unconstricted valve (UV) and internal water storage (IWS) outflow. Hess et al. (2016) reported that evapotranspiration made up 47, 43 and 70% of the water budget for sandy loam, sand and sand IWS systems, respectively.

4 Water Quality Improvement

One of the most common goals of storm management practice includes the removal and reduction of pollutants (McPherson et al. 2002; Hong et al. 2006). However, published field performance data on pollutant removal efficiency of rain garden is currently less but is rapidly growing. The results of pollutant removal may be satisfactory in some cases, but they can be misleading in a few cases. This cases mainly arises when the quality of influent water is relatively good, and concentration of pollutant is less, than the fractional removal may be low. But BMP cannot be elucidated unsatisfactory in this case.

4.1 Nitrogen and Phosphorus Removal

To determine the optimal characteristics for nutrient removal of a bioretention system, a series of column experiments were performed (Bratieres et al. 2008). Nitrogen removal by certain species of plants (*Melaleuca ericifolia and Carex appressa*) were significantly higher than other tested species. This clearly indicates that vegetation selection is an essential criterion in rain garden design for removal of nutrients. Around 85% of phosphorus removal was constantly observed in bioretention columns but if organic content of soil (in columns) is high than removal efficiency reduced (nearly 40%) due to the formation of orthophosphate in columns (Dietz and Clausen 2005).

4.2 *Removal of BOD and Pathogens*

Hunt et al. (2008) measured BOD_5 during 23 storm events of influent and effluent of a rain garden fabricated in North Carolina. He reported that a 63% reduction in concentration is observed, as average BOD_5 of influent is 8.54 mg/l which was reduced to 4.18 mg/l.

The counts of pathogens are considerably lesser in storm water than in wastewater, but the presence of pathogens is of considerable importance, because pathogens are hazardous for both human health and aquatic species. Rusciano and Obropta (2007) conducted column experiments and suggested that rain gardens have significant capacity to reduce the count of pathogens in storm water. He conducted 13 experiments containing a range of coliforms, a reduction of 91.6% was obtained in their count and removal efficiency ranges from 54.5 to 99.8%.

5 Implementation Issues

A major concern in the implementation of a rain garden is its poor construction practices. Since most of the contractors are unfamiliar with the construction of rain garden, this to the problems like the improper selection of soil, improper placement and poor establishment of vegetation (Cosgrove and Bergstrom 2004; Toronto and Region Conservation Authority 2008). Carpenter and Hallam (2010) concluded that infiltration rates of a rain garden are highly affected by construction method of the rain garden as well as soil mixtures used in rain garden. Preferential flow tracks developed in soil media of rain garden, if construction technique is poor (composed of 80% compost, 20% sand), this severely affects the hydrologic performance of garden. The effect of preferential flow paths can be observed as the storm water volume retention for a rainfall event of 2.3 cm, declined from an average of 92–40.2%. In similar conditions, rain garden constructed with a soil mixture of 20% compost, 50% sand and remaining 30% topsoil, exhibit a volume retention of about 82%. Morzaria-Luna et al. (2004) studied the process of integrating the rain gardens in storm water management plan of a new subdivision Wisconsin and concluded that if rain gardens are need to be used widely as storm water management practice, then the current regulatory approaches which discourage the use of non-traditional storm water management practices need to be revised.

6 Bioretention Modelling Research

It is necessary to accurately predict the hydrologic performance and water quality improvement characteristics of the rain garden to adopt it as integrated storm water management practice and verify the current guidelines available for the design of rain garden. This can be done with the help of modelling tools.

Researches at the University of Wisconsin developed the most comprehensive model for evaluating the hydrologic performance of a rain garden, the model is named as RECHARGE model (Dussaillant et al. 2004). Infiltration is simulated in this model using Richard's equation, the model includes soil layers (maximum three), surface ponding and underdrain (optional). RECARGA is an alternative version of RECHARGE, based on Green-Ampt equation to simulate infiltration. RECARGA has been adopted in the design guidelines of bioretention system in the state of Wisconsin (WDNR 2004). This model uses evapotranspiration and precipitation and computes the water balance terms including soil moisture, inflow, outflow, underdrain flow, no. of overflow events and evaporation.

HEC-HMS is a hydrologic model developed by Heasom et al. (2006) to determine the hydrologic impact of rain garden. This model introduced sub-basins (impermeable and permeable basin) to represent the drainage area of rain garden. With the help of sub-basin characteristics input rainfall is changed into runoff inflow. Diversion elements are also included in this model which simulates hydrologic conditions and effect of changes of soil moisture on infiltration rate. The model uses stage–discharge relationship to control the outflow and rain garden in this model is represented by a reservoir attached with a weir. This model was used to monitor data from rain gardens located at Villanova University to anticipate the water level in gardens. Storm Water Management Model (SWMM) modelling software also uses similar models but different programming languages (Lucas 2008; Li et al. 2016; Poresky et al. 2016).

7 Summary and Conclusion

In the past 20 years, rain garden has become one of the prominent storm water Best Management Practices (BMPs). Nowadays, the rain garden has become the key component of Low Impact Development (LID) storm water managing philosophy. Rain garden is a new breed of LID that integrates expertise from several disciplines including horticulture, botany, soil science, hydrology, hydraulics, surface and groundwater flow, landscape architecture and engineering. Consequently, the expansion of guidelines for bioretention system requires a team approach. The performance of rain garden has been studied extensively, through field studies and laboratory both and results have confirmed the potential of a rain garden for storm water treatment and storm water retention. Nevertheless, this technology is still immature and requires additional research to provide proper design specifications and performance information.

References

Aravena JE, Dussaillant A (2009) Storm-water infiltration and focused recharge modeling with finite-volume two-dimensional richards equation: application to an experimental rain garden. J Hydraul Eng 135(12):1073–1080

Bratieres K, Fletcher TD, Deletic A, Zinger Y (2008) Nutrient and sediment removal by storm water biofilters: a large-scale design optimization study. Water Res 4214:3930–3940

Carpenter DD, Hallam L (2010) Influence of planting soil mix characteristics on bioretention cell design and performance. J Hydrol Eng 15(6):404–416

Champagne P (2008) Wetlands. Natural processes and systems for hazardous waste treatment. ASCE, Reston, Va., pp 189–256

Coffman L, Green R, Clar M, Bitter S (1993a) Design considerations associated with bioretention practices. In: Proceedings of 20th anniversary conference on water management in the '90s. ASCE, Reston, Va., pp 130–133

Coffman L, Green R, Clar M, Bitter S (1993b) Development of bioretention practices for stormwater management. In: Proceedings 20th anniversary conference on water management in the '90s. ASCE, Reston, Va., pp 126–129

Cosgrove JFJ, Bergstrom JD (2004) Design and construction of biofiltration basins: lessons learned. In: Proceedings of 2003 world water and environmental resources congress. ASCE, Reston, Va., p 323

Dietz ME, Clausen JC (2005) A field evaluation of rain garden flow and pollutant treatment. Water Air Soil Pollut 167(1–4):123–138

Dussaillant AR, Wu CH, Potter KW (2004) Richard's equation model of a rain garden. J Hydrol Eng 93:219–225

Heasom W, Traver RG, Welker A (2006) Hydrologic modeling of a bioinfiltration best management practice. J Am Water Resour Assoc 425:1329–1347

Hess A, Wadzuk B, Walker A (2016) Evapotranspiration in rain gardens using weighing lysimeters. ASCE, J Irrig Drain Eng

Hong E, Seagren EA, Davis AP (2006) Sustainable oil and grease removal from synthetic storm water runoff using bench-scale bioretention studies. Water Environ Res 782:141–155

Hunt WF, Smith JT, Jadlocki SJ, Hathaway JM, Eubanks PR (2008) Pollutant removal and peak flow mitigation by a bioretention cell in urban Charlotte, N.C. J Environ Eng 1345:403–408

Ishimatsa K, Ito K, Mitani Y, Tanaka Y, Sugahara T, Naka Y (2016) Use of rain gardens for stormwater management in urban design and planning. Landscape Ecol. Springer

Le Coustumer S, Fletcher TD, Deletic A, Barraud S (2007) Hydraulic performance of biofilters for storm water management: first lessons from both laboratory and field studies. Water Sci Technol 56(10):93–100

Li J, Li Y, Li Y (2016) SWMM-based evaluation of the effect of rain gardens on urbanized areas. Environ Earth Sci 75:17

Li J, Li Y, Shen B, Li YJ (2014) Simulation of rain garden effects in urbanized area based on SWMM. J Hydroelectr Eng 33:60–67

Lucas WC (2008) Continuous simulation of integrated bioretention-infiltration systems for urban retrofits. In: International low impact development conference 2008. ASCE

McPherson TN, Burian SJ, Turin HJ, Stenstrom MK, Suffet IH (2002) Comparison of the pollutant loads in dry and wet weather runoff in a southern California urban watershed. Water Sci Technol 459:255–261

Morzaria-Luna HN, Schaepe KS, Cutforth LB, Veltman RL (2004) Implementation of bioretention systems: a Wisconsin case study. J Am Water Resour Assoc 40(4):1053–1061
Poresky AL, Allen VP, Reynolds SK (2016) Biofiltration equivalency: assessing the relative performance of innovative and conventional designs. In: World environmental and water resources congress 2016. ASCE
Prince George's County (1993) Design manual for use of bioretention in storm water management. PGC, Maryland, Department of Environmental Protection, Watershed Protection Branch, Landover, Md
Prince George's County PGC (2007) Bioretention manual. PGC, Maryland, Department of Environmental Resources, Environmental Services Div., Landover, Md
Rusciano GM, Obropta CC (2007) Bioretention column study: fecal coliform and total suspended solids reductions. Trans ASABE 504:1261–1269
Swain S, Nandi S, Patel P (2017) Application of SPI, EDI and PNPI using MSWEP precipitation data over Marathwada, India. In: 2017 IEEE International geoscience and remote sensing symposium (IGARSS), pp 355–357, IEEE
Swain S, Verma MK, Verma MK (2018a) Streamflow estimation using SWAT model over Seonath river basin, Chhattisgarh, India. In: Singh V, Yadav S, Yadava R (eds) Hydrologic modeling. Water Sci Tech Lib 81:659–665
Swain S, Nandi S, Patel P (2018b) Development of an ARIMA model for monthly rainfall forecasting over Khordha district, Odisha, India. In: Sa P, Bakshi S, Hatzilygeroudis I, Sahoo M (eds) Recent findings in intelligent computing techniques. Advances Int Sys Comput 708:325–331
Tang S, Luo W, Jia Z, Liu W, Li S, Wu Y (2016) Evaluating retention capacity of infiltration rain gardens and their potential effect on urban stormwater management in the sub-humid loess region of China. Water Resour Manage 30(3):983–1000
Toronto and Region Conservation Authority (2008) Performance evaluation of permeable pavement and a bioretention swale—Seneca College, King City, Ontario, Toronto and Region Conservation Authority, Toronto
USEPA (1999) Preliminary data summary of urban storm water best management practices. Rep. No. EPA-821-R-99–012, USEPA, Office of Water, Washington, D.C
USEPA (2003) National primary drinking water standards. Rep. No.EPA-816-F-03–016, USEPA, Office of Water, Washington, D.C
Verma M, Verma MK, Swain S (2016) Statistical analysis of precipitation over Seonath river basin, Chhattisgarh, India. Int J Appl Eng Res 11(4):2417–2423
WDNR (2004) Technical note for sizing infiltration basins and bioretention devices to meet state of Wisconsin storm water infiltration performance standards, Wisconsin Department of Natural Resources, Madison, Wis
Yuan J, Dunnett N, Stovin V (2017) The influence of vegetation on rain garden hydrological performance. Urban Water J 14(10):1083–1089

Al–Fe and Al–Ti Pillared Saponite Clay Catalysts: Preparation and Characterization

Sudha Minz, Renu Gupta and Sangeeta Garg

Abstract Saponite clay, a trioctahedral smectite clay was modified with Aluminum–Iron and Aluminum–Titanium solutions to form Al–Fe and Al–Ti pillared interlayer catalysts (PILC's). The PILC's were characterized using FTIR, XRD, N_2 adsorption–desorption isotherms, SEM, WD-XRF, and TGA/DTG techniques. The OH stretching region in IR spectra of saponite reflected the trioctahedral character with mainly $Mg(OH)_2$ units. XRD analysis showed the increase in basal spacing of Al–Fe PILC (18.37 Å) and Al–Ti PILC (16.99 Å) as compared to saponite clay (11.77 Å). The nitrogen adsorption indicated an increase in specific surface area and micropore volume of Al–Fe PILC (94.44 m^2/g) and Al–Ti PILC (73.78 m^2/g) as compared to smectite clay (33.96 m^2/g). SEM and WD-XRF analysis results indicated successful pillaring with oxides of Al–Fe and Al–Ti between the layers of the saponite clay. Thermal stability of the saponite clay and PILC's in nitrogen atmosphere was analyzed up to 900 °C. Al–Fe PILC and Al–Ti PILC were found to be thermally stable in the temperature range 200–600 °C with weight loss 6.7% and 3.5%, respectively, whereas for natural saponite clay it was 10.1%. The study revealed the increase in basal spacing, specific surface area and higher thermal stability of Al–Fe PILC and Al–Ti PILC over saponite clay.

Keywords Saponite clay · Al–Fe PILC · Al–Ti PILC · Characterization · Thermal analysis

S. Minz · R. Gupta (✉) · S. Garg
Department of Chemical Engineering, Dr. B. R. Ambedkar National Institute of Technology, Jalandhar 144011, Punjab, India
e-mail: bansalr@nitj.ac.in

S. Minz
e-mail: minz.sudha@gmail.com

S. Garg
e-mail: gargs@nitj.ac.in

A. K. Agnihotri et al. (eds.), *Sustainable Engineering*, Lecture Notes in Civil Engineering 30, https://doi.org/10.1007/978-981-13-6717-5_5

1 Introduction

Clay minerals are phyllosilicate minerals present on the earth's surface in tetrahedral and octahedral sheets in the ratio 2:1 and 1:1, respectively. The sheets either have no charge or have a net negative charge, which is balanced by Na^+ or K^+ cations (Madejová 2003). The clay minerals are extensively used to produce pottery, building materials, ceramics, foundry molds, paper coatings, polymer nanocomposites, pharmaceuticals, agrochemicals, adsorbents (Zhou et al. 2010). Nowadays, pillared interlayered clays (PILC's) constitute an important field of study for microporous materials developed by molecular engineering (Vicente et al. 2001).

Pillaring is a modification process of smectite clays by intercalation of active species into the interlayer clayed structure with a promising application in catalysis, adsorption, and separation processes (Cupic 2008). The PILC's are synthesized as: first, preparation of bulky inorganic polyoxocations by hydrolysis of multivalent cations that produce cationic polymeric species; second, constitution of intercalated clay by interchange of exchangeable charge that counterbalance the cations of swellable smectite by the polyoxocations prepared in first step; and third, formation of pillared clays by heating at higher temperature that transforms metastable polyoxocations to stable oxy-hydroxidic phases, which act as pillars between the layers of clay (Mata et al. 2007; Guerra et al. 2008; Toranzo et al. 1997). The metal oxide pillars introduced between the clay layers keep them separated, predominately increase the basal spacing in the clay thus averting their collapse and develops a porous network in the range of molecular dimensions. Montmorillonite is the commonly studied smectite clay mineral. Aluminum (Al^{III}) is the most common polycation $[Al_{13}O_4(OH)_{24}(H_2O)_{12}]^{7+}$ studied and reported by researchers that are easily synthesized and is stable (Belver et al. 2004). Other polycations studied and reported are Si^{IV}, Fe^{III}, Cr^{III}, Ti^{IV}, Zr^{IV}, and Ga^{III}. Studies of Aluminum cations combined with metal cations like Al–Fe, Al–Cu, Al–Si, Al–Zr has also been reported in the literature (Galeano et al. 2010; Li et al. 2017). Studies of PILC's with smectites other than montmorillonite like saponite, hectorite and beidellite are less reported in the literature. Saponite is a 2:1 trioctahedral phyllosilicate smectite mineral constituted of an octahedral sheet at the center with a brucite $[Mg(OH)_2]$ structure, in which four out of six OH-groups are exchanged by oxygen atoms. These O-atoms are linked to two tetrahedral sheets comprising of Si^{4+} and O^{2-} located on both sides of a central octahedral sheet. A limited amount of isomorphous substitution of Si^{4+} and Al^{3+} in the tetrahedral sheets leads in charge deficiency is compensated by interchangeable interlayer cations (Vogels et al. 2005a, b). Titanium and iron are interesting metals that have catalytic applications in the oxidation of various organic pollutants. Natural clays do not act as catalysts commonly, but they act as support material in the preparation of catalysts. So, clays with pillared metal oxides act as catalysts and have proved to be stable catalysts for oxidation of many organic pollutants (Fatimah 2014).

The purpose of the study is the preparation and characterization of PILC's, pillared with a combination of Aluminum–Titanium and Aluminum–Iron metal oxides into the layers of saponite clay. The PILC's were prepared by cation-exchange methods known as co-intercalation in which isomorphous substitution of Si^{4+} by Al^{3+}, Ti^{4+}, and Fe^{3+} in the tetrahedral sheets of saponite. The PILC's were characterized using several techniques to confirm the presence of pillared oxides of Al–Ti and Al–Fe between the interlayers of saponite clay. The thermal stability studies of Al–Fe PILC and Al–Ti PILC have also been made over smectite clay.

2 Materials and Method

2.1 Materials

Saponite clay was purchased from Nanoshel (USA), purity > 99%. Aluminum chloride hexahydrate ($AlCl_3.6H_2O$) (w/w, purity > 99%), ferric chloride hexahydrate ($FeCl_3.6H_2O$) (w/w, purity > 99%), titanium tetrachloride ($TiCl_4$) (purity 99.5%) was obtained from Loba Chemie Pvt. Ltd. (Mumbai, India). NaOH and HCl were obtained from S D Fine Chemical Limited (Mumbai, India). All the chemicals are of reagent grade with a purity greater than 99%. All the solutions were prepared using double distilled water.

2.2 Catalyst Preparation

2.2.1 Al–Fe PILC Preparation

Al–Fe PILC was prepared following the co-intercalation method described by Belver et al. (2004). The solution of aluminum chloride and ferric chloride ($AlCl_3.6H_2O$, 0.1 M and $FeCl_3.6H_2O$, 0.2 M) were mixed in the molar ratio of 5:1. The solution was maintained at 60 °C with continuous stirring in the magnetic stirrer. 0.2 M NaOH solution was added dropwise until the molar ratio of the titrated amount of OH^-/M^{3+} ($M^{3+} = Al^{3+} + Fe^{3+}$) reached to 2. The mixed solution was continuously stirred and maintained at 60 °C for 3 h and then cooled to room temperature. Dispersion of saponite in ethanol (2 wt%) was prepared by uninterruptedly stirring the solution for 24 h. The dispersed saponite solution was then added dropwise into the pillaring solution until the ratio of (M^{3+})/clay reached to 10 mmol/g. The mixture was continuously stirred for 3 h at 80 °C and then cooled for a day to room temperature. The solid mass was recovered by centrifugation and washed with double distilled water. It was then dried at 60 °C for 12 h and finally calcined at 500 °C for 2 h in the muffle furnace.

2.2.2 Al–Ti PILC Preparation

Al–Fe PILC was prepared similarly following the co-intercalation method described by Vicente et al. (2001). In the preparation of Al–Ti PILC, firstly, titanium tetrachloride ($TiCl_4$, 0.2 M) was dissolved in 5 M HCl. Then chloride solution of aluminum ($AlCl_3.6H_2O$, 0.1 M) $TiCl_4$ solution in 5 M HCl mixed in the molar ratio of Al^{3+}/Ti^{4+} = 5/1 was maintained at 60 °C with continuous vigorous stirring. The rest of the procedure remained the same as mentioned above in the preparation of Al–Fe PILC.

2.3 Characterization Techniques

The FTIR spectra were recorded for saponite clay and PILC's in the range of 4000–400 cm^{-1} with Perkin Elmer (Model RX-I, Tokyo, Japan) using a KBr pellet. The instrument was having a resolution of 1 cm^{-1}. X-ray diffraction diffractometer (X'PERT-PRO Model: PW3064, Philips Japan) with Cu–K-α radiation ($\lambda = 1.54$ Å) was employed to obtain XRD pattern of the saponite clay and PILC's. Surface area was determined by N_2 adsorption-desorption isotherms at −196 °C using the surface area analyzer (SMART SORB 93, Smart Instruments Co. Pvt. Ltd., India). The samples were degassed at 120 °C for 3 h. The surface morphology of the samples was observed using a field-emission scanning electron microscope (SEM) (JEOL Model: JSM 6510, Japan). The range of accelerating voltage was from 200 V to 30 kV with a resolution of 2 mm and magnification up to 5000 X. The elemental composition analysis of the saponite clay and PILC was carried out using wavelength dispersive X-ray fluorescence (WD-XRF S4 Pioneer, Bruker-AXS). Thermogravimetric analysis (TGA) and derivative thermogravimetric analysis (DTG) of the samples were performed using a TG/DTG instrument (EXSTAR TG-DTA 6300). A nitrogen stream was used at a heating rate of 10 °C.

3 Results and Discussion

3.1 Fourier Transform Infrared Spectroscopy (FTIR) Analysis

The FTIR spectra of saponite clay, Al–Fe, and Al–Ti pillared saponite clay is shown in Fig. 1. The absorption peak near 3632.91 cm^{-1} in saponite clay was assigned due to O–H stretching vibrations of water molecules weakly bonded to the Si–O surface. A broad absorption band near 3413.05 cm^{-1} observed in nearly all smectites was due to H–O–H vibrations of adsorbed water (Madejová 2003). The absorption peaks near 2922.47 and 2851.41 cm^{-1} were due to the aliphatic stretching vibration of C–H bond (Wang et al. 2009). The absorption band at 1039.95 cm^{-1} was due to the

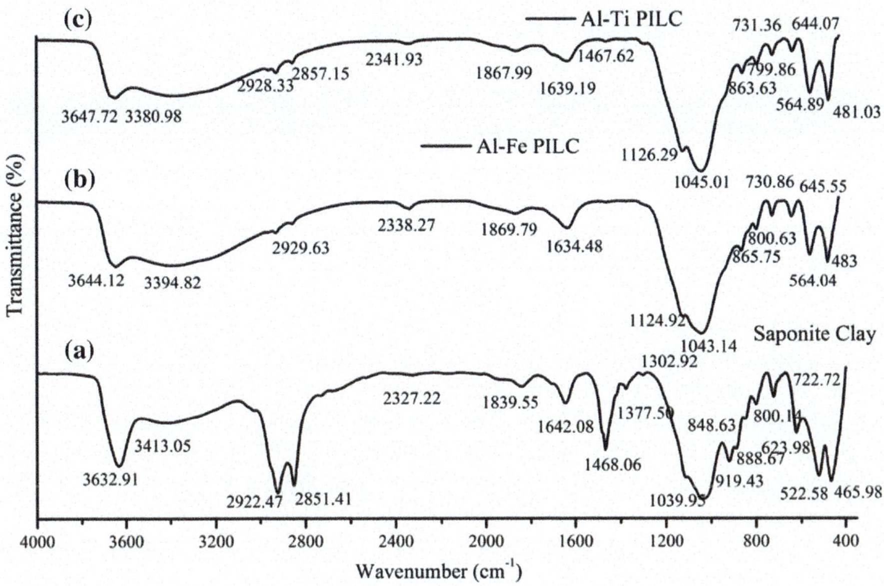

Fig. 1 FTIR spectra of **a** saponite clay, **b** Al–Fe PILC and **c** Al–Ti PILC

stretching vibration of Si–O bond in the tetrahedral sheet. The bands at 919.43 cm^{-1}, 888.67 cm^{-1}, 848.63 cm^{-1} corresponded to Al–Al–OH, Al–Fe–OH, and Al–Mg–OH bending vibrations, respectively. The sharp peak at 800.1 cm^{-1} with inflection near 722.72 cm^{-1} confirmed the presence of quartz admixture in the samples, the band at 623 cm^{-1} was due to coupled Al–O and Si–O out of plane vibration (Eren and Afsin 2008). The bands at 522.58 and 465.98 cm^{-1} were due to Al–O–Mg and Si–O–Fe bending vibrations, respectively (Zhou et al. 2014). The peak at 1642.08 cm^{-1} corresponded to an overtone of the bending vibration of water. The peak at 1468.06, 1377.50, and 1302.92 cm^{-1} was due to the presence of calcite and stretching of carbonate (Shillito et al. 2009). After pillaring of saponite clay with Al–Fe and Al–Ti metal cations, some of the absorption peaks disappeared and some new peaks were formed. A new peak at 1124.92 and 1126.29 cm^{-1} emerged after pillaring saponite was due to Si–O vibrations of amorphous silica with a three-dimensional frame network (Fig. 1b, c) (Eren and Afsin 2008). The peak at 1468.06 and 919.43 cm^{-1} in saponite almost disappeared after pillaring it with Al–Fe and Al–Ti. Also the intensity of the peaks at 3644.12, 3647.72 cm^{-1} and 1634.48, 1639.19 cm^{-1} decreased due to the reduction in –OH group after the pillaring process. The intensity of peaks was also decreased at 2929.63 and 2928.33 cm^{-1}, which was due to the reduction in stretching of C–H bonds.

3.2 X-Ray Diffraction (XRD) Analysis

The X-ray diffractograms at angle 2θ from 3° to 50° for the saponite clay and PILC with Al, Fe, and Ti metal cations are shown in Fig. 2. The 001 reflection at a 2θ angle of 7.5° in saponite presented a basal spacing of 11.77 Å. The different phases were identified in the saponite clay and PILC's samples: smectite, quartz, feldspar, and calcite (Mojovic et al. 2009). The reflections at a 2θ angle of 21 and 26.5 revealed the presence of quartz impurity, diffraction at 2θ value of 29° was due to calcite (Ayodele and Hameed 2013a; Kim and Lee 2004). After the pillaring process, the typical peaks corresponding to 001 reflection were sharp, distinct, and well defined, showing good crystallinity for the PILC's. Also, upon pillaring saponite clay with Al, Fe, and Ti metals the 001 reflection slightly shifted toward left indicating an increase in basal spacing. In Al–Fe PILC the refection was shown at 2θ angle of 4.8° (18.37 Å) and in Al–Ti PILC the refection was observed at a 2θ angle of 5.2° indicating basal spacing of 16.99 Å, respectively. A small peak around 33° and 36° in Al–Fe PILC plot attributed to the presence of Fe_2O_3 and FeO (Zhou et al. 2014; Ayodele and Hameed 2013a). In addition to d(001) reflections in PILC's d(002) reflections were also observed at 2θ angle of 9.46° with d(002) of 9.02 Å in Al–Fe PILC while Al–Ti PILC showed d(002) value of 9.38 Å at 2θ angle of 9.42° (Kaneko et al. 2001). In short, the XRD characterization results confirmed all PILC's (Al–Fe PILC and Al–Ti PILC) were successfully pillared as shown by d(001) reflection. After the pillaring process, the catalysts still maintained some structural features. However, various factors like intercalation of clay by different amount of metal species and calcination temperature influenced the crystalline behavior of PILC's and hence resulted in typical XRD patterns.

3.3 Surface Area (BET) Analysis

The N_2 adsorption/desorption isotherms of saponite clay and PILC's are shown in Fig. 3. According to IUPAC classification, the isotherm curves corresponded to type II and type IV with a hysteresis loop type H3 (Mata et al. 2007; Kooli and Jones 1998). The textural data obtained from the isotherms are represented in Table 1. The specific surface areas were obtained from BET method, the external surface area was found from t-method and the total pore volume was calculated from the amount of nitrogen adsorbed at a relative pressure of 0.99. It is observed that after pillaring saponite with different metal cations, there is a large increase in surface area in Al–Fe PILC (94.44 m^2/g) and Al–Ti PILC (73.78 m^2/g). Al–Fe PILC showed higher micropore surface area, total pore volume and average pore diameter than Al–Ti PILC.

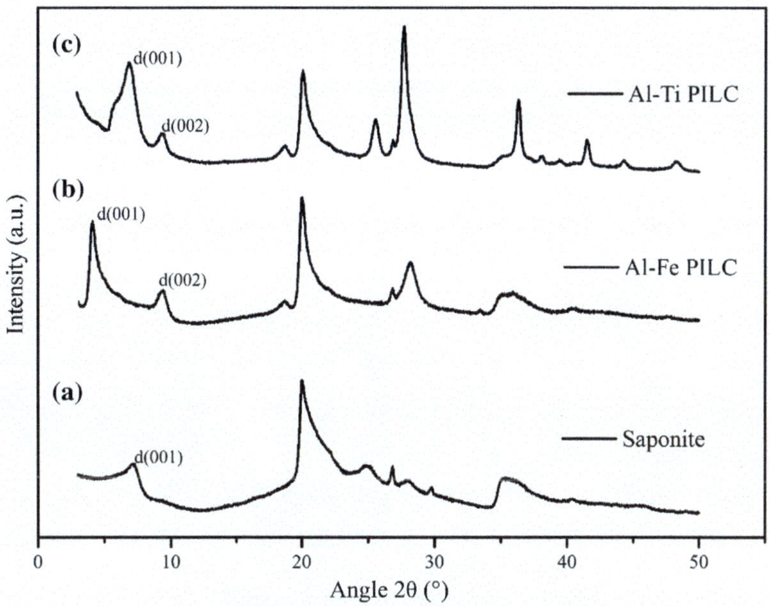

Fig. 2 X-ray diffraction patterns of **a** saponite clay, **b** Al–Fe PILC and **c** Al–Ti PILC

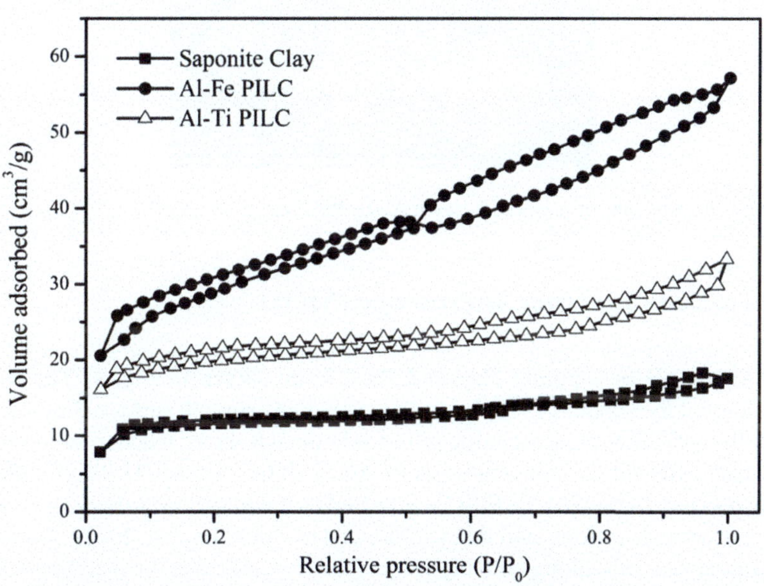

Fig. 3 Nitrogen adsorption/desorption curves for saponite clay and PILC's

Table 1 Specific surface area and total pore volumes of the saponite clay and PILC's calcined at 500 °C

Sample	S_{BET} (m^2/g)	$S_{\mu p}$ micropore surface area (m^2/g)	V_p (cm^3/g)	Average pore diameter (Å)
Saponite	33.96	8.39	0.004302	5.067
Al–Fe PILC	94.44	33.47	0.08614	36.48
Al–Ti PILC	73.78	13.52	0.05168	28.03

Fig. 4 SEM micrographs of **a** saponite clay **b** Al–Fe PILC and **c** Al–Ti PILC

3.4 *Scanning Electron Microscope (SEM)*

The SEM images of saponite clay and PILC's are shown in Fig. 4. The surface morphology of the saponite clay showed a layered and smooth surface (Fig. 4a). The intercalation of different metal oxide pillars into the interlayer space of saponite clay significantly affected the morphologies of Al–Fe PILC and Al–Ti PILC. After the pillaring process, the surface of PILC's appeared as flake-like crystals and the surface has got more rough, porous, and fluffy (Guerra et al. 2008; Li et al. 2017). This rough and porous appearance along with higher surface area indicated the increase in active sites over the surface of PILC's.

Table 2 Chemical composition analysis (wt%) of saponite clay, Al–Fe PILC, and Al–Ti PILC

Sample	SiO_2	Al_2O_3	Fe_2O_3	MgO	Na_2O	CaO	K_2O	TiO_2	MnO
Saponite clay	65.44	15.87	1.3	0.99	0.68	3.28	1.31	0.18	–
Al–Fe PILC	27.77	20.43	13.40	1.32	–	–	0.07	0.08	0.006
Al–Ti PILC	21.15	8.90	2.35	0.98	–	–	0.04	21.45	–

3.5 *Wavelength Dispersive X-Ray Fluorescence (WD-XRF)*

The chemical composition of various oxides present in the saponite clay and PILC's was analyzed using WD-XRF technique and is listed in Table 2. The values were consistent with smectite, and aluminum being the major clay mineral in the clay structure. The elemental analysis results revealed the decrease in metal oxide contents of Na, Ca, K, and in both Al–Fe PILC and Al–Ti PILC, which may be due to the replacement of these metal ions by Fe and Al ions to form iron and aluminum oxide pillars in between the silicate layers of saponite clay (Minz et al. 2018a; Sum et al. 2005). On pillaring saponite clay with Fe cations, the Fe_2O_3 content in the Al–Fe PILC was increased to 13.40 wt% from 1.30 wt% and TiO_2 was increased to 21.45 wt% from 0.18 wt% in Al–Ti PILC. Thus, the increase in the Fe and Ti in the Al–Fe PILC and Al–Ti PILC confirmed the successful pillaring of the metal cations.

3.6 *Thermogravimetric/Derivative Thermogravimetric (TG/DTG) Analysis*

The TG/DTG studies for saponite clay, Al–Fe PILC and Al–Ti PILC are made to analyze their stability and are shown in Fig. 5a–c. In the thermogram of saponite clay, the first weight loss was 25.6% from 22 to 200 °C due to a large amount of loosely bound water molecules present in the sample. During the second stage, the weight loss was 10.1% in the temperature range of 200–800 °C. The lower weight loss was may be due to the strong bonds between the water molecules and the interlayer ions of the clay sheets. The third stage weight loss (1.3%) was due to the dehydration of structural hydroxyl group of the clay above 800 °C (Minz et al. 2018b; Ayodele and Hameed 2013b). TG curves of Al–Fe PILC and Al–Ti PILC showed 2.9% and 1.6% weight loss in the first stage, respectively. The weight loss in both the PILC's was very low as compared to weight loss of saponite clay (25.6%) in the first stage. In the second stage, the Al–Fe PILC and Al–Ti PILC showed the weight loss of 6.7% and 3.5%, respectively, as compared to 10.1% in saponite clay. TG thermogram showed the phase change of Al–Fe PILC and Al–Ti PILC weight loss of 0.3 and 0.9% in the third stage above 800 °C. It is observed from TG studies that the thermal decomposition of saponite clay is more (38.7%) as compared to AL–Fe PILC (9.9%) and Al–Ti PILC (5.7%).

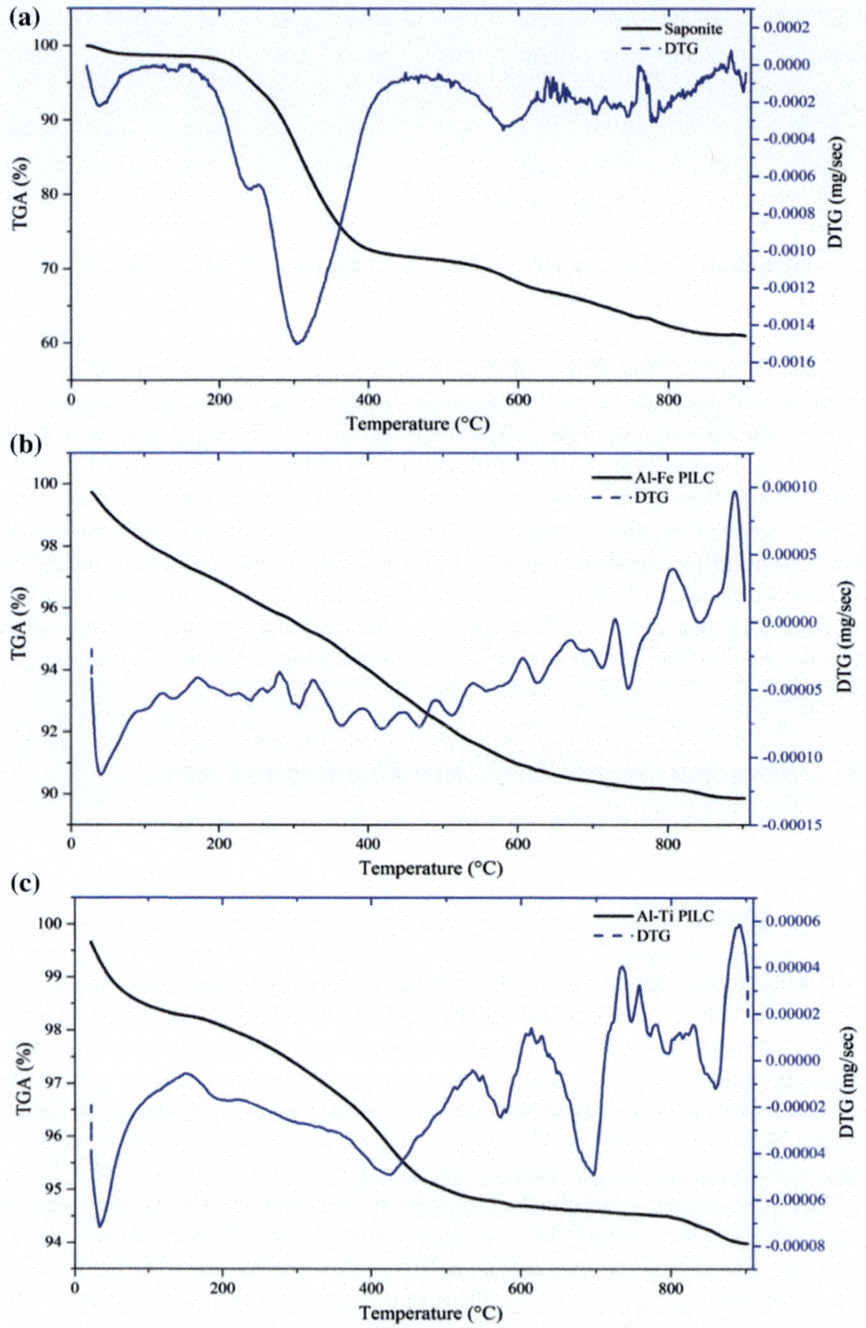

Fig. 5 TGA and DTG curves for **a** saponite clay, **b** Al–Fe PILC and Al–Ti PILC

The DTG analysis of the catalysts showed the rate of weight loss (mg per sec) with temperature. Saponite clay exhibited decomposition at 250 °C and 570 °C with the rate of weight loss 0.6×10^{-3} and 0.2×10^{-3} mg/s, respectively. However, the DTG curves of Al–Fe PILC shows the rate of weight loss of 0.1×10^{-3} mg/s at 370 °C and 0.1×10^{-4} mg/s at 720 °C. Al–Ti PILC also showed peaks with a lower rate of weight loss of 0.4×10^{-4} mg/s and 0.5×10^{-5} mg/s at 430 °C and 650 °C, respectively, as compared to saponite clay. Thus, the TG/DTG studies showed higher thermal stability of PILC's as compared to saponite clay.

4 Conclusions

The intercalation of saponite clay was done with the mixture of polycations of Al–Fe and Al–Ti. XRD analysis showed the increase in basal spacing of saponite (11.77 Å) to 18.37 Å in Al–Fe PILC and 16.99 Å in Al–Ti PILC after the pillaring process. BET surface area analysis showed an increase in surface area and pore volume from 33.96 m^2/g of saponite clay to 94.44 m^2/g in Al–Fe PILC and 73.78 m^2/g in Al–Ti PILC, respectively.

The elemental composition analysis results also confirmed the pillaring by showing increased contents of Fe_2O_3 and TiO_2. The TG/DTG characterization showed Al–Ti PILC as a more stable catalyst than Al–Fe PILC catalyst in spite of Al–Fe PILC showing greater surface area and basal spacing. Thus, FTIR, XRD, N_2 adsorption–desorption isotherms, SEM, WD-XRF characterization techniques confirmed the successful pillaring of the Al, Fe, and Ti metal cations into the layers of saponite clay.

References

Ayodele OB, Hameed BH (2013a) Synthesis of copper pillared bentonite ferrioxalate catalyst for degradation of 4-nitrophenol in visible light assisted Fenton process. J Ind Eng Chem 19:966–974

Ayodele OB, Hameed BH (2013b) Development of kaolinite supported ferric oxalate heterogeneous catalyst for degradation of 4-nitrophenol in photo-Fenton process. Appl Clay Sci 83:171–181

Belver C, Bañares-Muñoz MA, Vicente MA (2004) Fe-saponite pillared and impregnated catalysts: I. Preparation and characterization. Appl Catal B 50:101–112

Cupic ZD (2008) Synthesis, characterization and application of Al, Fe-pillared clays. In Proceedings of the tenth annual conference of the materials research society of Serbia

Eren E, Afsin B (2008) An investigation of Cu (II) adsorption by raw and acid-activated bentonite: a combined potentiometric, thermodynamic, XRD, IR, DTA study. J Hazard Mater 151:682–691

Fatimah I (2014) Preparation of ZrO2/Al2O3-montmorillonite composite as catalyst for phenol hydroxylation. J Adv Res 5:663–670

Galeano LA, Gil A, Vicente MA (2010) Effect of the atomic active metal ratio in Al/Fe-, Al/Cu-and Al/(Fe–Cu)-intercalating solutions on the physicochemical properties and catalytic activity of pillared clays in the CWPO of methyl orange. Appl Catal B 100:271–281

Guerra DL, Airoldi C, Lemos VP, Angélica RS (2008) Adsorptive, thermodynamic and kinetic performances of Al/Ti and Al/Zr-pillared clays from the Brazilian Amazon region for zinc cation removal. J Hazard Mater 155:230–242

Kaneko T, Shimotsuma H, Kajikawa M, TAG ME, Kodama T, Kitayama Y (2001) Synthesis and photocatalytic activity of titania pillared clays. J Porous Mater 8:295–301

Kim SC, Lee DK (2004) Preparation of Al–Cu pillared clay catalysts for the catalytic wet oxidation of reactive dyes. Catal Today 97:153–158

Kooli F, Jones W (1998) Al and Zr pillared acid-activated saponite clays: characterization and properties. J Mater Chem 8:2119–2124

Li K, Lei J, Yuan G, Weerachanchai P, Wang JY, Zhao J, Yang Y (2017) Fe, Ti, Zr and Al-pillared clays for efficient catalytic pyrolysis of mixed plastics. Chem Eng J 317:800–809

Madejová J (2003) FTIR techniques in clay mineral studies. Vib Spectrosc 31:1–10

Mata G, Trujillano R, Vicente MA, Belver C, Fernández-García M, Korili SA, Gil A (2007) Chromium–saponite clay catalysts: preparation, characterization and catalytic performance in propene oxidation. Appl Catal A 327:1–12

Minz S, Garg S, Gupta R (2018a) Catalytic wet peroxide oxidation of 4-nitrophenol over Al–Fe, Al–Cu and Al–Cu–Fe pillared clays. Indian Chem Eng 60(1):16–36

Minz S, Garg S, Gupta R (2018b) Catalytic wet peroxide oxidation of 4-Nitrophenol over Al–Fe PILC: kinetic study using Fermi's equation and mechanistic pathways based on TOC reduction. Chem Eng Commun 205(5):667–669

Mojovic Z, Bankovic P, Milutinovic-Nikolic A, Dostanic J, Jovic-Jovicic N, Jovanovic D (2009) Al, Cu pillared clays as catalysts in environmental protection. Chem Eng J 154:149–155

Shillito LM, Almond MJ, Wicks K, Marshall LJR, Matthews W (2009) The use of FT-IR as a screening technique for organic residue analysis of archaeological samples. Spectrochim Acta, Part A 72:120–125

Sum OS, Feng J, Hub X, Yue PL (2005) Photo-assisted fenton mineralization of an azo-dye acid black 1 using a modified laponite clay-based Fe nanocomposite as a heterogeneous catalyst. Top Catal 33:233–242

Toranzo R, Vicente MA, Bañares-Muñoz MA (1997) Pillaring of a saponite with aluminum–chromium oligomers. Characterization of the solids obtained. Chem Mater 9:1829–1836

Vicente MA, Bañares-Muñoz MA, Gandıa LM, Gil A (2001) On the structural changes of a saponite intercalated with various polycations upon thermal treatments. Appl Catal A 217:191–204

Vogels RJMJ, Kloprogge JT, Geus JW (2005a) Synthesis and characterisation of boron and gallium substituted saponite clays below 100 °C at one atmosphere. Microporous Mesoporous Mater 77:159–165

Vogels RJMJ, Kloprogge JT, Geus JW (2005b) Synthesis and characterization of saponite clays. Am Mineral 90:931–944

Wang S, Dong Y, He M, Chen L, Yu X (2009) Characterization of GMZ bentonite and its application in the adsorption of Pb (II) from aqueous solutions. Appl Clay Sci 43:164–171

Zhou CH, Tong D, Li X (2010) Synthetic hectorite: preparation, pillaring and applications in catalysis. Pillared clays and related catalysts. Springer, New York, pp 67–97

Zhou S, Zhang C, Hu X, Wang Y, Xu R, Xia C, Zhang H, Song Z (2014) Catalytic wet peroxide oxidation of 4-chlorophenol over Al–Fe, Al–Cu, and Al–Fe–Cu pillared clays: Sensitivity, kinetics and mechanism. Appl Clay Sci 95:275–283

Performance-Based Seismic Evaluation of Multi-storey R.C.C. Building with Addition of Shear Wall

Sumit and S. M. Gupta

Abstract It has been evident from the recent history of earthquake that the improper and inadequate design of structures may cause great damage. Many existing structures are seismically deficient as the engineers and designers are not well aware of the seismic behaviour of a structure. Hence, there is a need to determine the seismic response of such buildings for designing earthquake structures by conducting a seismic analysis of structures. Performance-based seismic design is an elastic design methodology which was carried out by considering the probable performance of a building under different ground motions. In this context, pushover analysis which is an iterative procedure should be looked upon as an alternative for the orthodox analysis procedures. The principal objective of this paper is to evaluate the performance of (G+10) and (G+20) storey RCC building with and without shear wall. Nonlinear static pushover analysis was performed by using SAP 2000, which is a product of computers and structures international. On the basis of the result, it was found that both buildings satisfy the drift criteria as recommended in IS 1893:2002 (part-1).

Keywords Performance-based seismic design · Pushover analysis · Base shear · SAP 2000

1 Introduction

Earthquakes are known as one of the most unpredictable and devastating of all natural disasters. Therefore, it becomes more challenging to prevent loss of human lives and destruction of properties (as structures are designed without considering earthquake forces) because of unpredictable nature of occurrence of the earthquakes (Monish and Karuna 2015). The existing research on Indian subcontinent shows that more than 60% of the country's land is vulnerable to all kinds of earthquakes (Pratap and Neelima 2015). A recent study (Chaudhari and Dhoot Gopal 2016) mentioned that the structure design criterion regarding earthquake forces is not as per

Sumit (✉) · S. M. Gupta
Department of Civil Engineering, NIT Kurukshetra, Kurukshetra, Haryana, India
e-mail: ekhatkar297@gmail.com

A. K. Agnihotri et al. (eds.), *Sustainable Engineering*, Lecture Notes in Civil Engineering 30, https://doi.org/10.1007/978-981-13-6717-5_6

the expectation. Seismic design codes are traditionally based on the force-based approach wherein structures are designed with a certain minimum lateral strength. However, it has been observed that such an approach, which relates to the elastic response, does not produce a consistent inelastic response in terms of the amount and distribution of damage in structural elements. Priestley (2000) concluded that the force-based approach provides widely different results if the performance of non-structural elements is considered. In view of the above, the displacement based approach (i.e. performance-based design approach) was identified as the suitable one (Satish Kumar and Venkateswarlu 2008a, b). A Performance-Based Seismic Design (PBSD) method is aimed at controlling the structural damage based on precise estimations of proper responded parameters. PBSD method evaluates the performance of a building frame for any seismic hazard (Khedkar 2014). PBSD demands the use of nonlinear analysis procedures to evaluate the response of structures under lateral loads. The nonlinear time history analysis is the most accurate but requires much computational effort, time and cost. Thus, the use of nonlinear static analysis procedure known as the pushover analysis has been proposed. The static pushover analysis not only provides information on strength capacity of the structure but also provides vital information on ductility as well as an insight on the progressive mode of failure of the structure (Akshara 2015). This paper focuses on the performance evaluation of (G+10) and (G+20) storey RCC building with and without shear wall by using nonlinear static pushover analysis. The target displacement is assumed as 4% of the height of the building. Modelling of RC shear walls has been done as mid-pier frame element. The shear wall was provided at central bay in X-direction of both buildings. Results of both the building are compared in terms of capacity curve, base shear, storey drift, etc., in X-direction only.

2 Methodology

Performance-based seismic evaluation of buildings is carried out using pushover analysis (Nikam and Kalurkar 2016). It is a static nonlinear procedure in which the magnitude of the structural loading is incrementally increased (Choudhary and Wadia 2014; Singh and Poluraju 2017). With the monotonic increment in the magnitude of the loading, weak links and the failure modes of the structure are found (Pednekar et al. 2015; Lakshmi et al. 2014). As the load and displacement increases, the element (beams, columns, etc.) begin to yield and deform elastically. Pushover analysis requires capacity curve, demand curve and performance objective (Pambhar 2012). Capacity curve, which is roof displacement versus base shear graph, is obtained from the nonlinear static analysis (Parishith and Preetha 2017). Capacity curve depends upon stiffness and deformation capacity of the structure. The demand curve is estimated by reducing 5% damped design spectrum by special reduction method. Capacity curve and demand curve is transformed into Acceleration Displacement Response Spectra (ADRS) which is acceleration versus displacement graph. The transformed

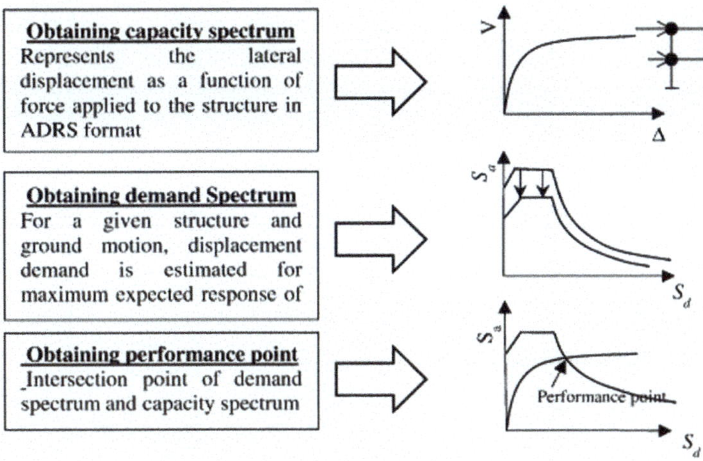

Fig. 1 Capacity curve, demand curve and ADRS

capacity curve is called the capacity spectrum. Capacity curve, demand curve and ADRS are shown in Fig. 1.

The nonlinear properties in beams and columns have to be introduced for analysis which can be introduced using predefined hinge properties in Sap 2000. The flexural default hinges (M3) to the beams at two ends was assigned. The interacting (P-M2-M3) frame hinges type a coupled hinge property was also assigned for all the columns at upper and lower ends. Sap 2000 takes auto-hinge property from ASCE 41-13 tables. All buildings are designed as per IS 456:2000, for adequate main and shear reinforcements, corresponding to the ultimate moment capacity level. Flexural plastic hinges will be developed along with the predicted values of ultimate moment capacity when there is no prior failure in shear. Two pushover loads push X and push Y were defined as displacement controlled. The maximum target displacement of the structure was kept at 4% of the height of the building. Both pushover loads were run from final conditions of dead load.

2.1 Description of Analysed Buildings

In this paper (G+10) and (G+20) storeys buildings are preliminary designed by Sap 2000v.19 for seismic zone V. Buildings were designed as Special Moment Resisting Frame (SMRF) as per IS 1893-2002. Plan dimensions of buildings kept as 20 m × 15 m with five bays in X-direction and three bays in Y-direction (Fig. 2).

The storey height for ground floor was kept as 3.5 m, while the height of a typical storey was kept uniform for all storeys, i.e. 3 m. Slab thickness was taken as 125 mm for all floors. Shear wall is modelled as mid-pier frame element by taking the thickness

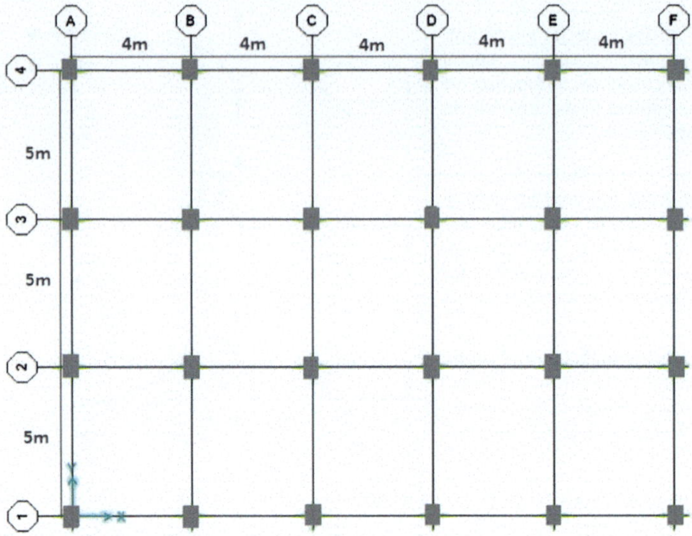

Fig. 2 Plan of building

of 200 mm. As the building is symmetric, the shear wall is provided in central bay of longer side of building frame. M25 grade of concrete and Fe415 grade of steel are used in all members.

2.2 *Loads Considered*

Following loads are considered for analysis of all buildings:

Dead Load:
The thickness of slab is taken as 125 mm; hence, the dead load of slab is taken as $125 \times 24 = 3$ kN/m^2 on all floors and 16.8 kN/m wall load on all beams.

Live load: 3.5 kN/m^2 on all floors. As per IS 875 (Part 2)-1987.

Earthquake load: as per IS 1893:2002 (Part1).

Type of Soil: Medium Soil (Type II) as per IS 1893:(2002).

Importance factor, I: 1.

The time period for both buildings is program calculated as per IS 1893:2002) was used to calculate the earthquake force for both buildings. For earthquake force analysis in Sap 2000, the following loads were considered in the mass source: dead load plus 0.5 times the live load.

The building is designed according to IS 456:2000 using SAP 2000 v.19 considering live load, dead load and earthquake load in both directions X and Y with minimum eccentricity. The dimensions of beams and column sand designed reinforcement details of columns are mentioned in Tables 1 and 2. Columns A1, A2, A3,

Table 1 (G+10) structural details as per analysis and design in SAP 2000

Storey	Height (m)	Beam size (B × D) (mm)	Column size (B × D) (mm)	Reinforcement percentage in column (%)					
				A1, A4, F1, F4	A2, A3, F2, F3	B1, B4, E1, E4	B2, B3, E2, E3	C1, C4, D1, D4	C2, C3, D2, D3
11th	33.5	230 × 375	375 × 375	1.59	0.95	1.29	0.8	1.13	0.8
10th	30.5	230 × 375	375 × 375	0.8	0.8	0.85	0.8	0.8	0.8
9th	27.5	275 × 425	425 × 425	0.8	0.89	0.92	0.98	0.8	1.04
8th	24.5	275 × 425	425 × 425	0.8	1.23	1.26	1.78	1.11	1.82
7th	21.5	275 × 425	425 × 425	0.8	1.8	1.75	2.72	1.67	2.78
6th	18.5	300 × 450	450 × 450	0.8	1.74	1.66	2.72	1.58	2.91
5th	15.5	300 × 450	450 × 450	1.03	2.21	2.01	3.65	2.00	4.05
4th	12.5	300 × 450	450 × 450	1.66	2.99	2.69	4.69	2.79	5.19
3rd	9.5	350 × 500	500 × 500	0.86	2.14	1.77	4.01	1.86	4.39
2nd	6.5	350 × 500	500 × 500	1.31	2.73	2.34	4.92	2.46	5.42
1st	3.5	350 × 500	500 × 500	2.26	3.77	3.23	5.85	3.37	4.66

Table 2 (G+20) details as per analysis and design in SAP 2009

Storey	Height (m)	Beam size (B × D) (mm)	Column size (B × D) (mm)	Reinforcement percentage in column (%)					
				A1, A4, F1, F4	A2, A3, F2, F3	B1, B4, E1, E4	B2, B3, E2, E3	C1, C4, D1, D4	C2, C3, D2, D3
21st	63.5	230 × 400	400 × 400	1.59	1.53	1.48	1.15	1.57	0.9
20th	60.5	230 × 400	400 × 400	1.04	1.49	1.32	1.41	1.36	1.35
19th	57.5	230 × 400	400 × 400	1.4	2.39	2.1	2.55	2.15	2.56
18th	54.5	300 × 450	450 × 450	0.97	2.23	1.83	2.43	1.89	2.43

(continued)

Table 2 (continued)

Storey	Height (m)	Beam size (B × D) (mm)	Column size (B × D) (mm)	Reinforcement percentage in column (%)					
				A1, A4, F1, F4	A2, A3, F2, F3	B1, B4, E1, E4	B2, B3, E2, E3	C1, C4, D1, D4	C2, C3, D2, D3
17th	51.5	300 × 450	450 × 450	1.39	2.64	2.31	2.93	2.36	3.00
16th	48.5	300 × 450	450 × 450	1.95	3.49	2.97	3.99	3.02	4.1
15th	45.5	325 × 500	500 × 500	1.2	2.65	2.25	3.05	2.32	3.19
14th	42.5	325 × 500	500 × 500	1.64	2.9	2.53	3.51	2.58	3.66
13th	39.5	325 × 500	500 × 500	2.51	3.79	3.24	4.36	3.31	4.54
12th	36.5	375 × 550	550 × 550	1.58	2.69	2.28	3.22	2.35	3.4
11th	33.5	375 × 550	550 × 550	2.06	2.96	2.63	3.61	2.78	3.8
10th	30.5	375 × 550	550 × 550	2.72	3.58	3.3	4.19	3.51	4.37
9th	27.5	400 × 600	600 × 600	1.91	2.59	2.48	3.15	2.65	3.38
8th	24.5	400 × 600	600 × 600	2.38	3.00	2.89	3.5	3.07	3.82
7th	21.5	400 × 600	600 × 600	2.88	3.66	3.51	4.00	3.73	4.33
6th	18.5	450 × 650	650 × 650	2.18	2.94	2.74	3.24	2.93	3.68
5th	15.5	450 × 650	650 × 650	2.57	3.47	3.17	3.75	3.43	4.16
4th	12.5	450 × 650	650 × 650	3.00	4.01	3.74	4.22	3.97	4.67
3rd	9.5	500 × 750	750 × 750	1.61	2.51	2.19	2.66	2.44	2.99
2nd	6.5	500 × 750	750 × 750	1.91	2.9	2.59	3.03	2.81	3.47
1st	3.5	500 × 750	750 × 750	2.56	3.47	3.12	3.55	3.35	3.93

A4, B1, B2, B3, B4, C1, C2, C3, C4, D1, D2, D3, D4, E1, E2, E3, E4, and F1, F2, F3, F4 can be identified from Fig. 2.

3 Results and Discussion

The seismic performance of a building is evaluated in terms of capacity curve, performance point and plastic hinge formation and storey drift. A detail discussion has been made below.

3.1 Capacity Curve

Pushover analysis gives an idea of the maximum base shear that the structure can resist at the time of the seismic load. For regular buildings, it can also give a rough idea about the global stiffness of the building. Capacity curve obtained after pushover analysis is plotted in Figs. 3 and 4. It is observed that buildings with shear wall are capable to resist larger base shear as compared to bare frame buildings. Maximum displacement was also observed in bare frame building.

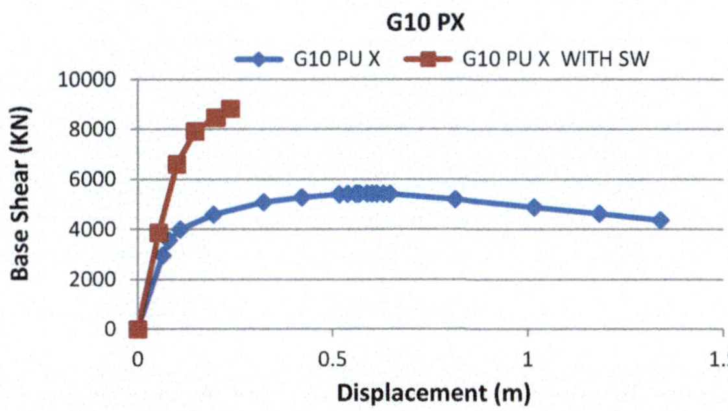

Fig. 3 Capacity curves of (G+10) storey buildings for pushover in X-direction

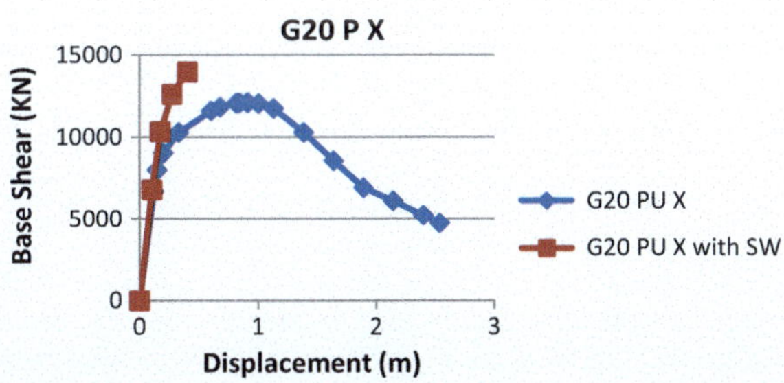

Fig. 4 Capacity curves of (G+20) storey buildings for pushover in X-direction

3.2 Status of Hinges at Ultimate Stage

Status of plastic hinges at the ultimate stage for (G+10) and (G+20) is shown in Figs. 5 and 6.

Table 3 is exhibiting the hinge status at the ultimate stage. It was observed that in (G+10) building without shear wall 98 hinges was formed between immediate occupancy and life safety range, while 242 hinges were formed beyond collapse range. On the other hand, around 14 hinges are formed between immediate occupancy and life safety range if the analysis is carried out by adding shear wall. So, a considerable decrement was observed in hinge formation by addition of shear wall and building lies between immediate occupancy and life safety range.

In (G+20) building without shear wall, 96 hinges are formed between immediate occupancy and life safety range and 424 hinges are formed beyond collapse prevention state, but with shear wall, addition hinges are not formed beyond immediate occupancy state. Building lies at immediate occupancy state; thus, both buildings performed well up to design seismic load, and it can be easily said that performance of high rise building is more affected than medium rise building with the addition of shear wall.

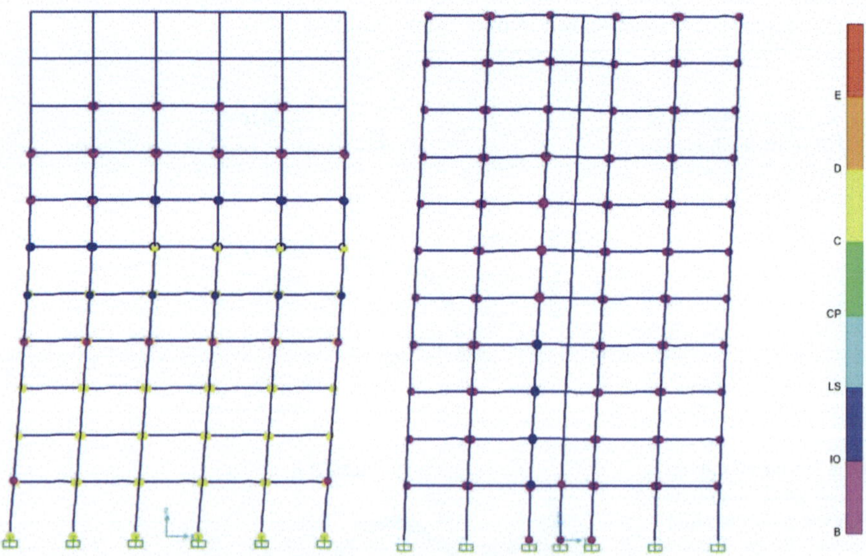

Fig. 5 Status of hinges at ultimate stage in pushover X case for (G+10) storey building

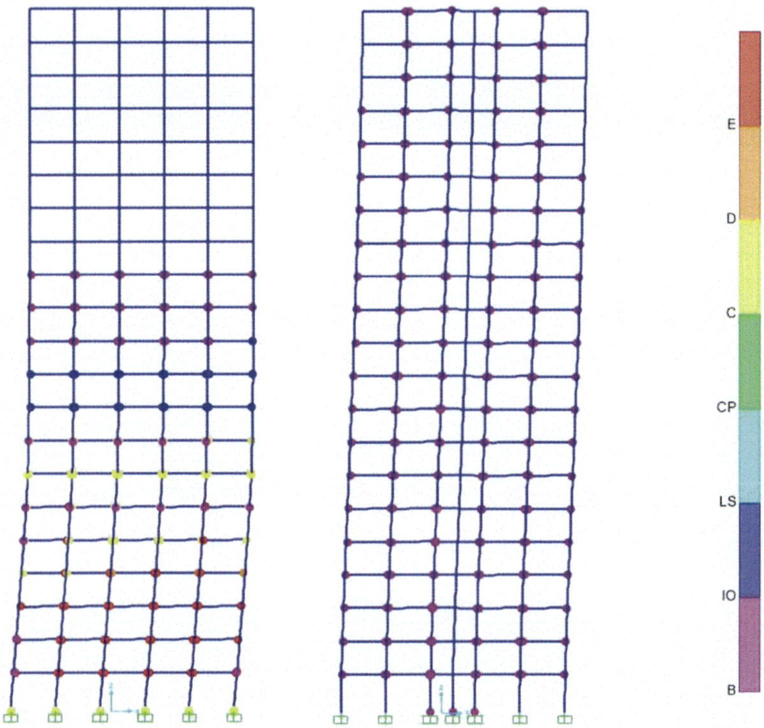

Fig. 6 Status of hinges at ultimate stage in pushover X case for (G+20) storey building

Table 3 Status of hinges at ultimate stage

Building	Monitored displacement (mm)	Base force (KN)	O-IO (Nos.)	IO-LS (Nos.)	LS-CP (Nos.)	>CP (Nos.)	Total (Nos.)
G10 P X	1340	4354.051	1024	98	0	242	1364
G10 P X SW	238.41	8822.869	1394	14	0	0	1408
G20 P X	2540	4998.35	2084	96	0	424	2604
G20 P X SW	395	11937.755	2688	0	0	0	2688

O Operational level, *IO* Immediate occupancy level, *LS* Life safety level, *CP* Collapse prevention level

3.3 *Capacity Demand Spectrum*

Capacity demand spectrum is the intersection of capacity curve and demand curve. It is the representation of structures ability to resist the seismic demand. The point of intersection of capacity spectrum and demand spectrum gives the required performance point of a particular structure. Figure 7 and Figure 8 show the capacity

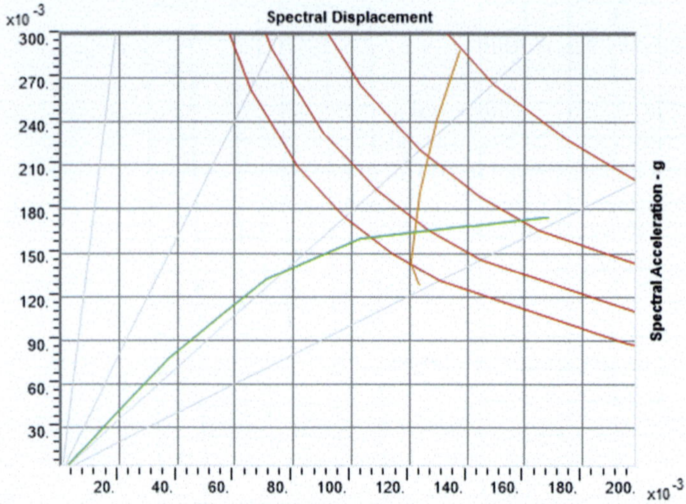

Fig. 7 Capacity spectrum curve as per ATC-40 of (G+10) story building

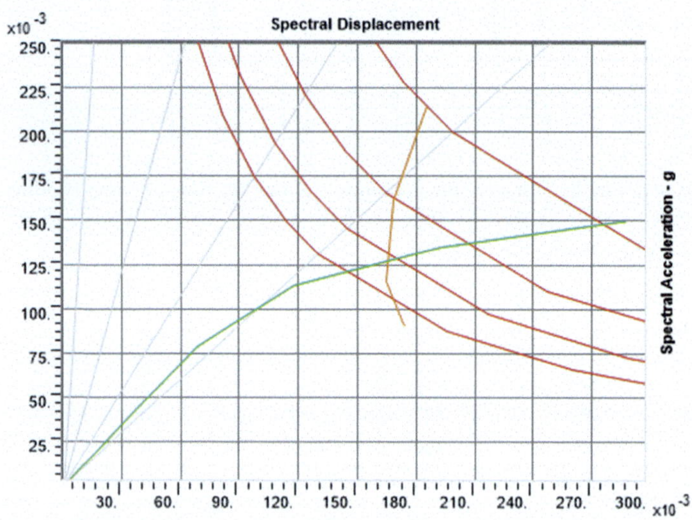

Fig. 8 Capacity spectrum curve as per ATC-40 of (G+20) story building

spectrum curve as per ATC-40 of (G+10) and (G+20) storey buildings including shear wall, respectively. Table 4 shows the performance point in terms of different structural parameter like base shear (V), displacement (D), spectral acceleration (Sa), spectral displacement (Sd), effective time period (Teff), and effective damping (βeff).

Table 4 Performance point in terms of (V, D) (Sa, Sd) (Teff, βeff)

Performance point	G10 P	G10 P SW	G20 P	G20 P SW
Performance point (V, D)	(4671.315, 0.216)	(8212.432, 0.173)	(9458.059, 0.245)	(11804.853, 0.237)
Performance point (Sa, Sd)	(0.083, 0.176)	(0.165, 0.122)	(0.094, 0.177)	(0.127, 0.166)
Performance point (Teff, βeff)	(2.906, 0.24)	(1.721, 0.16)	(2.717, 0.198)	(2.273, 0.147)

3.4 Storey Drift

Storey drift of (G+10) and (G+20) storey buildings for push x case with and without shear wall designed for dead load, live load and the earthquake load is presented in Figs. 9 and 10. All buildings were satisfying the strength requirements as well as satisfy inter-storey drift criteria as per IS 1893:2002 clause no. 7.11.1, maximum permissible inter-storey drift is 0.004 times the storey height.

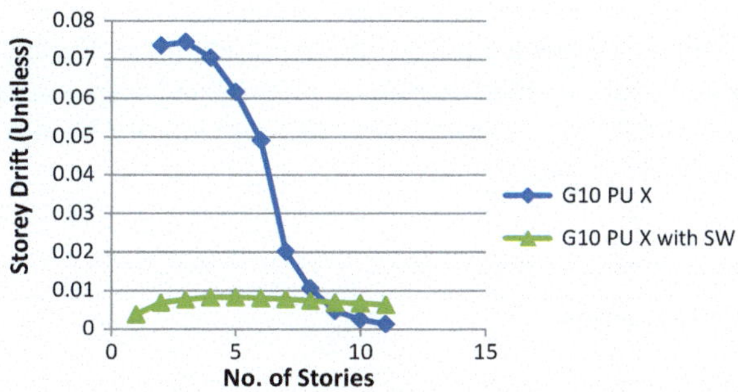

Fig. 9 Storey drift in X-direction for (G+10) storey building

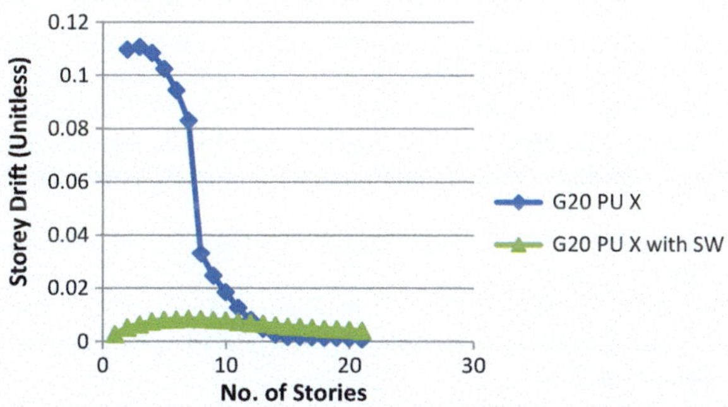

Fig. 10 Storey drift in X-direction for (G+20) storey building

4 Conclusions

In the present study, on the basis of several analyses, a number of conclusions have been made. For the present study, seismic performance evaluation was done with and without shear wall for two types of buildings, namely, (G+10) and (G+20) storeys buildings. Both buildings were designed for seismic zone V of India as per IS 1839:2002 and IS 456:2000. Pushover analysis of buildings was carried out using Sap 2000 nonlinear software tool. Based on the present study, the following conclusion can be drawn:

1. From pushover analysis, it is seen that the capacity of a structure increases with the introduction of shear walls in both G10 and G20 buildings.
2. There is a considerable reduction in hinge formation between immediate occupancy and life safety range, and there is no hinge formation above life safety range with the addition of shear wall.
3. As per IS 1893:2002, the allowable storey drift requirement is 0.004 times the storey height, which is satisfied by both G10 and G20 buildings.

It is concluded that pushover analysis gives a better view of building performance under seismic load and its failure mechanism. Further other analysis methods like time history method, pushover method and response spectrum analysis etc. Can be adopted for future study and other irregularity locations can be adopted to analyse the effect of same.

References

Akshara SP (2015) Performance based seismic evaluation of multi-storeyed reinforced concrete building using pushover analysis. IRJET 2(3). ISSN 2395-0056

ASCE (2000) Prestandard and commentary for the seismic rehabilitation of buildings. FEMA 356 report, prepared by the American Society of Civil Engineers for the Federal Emergency Management Agency, Washington, D.C. ATC, 1997a, NEHRP guidelines for the seismic rehabilitation of buildings, FEMA 273 report, prepared by the Applied Technology Council for the Building Seismic Safety Council, published by the Federal Emergency Management Agency, Washington, D.C

Chaudhari DJ, Dhoot Gopal O (2016) Performance based design of reinforced concrete building. OJCE 6, 188-194

Choudhary N, Wadia M (2014) Pushover analysis of R.C. frame building with shear wall. IOSR J Mech Civil Eng (IOSR-JMCE) 11(2):9–13. e-ISSN 2278-1684, p-ISSN 2320-334X. Ver. V (Mar–Apr 2014)

IS 1893 (Part 1):2002 Criteria for earthquake resisting design of structures—Indian Standard Code of Practice, Bureau of Indian Standards, New Delhi, India

IS 456:2000 Plain and reinforced concrete—Indian Standard Code of Practice, Bureau of Indian Standards, New Delhi, India

Kevadkar MD, Kodag PB (2013) (IJMER) Int J Modern Eng Res 3(3):1428–1434

Khedkar AS et al (2014) Performance based seismic design of reinforced concrete moment resistant frame with vertical setback. IJERT 3(2). ISSN 2278-0181

Lakshmi KO et al (2014) ISOI J Eng Comput Sci 1(1):7–17

Monish S, Karuna S (2015) A study on seismic performance of high rise irregular RC framed buildings. IJRET Int J Res Eng Technol. eISSN 2319-1163/pISSN 2321-7308

Nikam NM, Kalurkar LG (2016) Pushover analysis of building with shear wall. Int J Eng Sci Comput 6(8):2916–2918

Pambhar JD (2012) Performance based pushover analysis of R.C.C. frames. Int J Adv Eng Res Stud IJAERS 1/(3):329–333. e-ISSN 2249-8974

Parishith J, Preetha V (2017) Pushover analysis of RC frame buildings with shear wall a review. IJSRD Int J Sci Res Dev 4(12) (2017). ISSN (online) 2321-0613

Pednekar SC et al (2015) Pushover analysis of reinforced concrete structures. Int J Comput Appl 0975–8887. In: International conference on quality up gradation in engineering, science and technology (ICQUEST2015)

Poluraju P, Nageswara Rao P (2011) Pushover analysis of reinforced concrete frame structure using SAP 2000. Int J Earth Sci Eng 4(6):684–690. ISSN 0974-5904

Prakash V (2004) Whither performance-based engineering in India. ISET J. Earthq Technol 41(1):201–222. Paper no 447

Pratap Y, Neelima PVS (2015) Performance based design: case study. IJEMR 5. ISSN (Online) 2250-0758, ISSN (Print) 2394-6962

Priestely MJN (2000) Performance based seismic design. In: 12th world conference on earthquake engineering, paper no. 2831, New Zealand

Satish Kumar SR, Venkateswarlu G (2008) Performance based design of reinforced concrete plane frames. In: 14th world conference on earthquake engineering, Beijing, China

Satish Kumar SR, Venkateswarlu G (2008) Performance based seismic design of reinforced concrete plane frames. In: The 14th world conference on earth quake engineering, Beijing, China, 12–17 Oct 2008

Singh RR, Poluraju P (2017) Performance based seismic design of RC framed building. Int J Civ Eng Technol (IJCIET). ISSN (Print) 0976-6308, ISSN (Online) 0976-6316

Synthesis and Characterization of the Graft Copolymers of Starch for the Application in Packaging Films

Sangeeta Garg, Aanchal Mittal and Anshuman Premi

Abstract In this study, starch was grafted with oleic acid in the presence of the initiators such as potassium persulphate and ammonium persulphate. The mixture of water/dimethyl sulphoxide (DMSO) (1:3) was used as solvent for the reaction. Grafted starch was characterized by FT-IR, X-RD, SEM and TG/DTA. Analysis of the crystallinity of the samples was done by using the X-ray diffraction technique. The reduced area under the crystalline peaks of the grafted starch showed a reduction in the crystallinity of starch after grafting as compared to native starch. SEM images of native starch and grafted starch showed that grafted starch had a rougher surface than native starch. Thermal analysis of the starch and grafted starch was studied as thermogravimetric analysis (TGA), derivative thermogravimetric analysis (DTG) and differential thermal analysis (DTA). Grafted starch was more thermally stable as compared to the native starch. It was observed that graft copolymerization of starch with oleic acid led to decrease in the crystallinity of starch and increase in the roughness on the surface of starch which may improve the compatibility of grafted starch with other biopolymers. These grafted copolymers of starch can be used with other biopolymers for the application in packaging films.

Keywords Graft copolymerization · Initiators, blend films · Grafted starch

1 Introduction

Plastics are synthetic organic polymers of high molecular mass derived from petrochemical-based products. These synthetic petrochemical-based polymers are widely used in a variety of packaging materials and in various industries such as automotive, aerospace, etc. Dozens of different synthetic plastics, approximately 200

S. Garg (✉) · A. Mittal · A. Premi
Department of Chemical Engineering, Dr B. R. Ambedkar National Institute of Technology, Jalandhar 144011, Punjab, India
e-mail: gargs@nitj.ac.in

A. Mittal
e-mail: aanchalm.15.phd@nitj.ac.in

A. K. Agnihotri et al. (eds.), *Sustainable Engineering*, Lecture Notes in Civil Engineering 30, https://doi.org/10.1007/978-981-13-6717-5_7

million tonnes per year are produced throughout the world (Zhang and Sun 2004). But there are many problems associated with plastic such as toxicity, long life, costly, releases many toxic emissions during processing. These plastics become a major source of waste after their use because of their poor biodegradability (Rhim and Ng 2007). The impact of these discarded plastics on environment has directed research towards the use of natural-based raw materials such as starch, cellulosic materials and proteins for the development of plastics that degrade more rapidly in the environment. Biopolymers are sustainable, renewable, completely biodegradable, cheap and easily available as they are natural products. Among all biopolymers, starch is considered to be the most promising biopolymer as it is easily available, easily biodegradable and is of low cost. Starch is made up of two components linear amylase and a highly branched amylopectin; amylose mainly is responsible for the swelling of granules. Importance of starch is increasing day by day due to its completely biodegradable nature (Araújo et al. 2004), renewability (Zhang and Sun 2004), availability and low cost. Starch is completely biodegradable in a wide variety of environments and can be hydrolysed into glucose by microorganism or enzymes, and then metabolized into carbon dioxide and water (Primarini and Ohta 2000). Starch can be blended with other biopolymers or other synthetic polymers to produce fully biodegradable or partially degradable composite films with improved packaging properties (Chai et al. 2009; Azahari et al. 2011; Negim et al. 2014; Canché-Escamilla et al. 2011).

Polyvinyl alcohol (PVA) is one of the water-soluble and biodegradable synthetic polymers which has excellent mechanical properties and has good compatibility with starch (Chiellini et al. 2003). After blending of PVA with starch, the hydroxyl groups present on the surface of PVA and starch form intermolecular and intramolecular hydrogen bonding together (Lani et al. 2014). But the main disadvantage is their lower water resistance property which can be improved by further chemical modifications of starch (Tang and Alavi 2011).

Chemical modifications of starch such as etherification, esterification, oxidation of the available hydroxyl groups, etc., were done to improve the film forming capacity of the starch (Garg and Jana 2011). Grafting is an important technique to modify surface characteristics of polymers by imparting specific functional groups to the polymer backbone. The stress transfer between the two phases can be significantly improved by the graft copolymerization method. In this study, starch was graft copolymerized using oleic acid. The grafted starch was blended with PVA and cast films were prepared. The grafted was characterized using Fourier transform infrared spectroscopy (FT-IR), X-ray diffraction (X-RD), scanning electron microscopy (SEM) and thermal analysis (TG/DTA/DTG). Mechanical properties and water uptake (%) of the films before and after grafting were evaluated.

2 Methodology

2.1 Materials

Starch (corn starch) (S) and polyvinyl alcohol (PVA) used in the study were obtained from S.D. Fine Chemicals Limited, Mumbai, India. Oleic acid and dimethyl sulphoxide (DMSO) used for grafting of starch were obtained from Loba Chemie Pvt. Ltd., Mumbai, India. All the chemicals were used as received.

2.2 Methods

2.2.1 Grafting of Starch

Graft copolymerization of starch was performed by soaking starch overnight in the water/DMSO (1:3) solvent mixture. Initiators potassium persulphate and ferrous ammonium sulphate (KPS-FAS) were added in the reaction media followed by oleic acid which was added dropwise to the reaction flask. The reaction was carried out at 50 °C for 150 min with continuous stirring. The extraction of the graft copolymers of starch was done by soxhlet extractor using ethanol. The sample was dried in a hot air oven at 50 °C for 24 h. Percentage graft yield was calculated as per the method reported earlier (Mittal et al. 2016; Kohli et al. 2017)

$$Perecentage\, graft\ yield(\%) = \frac{W_g - W_i}{W_i} \times 100$$

where W_g is the weight of the starch after grafting and W_i is the weight of the starch initially.

2.2.2 Preparation of Blend Films

PVA/starch (PVA/S) films and PVA/grafted starch (PVA/GS) films were prepared by varying the PVA and starch ratio. PVA and starch were dissolved in hot water separately. Gelatinized starch was mixed with PVA solution in a flask and the reaction mixture was continuously stirred for 45 min using mechanical stirrer at 50 °C to form a homogeneous solution. Glycerol was added to the solution with continuous stirring and the total amount of polymeric mixture was 100 g. After 45 min, the suspension was poured onto the glass petri dish and dried in hot air for 24 h. The composition of the different blend films prepared is shown in Table 1.

Table 1 Composition of PVA/S and PVA/GS blend films

Films	PVA	Starch	Grafted starch
PVA	10	0	
PVA/S 3:1	7.5	2.5	
PVA/S 1:1	5	5	
PVA/S 1:3	2.5	7.5	
PVA/GS 3:1	7.5		2.5
PVA/GS 1:1	5		5
PVA/GS 1:3	2.5		7.5

2.3 *Characterization*

2.3.1 Fourier Transforms Infrared (FT-IR) Spectroscopy

Native starch and graft copolymers of starch with oleic acid were characterized by FT-IR spectrophotometer. FT-IR spectra were recorded by Perkin-Elmer Fourier transform infrared (FT-IR) spectrophotometer in the range from 4000 to 400 cm^{-1} at a resolution of 4 cm^{-1} using KBr pellets.

2.3.2 X-Ray Diffraction Studies

X-ray diffraction studies were carried out in order to know about the crystalline and amorphous regions of the starch after modification. X-ray diffractogram of starch and grafted starch were recorded by X-ray diffractometer (XPERT-PRO PW 3064). The radiation used is Cu K-alpha whereas nickel metal is used as beta filter. The range of diffraction angle was 5–40° (2θ).

2.3.3 Scanning Electron Microscopy

Surface morphology of starch and grafted starch was observed with a scanning electron microscope (LEO 435 VP). The range of accelerating voltage was up to 30 kV. The samples were first sputter coated with a thin layer of gold and then observed at a magnification of 500 X, 1000 X, 1500X and 2000 X.

2.3.4 Thermal Analysis

Thermal analysis was done to check the thermal stability of the samples. Thermal analysis of native starch and grafted starch was performed using EXSTAR TG-DTA 6300. Samples were heated up to 900 °C with a heating rate of 10 °C/min in an inert nitrogen atmosphere. Nitrogen was used as a purge gas at a flow rate of 200 mL/min.

The change in weight of sample against temperature (thermogravimetric analysis, TG), the rate of change of weight against temperature (derivative thermogravimetric analysis, DTG) and the heat evolved or absorbed in the sample (derivative thermal analysis, DTA) because of exothermic or endothermic activity in the sample were measured.

3 Results and Discussion

3.1 FT-IR Spectra

The FT-IR spectra of starch and grafted starch were shown in Fig. 1a, b. In the spectra of starch (Fig. 1a), a strong and broad absorption band appeared at 3253 cm^{-1} attributed to the vibrational stretching of inter and intramolecular bounded –OH groups on the starch. In the spectra of native starch, peaks at 1157, 1080 and 997 cm^{-1} were due to the –CO bond stretching (Namazi and Dadkhah 2010). The characteristic peaks of the anhydrous glucose ring (O–C stretching) were present at 1157 and 1080 cm^{-1} (Simi and Abraham 2007). The absorption band at 2932 cm^{-1} was present due to the C–H stretching in the glucose unit of starch.

In the FT-IR spectra of grafted starch (Fig. 1b), an increase in the intensity of adsorption peak present at 2927 cm^{-1} corresponding to the methyl and methylene C–H stretching associated with fatty acid substituents was observed on grafting

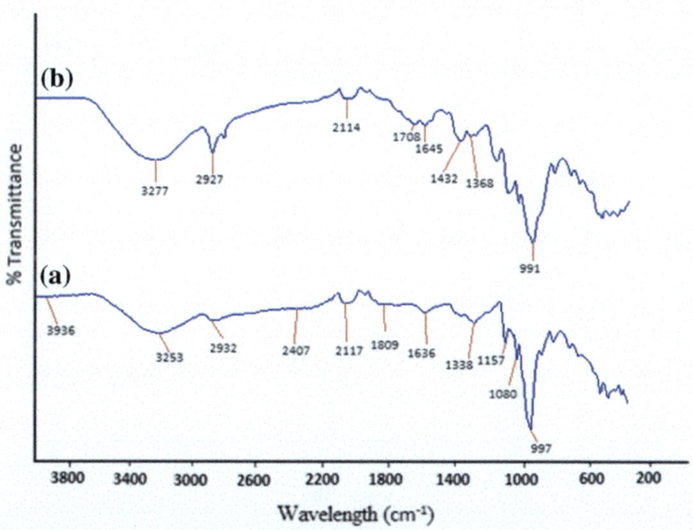

Fig. 1 FT-IR spectra of **a** starch and **b** grafted starch

(Namazi et al. 2011). Spectra of grafted starch showed a new peak at 1708 cm^{-1} attributed to the carbonyl groups, indicating the occurrence of grafting.

3.2 *X-Ray Diffraction of Starch and Grafted Starch*

X-ray diffraction was done to examine the change in crystalline and amorphous regions of starch after grafting. Figure 2a, b represents the X-ray diffractograms of starch and grafted starch, respectively. It is evident from the figure that starch showed peaks at 17.97° and 23.19° with relative intensities of 1757 (counts) and 1533 (counts), respectively, while in grafted starch, peaks were observed at 18.17° and 23.13° with relative intensities 1425 and 1169.5 (counts), respectively. The intensity of starch was lower as compared to the grafted starch. The value of d-spacing for starch were 4.77 and 3.79 (2θ: 17.97° and 23.19°) while the values of d-spacing for grafted starch increased to 4.88 and 3.85 (2θ: 18.17° and 23.13°) which showed that grafting of starch led to amorphous structure (Mittal et al. 2016).

3.3 *Scanning Electron Microscopy (SEM)*

Scanning electron micrographs of starch and grafted starch were shown in Fig. 3a, b.

Figure 3a shows the cellular and porous structure of the starch. After modification, it was observed that the surface of the starch became rough due to grafting of oleic acid on starch which led to the change in the surface topology. After grafting, the granular structure of the starch was completely destroyed and amorphous particles of gelatinized starch were clearly visible (Simi and Abraham 2007).

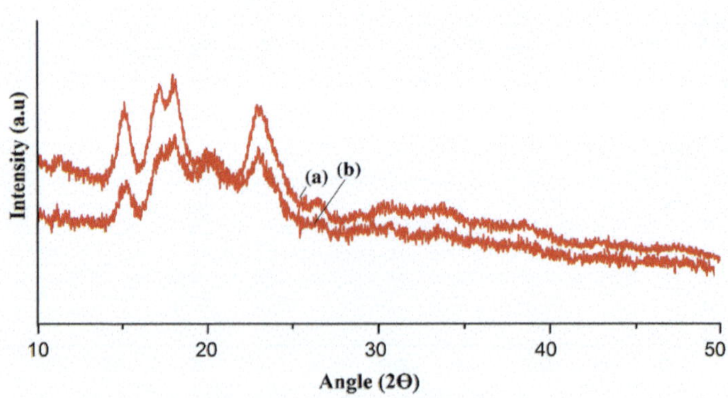

Fig. 2 X-ray diffractograms of **a** starch and **b** grafted starch

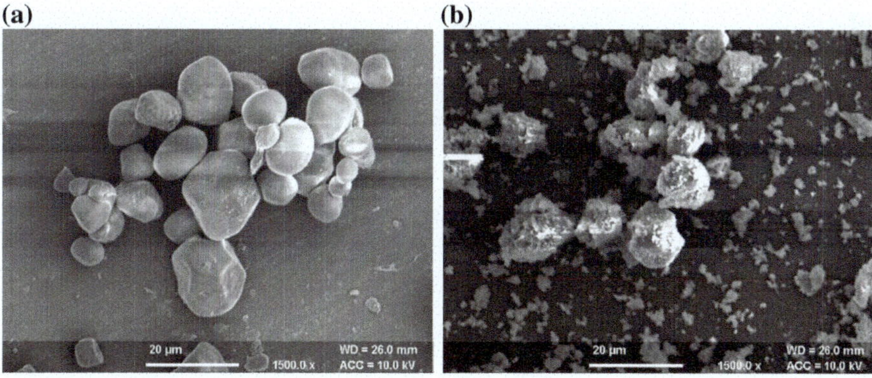

Fig. 3 SEM images of **a** starch and **b** grafted starch

3.4 *Thermal Analysis*

Thermogravimetric analysis (TGA), derivative thermogravimetric analysis (DTG) and differential thermal analysis (DTA) curves of starch and grafted starch are shown in Fig. 4a and b, respectively. TG analysis (change in weight %) of starch showed two-stage weight losses. Initially, there was 9.4% weight loss up to 250 °C due to the presence of free moisture. The first-stage weight loss was 57.5% between 250 and 325 °C while the second stage weight loss was 33% between 325 and 497 °C. The first-stage weight loss was due to the breakdown of pyranose glycosidic linkage present in starch and second stage weight loss was because of the thermal decomposition of starch and volatilization of the polymer products. TG analysis of grafted starch also showed two-stage weight losses. Initially, 13.5% weight loss was observed due to free moisture while first-stage weight loss was 39.76% between 230 and 300 °C and second stage weight loss was 27.84% from 300 to 442 °C. The first-stage weight loss was due to the breakdown of pyranose glycosidic linkage present in starch and second stage weight loss was due to the breakdown of the C=C bonds present in grafted starch and decomposition of starch (Namazi and Dadkhah 2010). Results showed that the decomposition temperature of grafted starch was higher compared to native starch. DTA analysis of starch shows two endothermic peaks at 349 and 493 °C. However, DTA analysis of grafted starch showed an endothermic peak at 317 and 400 °C which is due to the higher energy needed to burn and cause decomposition of grafted starch.

DTG analysis described the rate of weight loss with respect to time. DTG curve of starch showed two peaks at 306 °C and 479 °C with a rate of weight loss of 92 μg/°C and 17 μg/°C, respectively. DTG curve of grafted starch showed two peaks at 279 °C and 403 °C with a rate of weight loss of 150.5 μg/°C and 20.6 μg/°C, respectively.

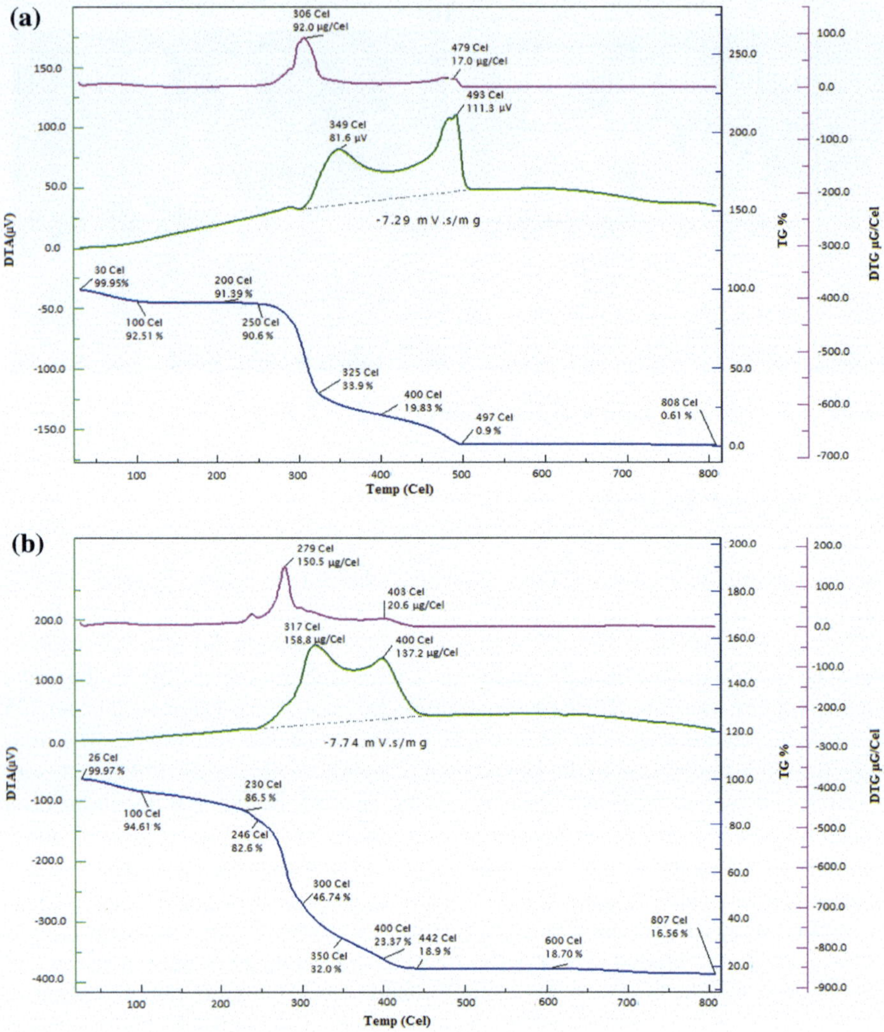

Fig. 4 DTA, DTG and TG curve of **a** starch and **b** grafted starch

3.5 Mechanical Properties

TS and % elongation of PVA and PVA/S films are shown in Fig. 5.

TS of PVA film was 42.41 MPa which decreased in the PVA/S films with an increase in the content of starch in the matrix. TS of PVA/S film with PVA/starch ratio 1:1 was 10.15 MPa. The elongation (%) of pure PVA film was 209.33% and it decreased to 61.76% with the addition of 50% starch in the film. However, the TS and % elongation of the PVA/GS blend film with PVA/grafted starch ratio 1:1 was

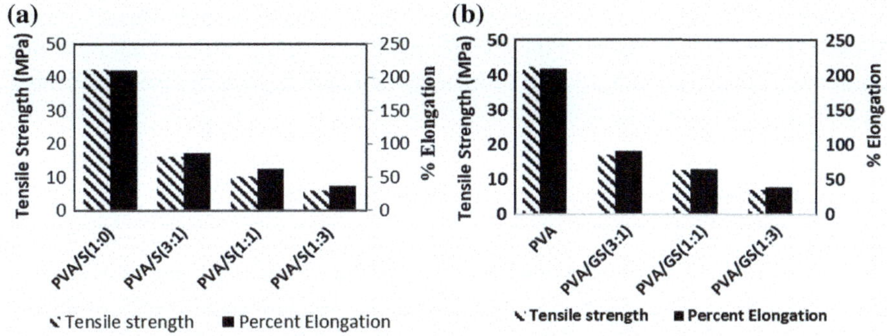

Fig. 5 Tensile strength and % elongation of **a** PVA/S blend films and **b** PVA/GS blend films

12.6 MPa and 65.2%, respectively (Fig. 5b). There was an increase in the mechanical properties of the PVA/grafted starch blend films as compared to the PVA/starch blend films due to improved compatibility between PVA/starch after grafting of starch.

3.6 Water Uptake (%)

Water uptake (%) of blend films as a function of time (days) is shown in Fig. 6.

In the pure PVA film, water uptake (%) was found to be around 193% after 1 day which increased up to 202.45% after 6 days. However, it was observed that water uptake decreased in the case of PVA/S films. Water uptake (%) of PVA/starch film with PVA/S ratio 1:1 was observed 141.82%, which was lesser as compared to pure PVA films. Starch and PVA both are hydrophilic materials, and the water uptake (%)

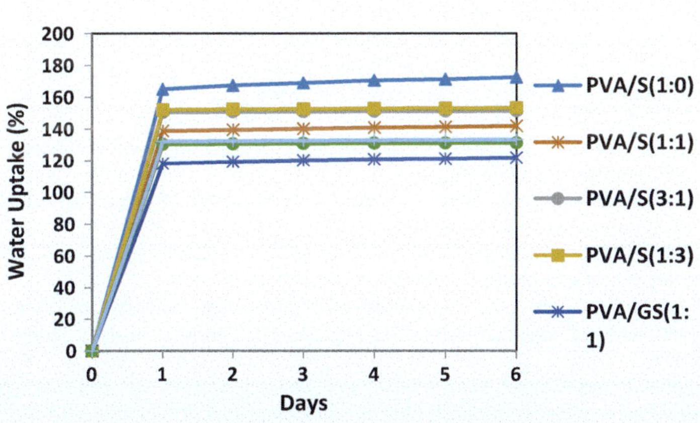

Fig. 6 Water uptake (%) of PVA/S and PVA/GS blend films

was lesser in PVA/S (1:1) films due to the presence of intermolecular bonds between PVA and starch. Water uptake of the PVA/starch blend films decreased after grafting of starch with oleic acid. PVA/GS 1:1 film showed 121.8% water uptake after 6 days.

4 Conclusions

Starch was graft copolymerized by oleic acid in the presence of initiators such as ammonium persulphate and potassium persulphate. The modification of starch with oleic acid induced the physiochemical and morphological changes on the surface of starch. FT-IR spectra confirmed the grafting of starch by showing the peaks at 2927 and 1708 cm^{-1} corresponding to the stretching of C–H bond and carbonyl groups present in the grafted starch, respectively. X-ray diffractograms of grafted starch showed the lower crystallinity as compared to the starch by showing reduced area under the crystalline peaks. SEM analysis showed the rough surface of the grafted starch as compared to the starch which was favourable for the preparation of PVA/starch polymer blend films. Thermal analysis showed that the thermal decomposition temperature of grafted starch was higher as compared to the starch. Mechanical properties and water-resistant properties of the PVA/grafted starch blend films were higher as compared to the PVA/starch blend films.

References

Araújo MA, Cunha A, Mota M (2004) Enzymatic degradation of starch-based thermoplastic compounds used in protheses: identification of the degradation products in solution. Biomaterials 25:2687–2693

Azahari NA, Othman N, Ismail H (2011) Biodegradation studies of polyvinyl alcohol/corn starch blend films in solid and solution media. J Phys Sci 22(2):15–31

Canché-Escamilla G, Canché-Canché M, Duarte-Aranda S, Cáceres-Farfán M, Borges-Argáez R (2011) Mechanical properties and biodegradation of thermoplastic starches obtained from grafted starches with acrylics. Carbohyd Polym 86(4):1501–1508

Chai WL, Chow JD, Chen CC, Chuang FS, Lu WC (2009) Evaluation of the biodegradability of polyvinyl alcohol/starch blends: a methodological comparison of environmentally friendly materials. J Polym Environ 17:71–82

Chiellini E, Corti A, D'Antone S, Solaro R (2003) Biodegradation of poly (vinyl alcohol) based materials. Prog Polym Sci 28(6):963–1014

Garg S, Jana AK (2011) Characterization and evaluation of acylated starch with different acyl groups and degrees of substitution. Carbohyd Polym 83:1623–1630

Kohli D, Garg S, Jana AK, Maiti M (2017) Synthesis of graft copolymers for green composite films and optimization of reaction parameters using Taguchi (L16) orthogonal array. Indian Chem Eng 59(2):136–158

Lani NS, Ngadi N, Johari A, Jusoh M (2014) Isolation, characterization, and application of nanocellulose from oil palm empty fruit bunch fiber as nanocomposites. J Nanomater 13

Mittal A, Garg S, Kohli D, Maiti M, Jana AK, Bajpai S (2016) Effect of cross linking of PVA/starch and reinforcement of modified barley husk on the properties of composite films. Carbohyd Polym 151:926–938
Namazi H, Dadkhah A (2010) Convenient method for preparation of hydrophobically modified starch nanocrystals with using fatty acids. Carbohyd Polym 79:731–737
Namazi H, Fathi F, Dadkhah A (2011) Hydrophobically modified starch using long-chain fatty acids for preparation of nanosized starch particles. Sci Iran C 18(3):439–445
Negim ESM, Rakhmetullayeva RK, Yeligbayeva G, Ukrimbaeva PI, Primzharova ST, Kaldybekov DB, Khatib JM, Mun GA, Craig W (2014) Improving biodegradability of polyvinyl alcohol/starch blend films for packaging applications. Int J Basic Appl Sci 3(3):263–273
Primarini D, Ohta Y (2000) Some enzyme properties of raw starch digesting amylases from streptomyces sp. no. 4. Starch 52:28–32
Rhim JW, Ng PK (2007) Natural biopolymer-based nanocomposite films for packaging applications. Crit Rev Food Sci Nutr 47(4):411–433
Simi CK, Abraham TE (2007) Hydrophobic grafted and cross-linked starch nanoparticles for drug delivery. Bioprocess Biosyst Eng 30:173–180
Tang X, Alavi S (2011) Recent advances in starch, polyvinyl alcohol based polymer blends, nanocomposites and their biodegradability. Carbohyd Polym 85:7–16
Zhang JF, Sun XZ (2004) Mechanical properties of PLA/starch composites compatibilized by maleic anhydride. Biomacromol 5:1446–1451

Application of RSM for Optimizing the Biodiesel Production Catalyzed by Calcium Methoxide

Kashish Mehta, Neetu Divya and Mithilesh Kumar Jha

Abstract In the present study, calcium methoxide ($Ca(OCH_3)_2$) catalyst was synthesized by the methanolysis of calcined calcium carbonate. The prepared catalyst was then characterized by different physicochemical techniques. The optimization of various reaction parameters for the transesterification of Jatropha curcas oil (JCO) was done by using response surface methodology (RSM) coupled with central composite design (CCD) by using Design Expert software. The polynomial equations based on the second-order model were obtained for transesterification reaction. Independent variables, i.e., methanol to oil molar ratio, catalyst concentration, reaction temperature, and reaction time, were investigated in order to find the maximum value of % biodiesel yield. The adequacy of regression modeling was tested by using analysis of variance (ANOVA). It was observed that biodiesel yield of 90.76% was predicted at the optimal level of catalyst concentration (1 wt%), 9:1 methanol to oil molar ratio, reaction temperature (60 °C), and time (210 min).

Keywords Heterogeneous catalyst · JCO · Transesterification · Biodiesel · RSM

1 Introduction

In this day and age, the search of newer energy sources has been boosted by several factors such as depleting fossil fuel reserves due to its continue unsustainable exploitation, the environmental impact associated to their consumption, limitations to its use, and technological changes (Blanco-Marigorta et al. 2013). In this respect, biodiesel derived from renewable sources of fats or oils (mainly nonedible oils) is an emerging alternative to diesel fuel (Konwar et al. 2014). Moreover, it has similar physical properties to petroleum diesel and has environment-friendly emission profile which makes it more interesting (Canakci and Van Gerpen 2001).

K. Mehta · N. Divya (✉) · M. K. Jha
Department of Chemical Engineering, Dr B. R. Ambedkar National Institute of Technology, Jalandhar, India
e-mail: divyan@nitj.ac.in

A. K. Agnihotri et al. (eds.), *Sustainable Engineering*, Lecture Notes in Civil Engineering 30, https://doi.org/10.1007/978-981-13-6717-5_8

Biodiesel also called fatty acid alkyl ester (FAAE) is conventionally synthesized by the alcoholysis of triglyceride molecule such as fat or oil in the presence of the catalyst. The esterification reaction of oil with methanol can be carried out using homogeneous (acidic or basic) catalyst resulting in high FAME yield under moderate reaction conditions. However, the major drawback of the homogeneous catalyst is its non-reusability because the catalyst gets consumed during the reaction. Besides this, the catalyst segregation from the reaction products is strenuous followed by immense wastewater generation (Boro et al. 2014).

Recently, a wide range of solid catalysts have been introduced in the field of biodiesel production such as zeolites, modified zeolites, alkaline earth metal oxides, and supported alkaline earth metal oxides, hydrotalcites, zirconia-based, and various others (Deshmane and Adewuyi 2013). However, for higher fatty acid to methyl ester conversion, most of the heterogeneous catalysts require high temperature and pressure conditions and longer reaction time. All these facets lead to an increment in the production cost (Fukuda et al. 2001).

Extensive research is being carried out on many alkaline metal oxides especially calcium oxide owing to its low methanol solubility and high basic character. Calcium methoxide ($Ca(OCH_3)_2$), a solid heterogeneous catalyst appear to be a promising catalyst for base transesterification (Teo et al. 2015). Liu and his co-workers have reported in their study that $Ca(OCH_3)_2$ can be reused up to 20 times.

Thus, the present study aims at investigating the effect of various process parameters on biodiesel production from JCO using $Ca(OCH_3)_2$ as a basic catalyst. Response surface methodology (RSM) coupled with central composite design (CCD) is used to determine the optimum reaction conditions for maximum biodiesel yield.

2 Materials and Methods

2.1 Materials

Calcium carbonate (Qualigens Fine Chemicals, 99%), methanol extra pure (Loba Chemie, 99.5%), $CHCl_3$ (Loba Chemie), potassium hydroxide pellets GR (SDFCL, 85%) and all necessary solvents were procured from Merck, India Ltd. Jatropha curcas oil was purchased from an oil marketing company in Rajasthan, India. The physical properties of JCO determined as per AOCS official method are listed in Table 1.

Table 1 Fuel properties of JCO

Property	JCO
Acid value (mg KOH/g)	22
Viscosity (cSt, @40°C)	51.76
Iodine value (g/100 g)	92
Saponification value (mg/g)	191.4

2.2 *Methods*

2.2.1 Pretreatment and Acid Esterification

In order to remove water traces and other impurities as purchased JCO was pretreated. The free fatty acid profile of treated oil was analyzed, and it is given in Table 1. The oil had an initial acid value of about 22 mg KOH/g corresponding to the FFA level of 11% which is far behind the limit for transesterification reaction (1 and 2%). Single step acid pretreatment was carried out to reduce the FFA of the oil. The reaction was carried out in a 250 ml three-neck flask using magnetic hot plate. The methanol (0.40 v/v oil) and sulphuric acid (1% w/w oil) were added to the flask by constantly stirring. The acid was thoroughly mixed with methanol and then preheated oil was added. The mixture was stirred at a constant speed at 60 °C for 2 h. The reaction mixture was then poured in separating funnel and allowed to settle for 2 days. The top layer had unreacted methanol and water whereas the bottom layer had oil and fatty acid methyl esters (FAMEs) (Helwani et al. 2009).

2.2.2 Catalyst Preparation

Calcium methoxide was synthesized as described by Theam et al. (2015). First, 10 g of calcium carbonate ($CaCO_3$) was calcined at 900 °C for 2 h. The calcined calcium carbonate was then dispersed in 100 ml of methanol in a glass reactor attached with a reflux condenser at 65 °C for 1 h under continuous stirring. After the reaction excess methanol was distilled off under vacuum. Then, the synthesized catalyst was dried overnight at 100 °C.

2.2.3 Characterization of Catalyst

The synthesized catalyst was characterized using different techniques such as XRD and FT-IR to confirm its formation. The powder X-ray diffractograms were recorded by X'PERT-PRO using Cu–Kα radiation as source ($\lambda = 1.54$ Å) and Ni as filter. The range of scanning angle (2θ) was kept at 3–60°. Measurement of IR spectra was performed using KBr pellets on Perkin Elmer Spectrum RX-I with a resolution of 4 cm^{-1}. The BET surface area and total pore volume were measured with SMART

SORB 93. A weighed sample was prepared for analysis by outgassing it at 150 °C for 5 h on the degas port. The adsorption isotherms were generated by dosing nitrogen onto the catalyst in liquid nitrogen bath at 77 K.

2.2.4 Transesterification Reaction Procedure

The transesterification of JCO to biodiesel was carried out in round bottom three-neck flask with jacketed water condenser placed on a hot magnetic plate. Based on oil weight %, catalyst was added to the flask. Then methanol and oil were added and the mixture was vigorously stirred for a definite time period. A small aliquot was collected from the reaction mixture at regular intervals. The reaction mixture was quenched to stop the reaction as the reaction period over. The catalyst was separated from the product by centrifugation. Biodiesel yield was determined by 1H NMR technique using $CDCl_3$ as a solvent by the given formula:

$$\% \text{ Yield} = \frac{2A_{Me}}{3A_{CH_2}} \times 100 \qquad (1)$$

where, A_{Me} is the area under methoxy protons of methyl esters, and A_{CH_2} is the area under α-methylene protons.

2.2.5 Batch Study of Biodiesel Synthesis

For the batch study of transesterification reaction, the following variables were considered: reaction temperature of 30–70 °C, reaction time of 50–210 min, catalyst concentration of 0.5–1.5 wt%, and methanol to oil molar ratio of 6:1–15:1.

2.2.6 Response Surface Methodology (RSM)

For the optimization of process variables, different statistical methodologies are available such as Taguchi, iterative mathematical search technique, heuristic search technique, metaheuristic search technique, genetic algorithm, simulated annealing, and response surface methodology (RSM) (Mukherjee and Ray 2006).

In this study, RSM was used to obtain the optimum reaction conditions for the transesterification of JCO with central composite design (CCD). CCD provides a quadratic surface model for sequential experimentation. The Design Expert v7.0.0 software (Stat-Ease, USA) was used for the regression and statistical data analysis.

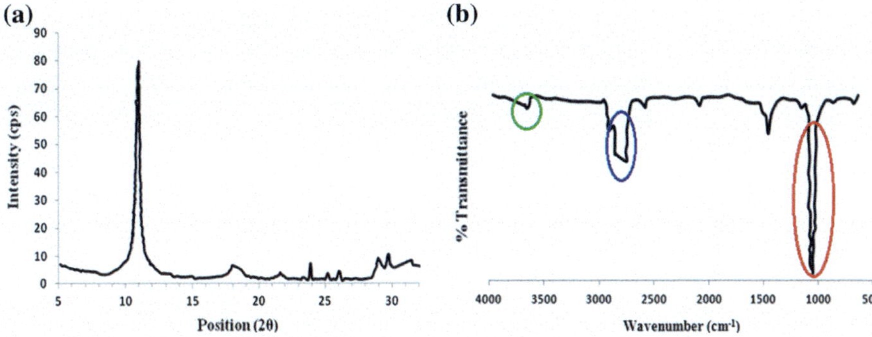

Fig. 1 **a** XRD pattern and **b** FT-IR spectrum of calcium methoxide catalyst

3 Results and Discussions

3.1 Catalyst Characterization

The X-ray diffractogram of the catalyst is shown in Fig. 1a. After reacting with methanol under reflux, the calcined calcium carbonate gets modified into calcium methoxide as evident from the characteristic peak of calcium methoxide at 12°. Similar results have already been reported in by other researchers confirming the synthesized catalyst to be calcium methoxide (Kouzu et al. 2008).

Figure 1b shows the IR spectra of $Ca(OCH_3)_2$ in which intense absorption bands were observed at 1057 and 1500 cm^{-1} attributed to the stretching and bending vibration of –C–H bond, respectively. Another peak around 3650 cm^{-1} was attributed to –OH stretching vibration. A series of peaks centered at 2870 cm^{-1} was assigned to CH_3 stretching vibrations as indicated by Teo et al. (2015) during their study. Lutz et al. (1994) reported that peak at 3650 cm^{-1} is due to adsorption deriving from hydration of the surface which corresponds to calcium methoxide.

The surface area of the catalyst affects the reactivity of the catalyst during the reaction. Hence, BET surface area is an important catalyst characterization. The BET surface area of the synthesized catalyst was found to be 34.7 $m^2\ g^{-1}$ with a pore volume of 0.197 cc g^{-1} Due to high surface area, the catalyst can provide a sufficient number of active sites for use in liquid phase reactions (Kouzu et al. 2008).

3.2 Experimental Design

Batch experiments were performed for the transesterification of JCO to produce biodiesel by using RSM based CCD. A quadratic regression model in terms of process variables in coded form was suggested by the software. The positive sign of the coefficients in Eq. (2) indicates the synergistic effect, whereas negative sign suggests antagonistic effect. It is clear from Eq. (2) that individual operating variables have a net positive effect. The factors chosen were catalyst concentration (A), methanol to oil molar ratio (B), reaction temperature (C), and reaction duration (D).

$$\begin{aligned}\%\,Yield &= (+33.95) + (6.21A) + (9.31B) + (12.74C) + (14.46D) \\ &- (0.59AB) + (4.69AC) - (1.12AD) + (1.59BC) - (2.17BD) \\ &+ (5.15CD) - \left(1.85A^2\right) + \left(0.18B^2\right) + \left(2.88C^2\right) \\ &+ \left(1.77D^2\right)\end{aligned} \tag{2}$$

In the above equation, the individual operating parameters which have positive coefficients will have a positive effect on biodiesel yield, whereas the negative coefficients indicate a decrease in biodiesel yield. The significance of linear coefficients was found to be more as compared to quadratic and interacting coefficients. It is clear from Table 2 that model F-value is 739.33 with a corresponding p-value of 0.0001 and high Sum of squares (13,277.84). It was observed that the developed model is significant and can appropriately explain the relationship between response and independent variables. ANOVA study suggested that reaction time ($p = 0.0001$, $SS = 5015$, $F = 3909.84$) has the most significant effect on biodiesel yield followed by reaction temperature ($p = 0.0001$, $SS = 3893.09$, $F = 3034.81$).

3.3 Combined Effect of Process Variables on Biodiesel Yield

In order to investigate the effect of different process parameters on biodiesel yield, three-dimensional (3D) response surface plots were generated. The highest predicted values were shown by smallest ellipse in the contour diagram. Elliptical contours were obtained when there was a significant interaction between the independent variables. The graphs were plotted as a function of two variables and the other two variables were taken at center points.

The interactive effect of catalyst concentration and methanol to oil molar ratio on percent yield of biodiesel using calcium methoxide as the catalyst is shown in Fig. 2a. More than 70% yield has been obtained at high catalyst dose and molar ratios. It is observed that at high catalyst dose, the yield of methyl esters increases significantly with an increase in methanol amount. This is due to the fact that the catalyst solubility increases with rise in methanol amount, and it favors the reaction toward biodiesel formation with higher rate of reaction and thus results in high yield (Encinar et al. 2012).

The 3D plot of interactive effect of catalyst concentration and reaction temperature has been shown in Fig. 2b where the actual factors such as reaction time and methanol amount are kept constant at 210 min and 9:1, respectively. When the catalyst dose was kept high, the biodiesel yield increases exponentially and reaches the maximum value

Table 2 ANOVA statistics for biodiesel production from JCO

Source	Sum of squares	df	Mean square	F-value	p-value prob > F	
Model	13277.84	14	948.42	739.33	<0.0001	Significant
A–A	925.66	1	925.66	721.59	<0.0001	
B–B	2078.55	1	2078.55	1620.31	<0.0001	
C–C	3893.09	1	3893.09	3034.81	<0.0001	
D–D	5015.60	1	5015.60	3909.84	<0.0001	
AB	5.58	1	5.58	4.35	0.0545	
AC	351.84	1	351.84	274.28	<0.0001	
AD	20.00	1	20.00	15.59	0.0013	
BC	40.23	1	40.23	31.36	<0.0001	
BD	75.13	1	75.13	58.56	<0.0001	
CD	424.05	1	424.05	330.56	<0.0001	
A^2	93.44	1	93.44	72.84	<0.0001	
B^2	0.88	1	0.88	0.69	0.4201	
C^2	228.18	1	228.18	177.87	<0.0001	
D^2	86.10	1	86.10	67.12	<0.0001	
Residual	19.24	15	1.28			
Lack of fit	17.30	10	1.73	4.46	0.0564	Not significant
Pure error	1.94	5	0.39			
Cor total	13297.08	29				

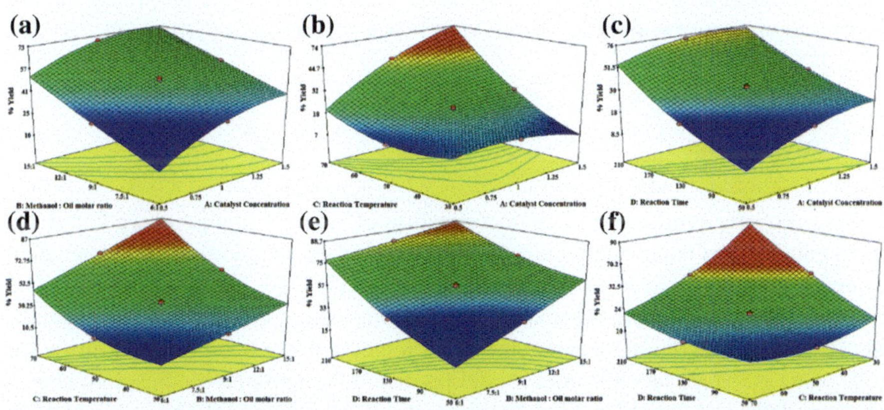

Fig. 2 3D surface plots explaining the interaction effects of **a** catalyst concentration and methanol: oil molar ratio, **b** catalyst concentration and temperature, **c** catalyst concentration and time, **d** methanol: oil molar ratio and temperature, **e** methanol amount and time, and **f** temperature and time on biodiesel production

of 75.3%. This can be attributed to the fact that as the catalyst amount increases the reaction attains its equilibrium faster as more catalytic active sites become available for interaction with reactants (Park et al. 2008).

The combined effect of catalyst concentration and reaction time on percent yield keeping the other parameters constant has been shown in Fig. 2c. It was observed that at low catalyst doses, the yield increases significantly as the reaction time increases. The reason offered is that the increase in reaction time enhances the diffusivity between the reaction mixtures and improves the yield of biodiesel content. Therefore, biodiesel yield is greatly influenced by the interaction of catalyst dosage and reaction time.

Figure 2d shows the 3D surface plot of for biodiesel production as a function of methanol to oil molar ratio and reaction temperature under experimental condition defined by CCD. About 87% of methyl esters yield was obtained when methanol to oil molar ratio was kept 12:1 at 60 °C while the catalyst dose was 1.5 wt% and reaction time 210 min. This happens because a higher reaction temperature can enhance the biodiesel yield by improving the transesterification process. The similar trend has been reported by Hanh et al (2009) during their study.

As discussed above, with an increase in the molar ratio of methanol, the conversion of oil to methyl esters increases. Figure 2e clearly depicts that high methanol amount and 210 min of time significantly improves the biodiesel yield to a maximum of 88.7%. As the reaction time increases, the excess amount of methanol increases the product solubility (glycerol) and which further leads to an increase in the yield of fatty acid methyl esters (Tan and Lee 2009).

Figure 2f represents the 3D surface plot, depicting the interaction effects of time and temperature on biodiesel yield while keeping the catalyst concentration and methanol amount constant. The maximum biodiesel yield of 90% was achieved at a reaction temperature of 60 °C and 210 min. It was observed that on increasing the reaction temperature with time, the yield of biodiesel increases (Silva et al. 2011).

4 Conclusions

In this study, the synthesis and characterization of calcium methoxide, a heterogeneous catalyst has been carried out. The efficacy of the prepared catalyst was then studied for transesterification of Jatropha curcas oil (JCO) to produce biodiesel. The optimization of the process variables has been done using a conventional and statistical method. Optimization of transesterification reaction was achieved by four-factor CCD using RSM. A second-order model predicted the jatropha biodiesel yield as a function of process variables. The maximum biodiesel yield of 90.76% was obtained at the following condition: catalyst concentration of 1 wt%, methanol to oil molar ratio of 9:1, temperature of 60 °C, and time of 210 min. Jatropha biodiesel has similar characteristics of petroleum fuel, and it also proves to be cost-effective.

References

Blanco-Marigorta AM, Suárez-Medina J, Vera-Castellano A (2013) Exergetic analysis of a biodiesel production process from Jatropha curcas. Appl Energ 101:218–225

Boro J, Konwar LJ, Thakur AJ, Deka D (2014) Ba doped CaO derived from waste shells of Tstriatula (TS–CaO) as heterogeneous catalyst for biodiesel production. Fuel 129:182–187

Canakci M, Van Gerpen J (2001) Biodiesel production from oils and fats with high free fatty acids. Trans-ASAE 44(6):1429–1436

Deshmane VG, Adewuyi YG (2013) Synthesis and kinetics of biodiesel formation via calcium methoxide base catalyzed transesterification reaction in the absence and presence of ultrasound. Fuel 107:474–482

Encinar JM, Pardal A, Martínez G (2012) Transesterification of rapeseed oil in subcritical methanol conditions. Fuel Process Technol 94:40–46

Fukuda H, Kondo A, Noda H (2001) Biodiesel fuel production by transesterification of oils. J Biosci Bioeng 92(5):405–416

Hanh HD, Dong NT, Okitsu K et al (2009) Biodiesel production through transesterification of triolein with various alcohols in an ultrasonic field. Renew Energ 34:780–783

Helwani Z, Othman MR, Aziz N et al (2009) Solid heterogeneous catalysts for transesterification of triglycerides with methanol: a review. Appl Catal A 363:1–10

Konwar LJ, Boro J, Deka D (2014) Review on latest developments in biodiesel production using carbon-based catalysts. Renew Sust Energ Revi 29:546–564

Kouzu M, Kasuno T, Tajika M et al (2008) Active phase of calcium oxide used as solid base catalyst for transesterification of soybean oil with refluxing methanol. Appl Catal A: Gen 334(1):357–365

Lutz HD, Möller H, Schmidt M (1994) Lattice vibration spectra. Part LXXXII. Brucite-type hydroxides $M(OH)_2$ (M = Ca, Mn Co, Fe, Cd)—IR and Raman spectra, neutron diffraction of $Fe(OH)_2$. J Mol Struct 328:121–132

Mukherjee I, Ray PK (2006) A review of optimization techniques in metal cutting processes. Comp Ind Eng 50:15–34

Park YM, Lee DW, Kim DK et al (2008) The heterogeneous catalyst system for the continuous conversion of free fatty acids in used vegetable oils for the production of biodiesel. Catal Today 131(1):238–243

Silva GF, Camargo FL, Ferreira ALO (2011) Application of response surface methodology for optimization of biodiesel production by transesterification of soybean oil with ethanol. Fuel Process Technol 92:407–413

Tan KT, Lee KT (2009) Production of FAME by palm oil transesterification via supercritical methanol technology. Biomass Bioenerg 33:1096–1099

Theam KL, Islam A, Choo YM et al (2015) Biodiesel from low cost palm stearin using metal doped methoxide solid catalyst. Ind Crop Prod 76:281–289

Teo SH, Taufiq-Yap YH, Rashid U et al (2015) Hydrothermal effect on synthesis, characterization and catalytic properties of calcium methoxide for biodiesel production from crude Jatropha curcas. RSC Adv 5(6):4266–4276

Synthesis and Characterization of Chemically Derived Graphene Oxide from Graphite

Pankaj Kumar, Neetu Divya and J. K. Ratan

Abstract Graphene oxide (GO) has unique physical, chemical, and mechanical properties. It has also acquired great interest for its unique mechanical, electrical, electronic, and optical properties. GO possesses different types of functional groups, which allows it to interact with organic and inorganic materials in covalent, non-covalent, or ionic manner. Therefore, GO behaves like a semiconductor due the presence of impurity such as hydroxyl, carbonyl, and carboxyl groups as a functional group. The present paper deals with the synthesis of ultrathin layer of GO from commercially available graphite powder. The synthesis process starts with the chemical oxidation of commercial graphite powder into GO by modified Hummer's and Offeman method. Modified Hummer's method is simple and easily adoptable in laboratory. The obtained GO material was characterized by X-ray diffraction (XRD), scanning electron microscope (SEM), and Fourier transform infrared spectroscopy (FTIR).

Keywords Modified hummer's method · Graphite · Characterization

1 Introduction

In recent years, many researchers have been attracted enormous attention to synthesis of ultrathin layer of graphene oxide (GO) from graphite due to low cost and easy access. Moreover, by using chemicals methods we have widespread ability to synthesize GO from graphite powder. Graphite is a multilayer of carbon atoms and GO is a single layer of carbon atoms (Stankovich et al. 2006). Structurally, GO is similar to graphene, which is an ultrathin layer of sp^2 hybridization carbon atoms. It has a hexagonal lattice structure. GO thin layer consists partly of tetrahedrally bonded sp^3 carbon atoms, which are slightly displaced above or below the graphene sheet (Yu et al. 2016), whereas the edges of GO sheet are mostly decorated by different

P. Kumar (✉) · N. Divya · J. K. Ratan
Department of Chemical Engineering, Dr. B. R. Ambedkar National Institute of Technology, Jalandhar, India
e-mail: pkbhogal03@gmail.com

A. K. Agnihotri et al. (eds.), *Sustainable Engineering*, Lecture Notes in Civil Engineering 30, https://doi.org/10.1007/978-981-13-6717-5_9

types of functional groups (Suk et al. 2010). Four types of oxygen functionalities are known to exist in GO sheet which are epoxide (–O–), hydroxyl (−OH), carbonyl (−C = O), and carboxyl (−COOH). The major components, epoxide and hydroxyl groups, are located on the basal plane of GO, while carbonyl and carboxyl groups are distributed at the edges of GO sheet (Szabó et al. 2006). The availability of such types of functional groups on the basal plane edge of GO sheet assent to interact with a broad range of organic and inorganic materials in covalent, non-covalent, and ionic manner (Park and Ruoff 2009). Moreover, due to strong bonding between GO and graphite, it is very difficult to strip into a single layer (Singh et al. 2011). GO is an oxygenated form of monolayer graphene platelet, it has attracted great interest because of its unique and extraordinary physical, chemical, optical, electronic, and mechanical properties. GO material can be used in several applications like composite materials, energy storage, solar cell, catalyst, thermal, electrical, and electronics (Titelman et al. 2005).

GO can be synthesized by various methods such as scotch tape, mechanical exfoliation and cleavage (Choucair et al. 2008), chemical vapor deposition (CVD) (Kim et al. 2009), and annealing a single crystal (SiC) under ultrahigh vacuum (Park and Ruoff 2018), etc. These methods have various types of disadvantages including low yield, high-energy requirement, high cost, low purity, and limitation of instruments. Chemical reduction of graphite powder has been considered as an effective route to synthesize GO sheet (Hummers and Offeman 1958). Therefore, in the present study, conventional Hummer's method and its modification have been used for preparing GO (Stankovich et al. 2006). It is a convenient method and provides GO in large amount as compared to other methods. The obtained samples were analyzed by XRD, SEM, and FTIR spectroscopy methods.

2 Materials and Methods

Graphite (fine powder), sulphuric acid (H_2SO_4) (98%), and sodium nitrate (99.5%) grade AR were purchased from S D Fine-Chem Ltd. Hydrochloric acid (98%) and hydrazine hydrate were procured from RFCL Ltd. Potassium permanganate ($KMnO_4$) (99.5%) was obtained from Central Drug House (P) Ltd. Hydrogen peroxide (30%) was purchased from Merck Ltd.

2.1 Synthesis of Graphene Oxide

Graphene oxide was synthesized from the natural graphite or pure graphite powder by using Hummer's method and its modification (Zaaba et al. 2017). Sample preparation of GO includes 2.0 g of graphite powder and 1.0 g of sodium nitrate mixture was added into a flask. 46 ml of H_2SO_4 was slowly added at normal room temperature. The temperature of the mixture was kept up to 10 °C for 30 min under stirring

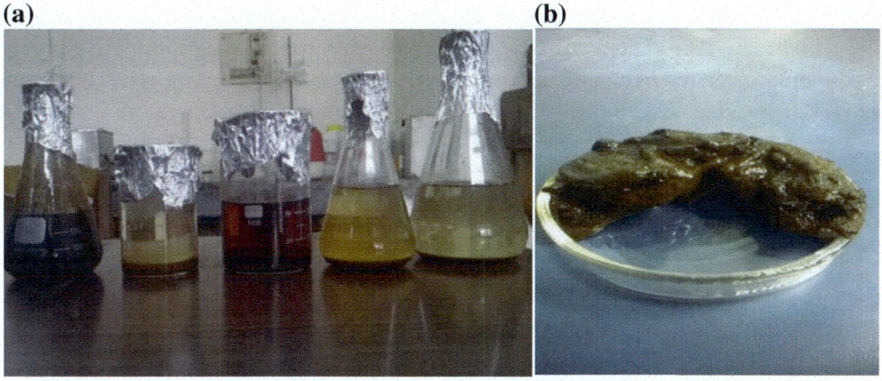

Fig. 1 **a** Systematic photographic image of chemically derived GO after completing the reaction. **b** Systematic photographic image of collected GO after centrifuge

condition. 8.0 g $KMnO_4$ was added dropwise during stirring, and a greenish colored suspension was obtained. The temperature of the mixture was kept below 8–10 °C. Then, 92 ml of distilled water was added dropwise into the mixture, stirred at 35 °C for 30 min, and further diluted with 80 ml distilled water. Thereafter, mixed solution was placed at 98 °C for 15 min until a homogenous suspension mixture was obtained which was light yellow in color (as shown in Fig. 1a–e).

Only variation in stirrer time from (10–20) mint, and remaining other reaction parameters remain constant which results in the change in color of sample and denoted as GO1–GO5.

Then, 25 ml of H_2O_2 with concentration of 30% was slowly added into the obtained mixture to reduce the residual $KMnO_4$ and left it for overnight. The slurry solution was then centrifuged at 10,000 rpm for 5 mins and washed with 5% HCl twice to remove metal ion, proceeded 4–5 times distilled water wash until the pH of the solution reaches 5–7. Finally, obtained graphene oxide (GO) paste, Fig. 1b, was dried at 60 °C in vacuum oven for further characterization.

3 Dispersion of Graphene Oxide

Graphene oxide (GO) is easily dispersed in water as compared to other solvents like acetone, alcohol, and methanol. By dispersion of GO in basic media or water, the exfoliation of graphite oxide produces atomically thin layer of graphene oxide sheet. As a result, GO forms a colloidal suspension of thin layer atomic sheets in water. After an ultrasonic treatment for 20 min, exfoliation process can produce stable dispersions of very thin GO layer in water. However, in alcohol and methanol solution, GO flakes get randomly aggregated, closely associated with each other, and form a disordered solid. Thin GO materials have different types of properties from graphitic platelets due to electrical conductivity. The variation of properties between thin GO materials and graphite platelets are due to its electrical conductivity.

4 Characterizations

The X-ray diffraction (XRD) analyzer used was REKAGU MINIFLEX II diffractometer. The X-rays have a wavelength of 1.5418 Å. The settings for all analyses were as follows: 2θ was measured from 5° to 80° in step increments of 0.025°. Diffractometer was set at a voltage of 30–40 kV and current of 10–15 mA. Scanning electron microscopy (SEM) was done using JEOL field emission SEM, with focused ion beam. SEM is an analytical technique that produces high-magnification images of the given sample. SHIMAZDU Fourier transform infrared spectroscopy (FTIR) analysis is a technique that is used to obtain an infrared spectrum of absorption, photoconductivity, emission or Raman scattering of a liquid, solid, or gas. FTIR spectrometer collects spectral data for a wide range of wavelengths.

5 Results and Discussion

5.1 X-Ray Diffraction (XRD)

Hummer method and its modified method are used to prepare GO samples, and XRD image of obtained GO is shown in Fig. 1a. The oxidizing agent used for GO-1 is sodium nitrate ($NaNO_3$) while for the remaining GO samples potassium permanganate ($KMnO_4$) + sodium nitrate ($NaNO_3$) is used as an oxidizing agent. From the graph, characteristic peak of GO is obtained at $2\theta = 26.2^o$ which is shown in Fig. 2a and the unknown peak was obtained at $2\theta = 43.18^o$ while the peak occurred at 54.4° due to the presence of small amount of graphite potassium in it. After the chemical oxidation and exfoliation into GO-2, the intercalation of oxygen distance functional groups and water molecules into GO sheet was attributed to an increased interlayer between successive GO basal planes (El Achaby et al. 2012 and Wojtoniszak et al. 2012).

The peak obtained at $2\theta = 26.36^o$ shows the presence of graphite nitrate and last peak at $2\theta = 42.7^o$ due to the presence of graphite potassium. First peak around 26° disappears and the new peak at an angle of 2θ around 10° appears for oxidized GO sample. The main differences occurred between peaks are due to the change in oxidizer.

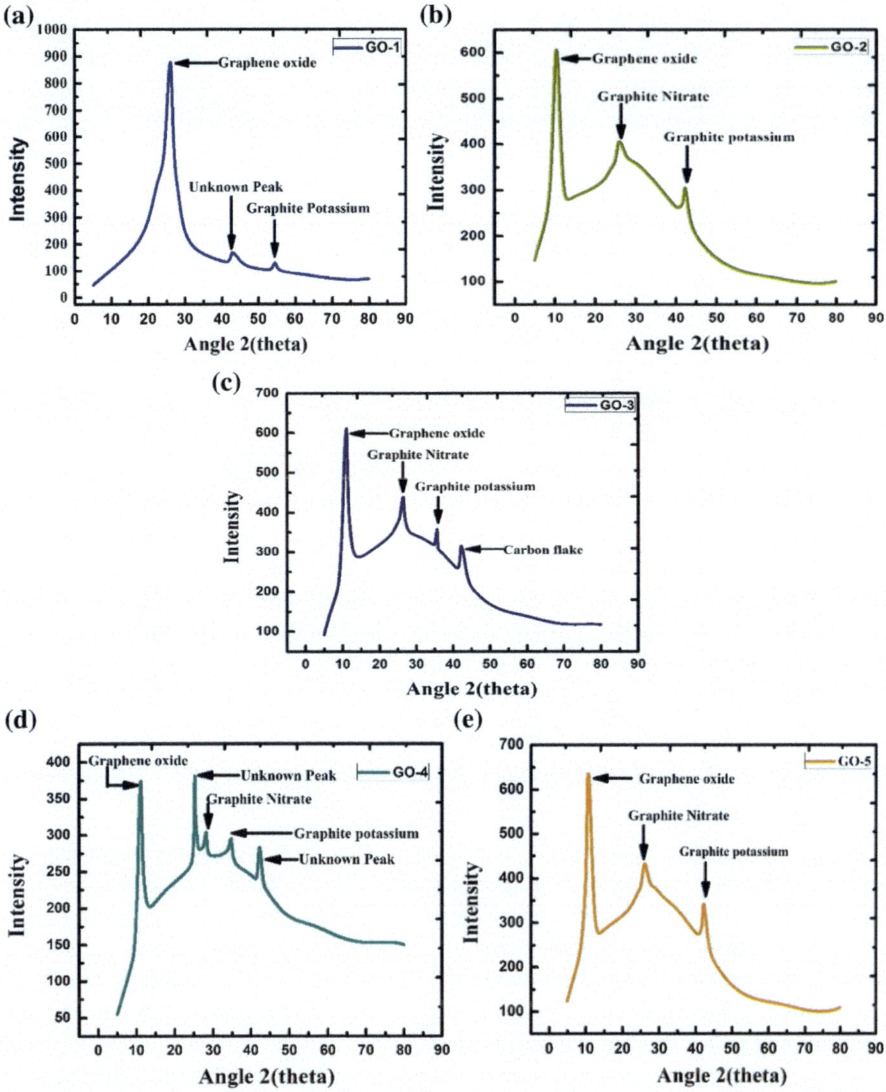

Fig. 2 **a** XRD analysis image from chemically expanded GO-1 from graphite. **b** XRD analysis image from chemically expanded GO-2 from graphite. **c** XRD analysis image from chemically expanded GO-3 from graphite. **d** XRD analysis image from chemically expanded GO-4 from graphite. **e** XRD analysis image from chemically expanded GO-5 from graphite

Figure 2c represents the XRD analysis of chemically expanded GO-3 from graphite. The change of peak was observed at $2\theta = 41.1^{\circ}$ which shows the presence of carbon flakes in it.

Graph of GO-4 shows five major peaks in which second and fifth peaks are unknown while first peak is graphene oxide, third is graphite nitrate, and fourth is graphite potassium as shown in Fig. 2d. From the graph of GO-5 and GO-2, both show similar properties (referred from JCPDS database). All these variations occur due to change in the reaction parameters like temperature, time variation, and many other parameters but mainly depend on pH value.

5.2 *Scanning Electron Microscope (SEM)*

The morphology of GO was investigated by scanning electron microscopy (SEM). To analyze GO samples, SEM is obviously the most durable microscopic method.

The images of GO samples were generated by detecting and collecting the secondary electrons emitted by specimen atoms excited by the incident electron beam. High-quality GO images can be obtained with varying the magnification from 500 to 100,000. SEM was used as the main instrumental method in the study contesting the two-component structural model of GO as shown in Fig. 3a–e.

5.3 *Fourier Transform Infrared Spectroscopy (FTIR)*

FTIR spectrum for the obtained GO in full range of wavenumbers (400–4000 cm^{-1}) is shown in Fig. 4a–e. A typical FTIR spectrum of GO can be arbitrarily divided into three characteristic regions: (a) stretching of O–H bands an intense and very broad adsorption band in the 3600–2400 cm^{-1} region, (b) in the middle of the spectrum the most recognizable absorption bands at 1723 and 1619 cm^{-1}, and (c) a bunch of overlapping signals in the region of 1051–1060 cm^{-1}. The peaks shown in Fig. 4a represent that GO-1 sample is not fully oxidized.

In Fig. 4b, the water molecule functional group was observed peak at 3600 cm^{-1} due to O–H stretching. And the presence of different types of oxygen functionalities in GO was confirmed at 1716 and 1619 cm^{-1}. The stretching vibrations of C=O were observed at 1716 cm^{-1} and C=C was observed at 1619 cm^{-1}. In fingerprint region, the bands assigning were very difficult due to overlapping of multiple bands while the most intense band in this region was observed at 1023 cm^{-1} which was due to C–O stretching. In Fig. 4b, the water molecule functional group had observed peak at 3585 cm^{-1} due to O–H stretching. Remaining same peak had been observed in Fig. 4b–e (Dimiev et al. 2012; Choi et al. 2010).

In Fig. 4c, d, the peak 3400 cm^{-1} was not observed. Like Fig. 4c, the peaks at 1712 cm^{-1} and 1618 cm^{-1} were due to C=O and C=C, respectively. The stretching

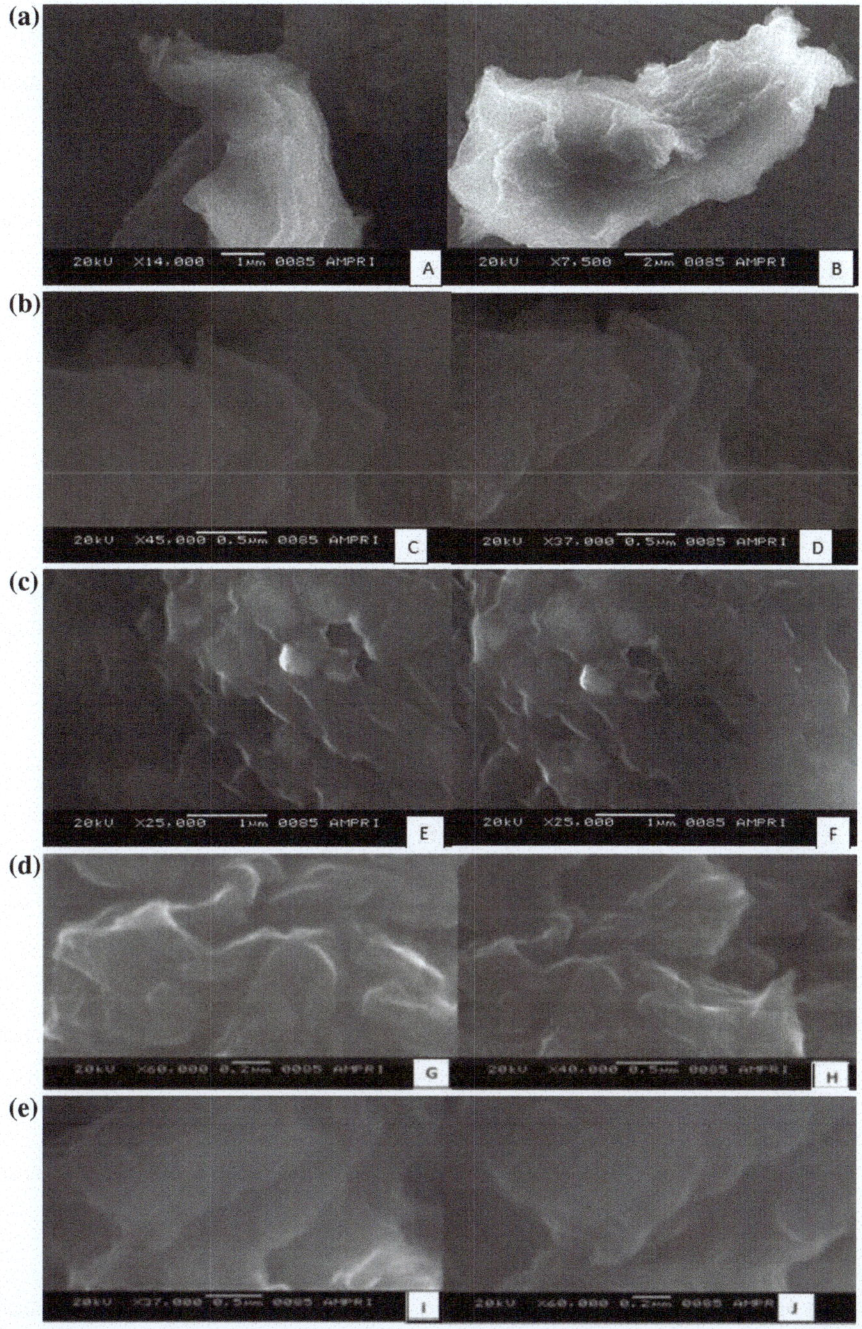

Fig. 3 **a** SEM images of graphene oxide (GO-1): (A) at high magnification (B) at low magnification. **b** SEM images of graphene oxide (GO-2): Both (C, D) are at same magnification. **c** SEM images of graphene oxide (GO-3): Both (E, F) are at same magnification. **d** SEM images of graphene oxide (GO-4): (G) at low magnification (H) at high magnification. **e** SEM images of graphene oxide (GO-5): (I) at high magnification and (J) at low magnification

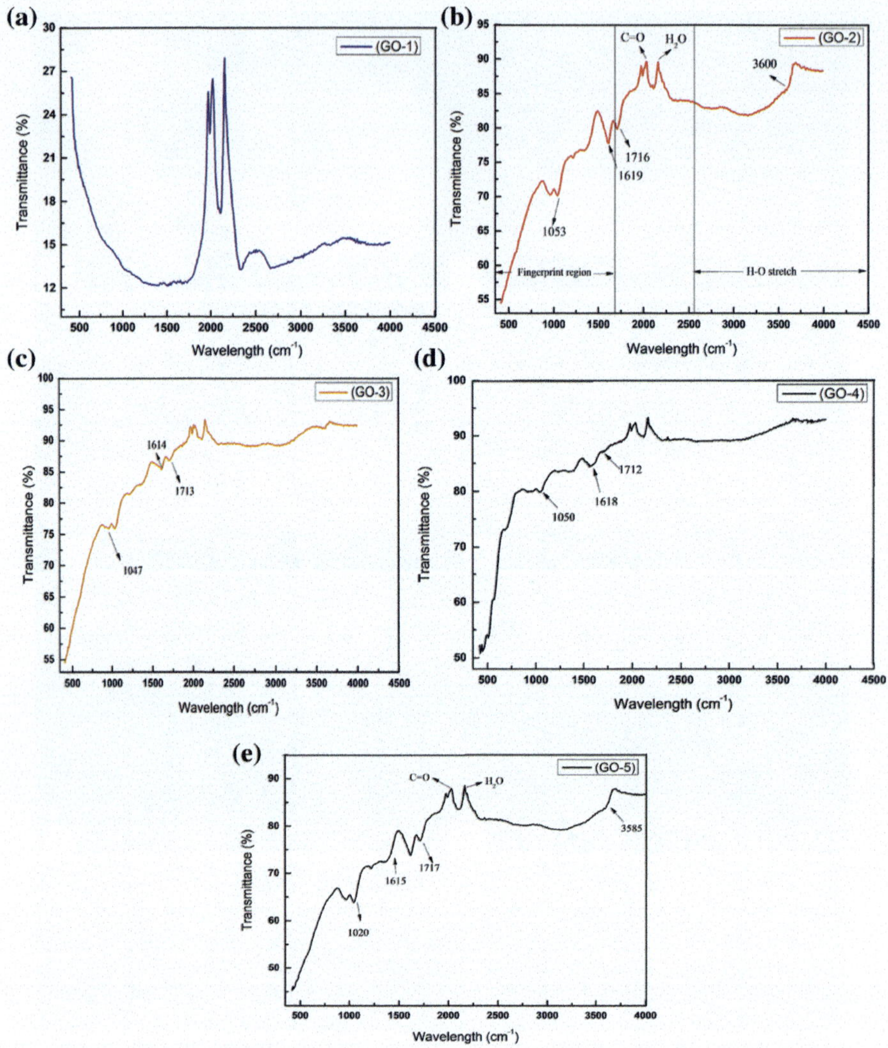

Fig. 4 **a** FTIR analysis image of graphene oxide (GO-1). **b** FTIR analysis image of graphene oxide (GO-2). **c** FTIR analysis image of graphene oxide (GO-3). **d** FTIR analysis image of graphene oxide (GO-4). **e** FTIR analysis image of graphene oxide (GO-5)

vibration observed at 1050 cm^{-1} was due to presence of C–O from the unoxidized graphitic powder (Lian et al. 2010).

These observations indicate that oxygen-containing functional groups existed on the surface of graphene oxide, during the oxidation process of the graphite powder in concentrated sulphuric acid with potassium permanganate, and the original extended conjugated p-orbital system of the graphite was destroyed and oxygen-containing functional groups such as hydroxyl, carbonyl, and carboxyl were inserted into the carbon skeleton (Oh et al. 2014). Moreover, only the broad O–H and the C–O stretching bands indicate that after a hydrazine reduction process, the GO had been successfully restored to graphene.

6 Conclusion

In this study, graphene oxide was synthesized from graphite powder by using modified Hummer's method. Synthesized GO materials from graphite flakes in a cost-effective and easy manner process used and it has various applications. However, using Hummer's method, huge amounts of impurities are present in the GO. The impurities are epoxide, hydroxide, and carbonyl groups found at the edge of GO sheet. Structurally graphene is like GO but electrical conductivity of graphene oxide is low due to the presence of oxygen as functional groups at its base. The papered GO sample was analyzed by XRD, SEM, and FTIR for the existence of GO in it. It was concluded from XRD, SEM, and FTIR that GO is present in the form of sheet.

References

Choi W, Lahiri I, Seelaboyina R, Kang YS (2010) Synthesis of graphene and its applications: a review. Crit Rev Solid State Mater Sci 35(1):52–71

Choucair M, Thordarson P, Stride J (2008) Gram-scale production of graphene based on solvothermal synthesis and sonication. Nat Nanotechnol 4:30–33. https://doi.org/10.1038/nnano.2008.365

Dimiev A, Alemany L, Tour J (2012) Graphene oxide. Origin of acidity, its instability in water, and a new dynamic structural model. ACS Nano 7:576–588. https://doi.org/10.1021/nn3047378

El Achaby M, Arrakhiz F, Vaudreuil S et al (2012) Piezoelectric β-polymorph formation and properties enhancement in graphene oxide—PVDF nanocomposite films. Appl Surf Sci 258:7668–7677. https://doi.org/10.1016/j.apsusc.2012.04.118

Hummers W, Offeman R (1958) Preparation of graphitic oxide. J Am Chem Soc 80:1339–1339. https://doi.org/10.1021/ja01539a017

Kim K, Zhao Y, Jang H et al (2009) Large-scale pattern growth of graphene films for stretchable transparent electrodes. Nature 457:706–710. https://doi.org/10.1038/nature07719

Lian P, Zhu X, Liang S et al (2010) Large reversible capacity of high quality graphene sheets as an anode material for lithium-ion batteries. Electrochim Acta 55:3909–3914. https://doi.org/10.1016/j.electacta.2010.02.025

Oh Y, Yoo J, Kim Y et al (2014) Oxygen functional groups and electrochemical capacitive behavior of incompletely reduced graphene oxides as a thin-film electrode of supercapacitor. Electrochim Acta 116:118–128. https://doi.org/10.1016/j.electacta.2013.11.04

Park S, Ruoff R (2009) Chemical methods for the production of graphenes. Nat Nanotechnol 4:217–224. https://doi.org/10.1038/nnano.2009.58

Park S, Ruoff R (2018) Chemical methods for the production of graphenes. Accessed 14 Mar 2018

Singh V, Joung D, Zhai L et al (2011) Graphene based materials: past, present and future. Prog Mater Sci 56:1178–1271. https://doi.org/10.1016/j.pmatsci.2011.03.003

Stankovich S, Dikin D, Dommett G et al (2006) Graphene-based composite materials. Nature 442:282–286. https://doi.org/10.1038/nature0496

Suk J, Piner R, An J, Ruoff R (2010) Mechanical properties of monolayer graphene oxide. ACS Nano 4:6557–6564. https://doi.org/10.1021/nn101781v

Szabó T, Berkesi O, Forgó P et al (2006) Evolution of surface functional groups in a series of progressively oxidized graphite oxides. Chem Mater 18:2740–2749. https://doi.org/10.1021/cm060258+

Titelman G, Gelman V, Bron S et al (2005) Characteristics and microstructure of aqueous colloidal dispersions of graphite oxide. Carbon 43:641–649. https://doi.org/10.1016/j.carbon.2004.10.035

Wojtoniszak M, Chen X, Kalenczuk R et al (2012) Synthesis, dispersion, and cytocompatibility of graphene oxide and reduced graphene oxide. Colloids Surf B 89:79–85. https://doi.org/10.1016/j.colsurfb.2011.08.026

Yu H, Zhang B, Bulin C et al (2016) High-efficient synthesis of graphene oxide based on improved hummers method. Sci Rep. https://doi.org/10.1038/srep36143

Zaaba N, Foo K, Hashim U et al (2017) Synthesis of graphene oxide using modified hummers method: solvent influence. Procedia Engineering 184:469–477. https://doi.org/10.1016/j.proeng.2017.04.118

Rainfall Trend Analysis of Various Districts of Haryana, India

Deeksha Malik and K. K. Singh

Abstract The study attempts to explore rainfall pattern characteristics in the Haryana region during (1997–2014) Kharif season. In this work, daily and seasonal variations of rainfall along with the determination of dry spells (interval between two wet spells of 7 days magnitude with at least 25 mm of rain) and wet spells (a period of number of consecutive days on each of which precipitation exceeding a specific minimum amount has occurred) during the specified time period have been studied. Mann–Kendall test is applied to detect trend and Sen's slope estimator is for the determination of slope. Results demonstrate that monthly maximum and total rainfall have positive trend and there is a strong spatial relationship in their variability. The increase of monthly precipitation is mainly associated with the increase of frequency and intensity of heavy precipitation during Kharif season. The variation of precipitation is likely to increase flood and drought risk.

Keywords Rainfall · Trends · Dry spells · Wet spells · MK test · Sen's slope estimator

1 Introduction

Variability of rainfall is an important feature of every climate. Change in climate is very likely to increase the magnitude, frequency, and variability of extreme weather events such as droughts, floods and storms. The economic condition of Haryana mostly depends on agriculture. In spite of increased grain yield through various methods, the agricultural scenario of the state is still heavily dependent on annual rainfall and rainfall distribution patterns. Rainfall occurrence and distribution are erratic, temporal and spatial variations in nature. The spatial difference in trends occurs as a result of spatial difference in the changes in the temperature, rainfall and catchment characteristics. For sound crop planning, knowledge of rainfall in any particular region is very helpful. Many researchers have studied the rainfall analysis

D. Malik (✉) · K. K. Singh
NIT Kurukshatra, Kurukshetra 136119, Haryana, India
e-mail: deeksha.malik08@gmail.com

A. K. Agnihotri et al. (eds.), *Sustainable Engineering*, Lecture Notes in Civil Engineering 30, https://doi.org/10.1007/978-981-13-6717-5_10

for crop planning at different temporals as well as spatial scales. Variations in the hydro-metrological series can take place in many different ways. A change can occur suddenly (steep changes) or gradually (trend) or in more complex forms.

Impact of changing climate is quite severe as given by International Protocol for Climate Change (IPCC) reports that there will be reduction in the freshwater availability because of climate change. By the middle of twenty-first century, decrease in annual average run-off and availability of water will project up to 10–30% (IPCC 2007). By the study of different time series data, it has been proved that trend is either increasing or decreasing, both in case of temperature and in case of rainfall. It has been reported that annual average precipitation received by India changed from 4000 (CWC 2005) to 3882.07 (CWC 2008) billion cubic metre (BCM), out of that utilised surface water and groundwater resources are approximated to be only 690 BCM and 433 BCM, respectively, (CWC 2008). Mondal et al. (2012) examined that there are rising rates of precipitation in some months and decreasing trend in some other months in the north-eastern part of Cuttack district, Orissa obtained by the statistical tests suggesting overall insignificant changes in the area. Dubey et al. (2016) examined the significant decreasing trend in annual monsoon and winter rainfall.

The study of trend analysis in rainfall finds use in many applications like irrigation engineering, water harvesting, public health engineering, etc. Also, rainfall trend analysis is very useful in rainfall forecasting. Researchers have utilised various methods and techniques to identify trends and shifts in hydrological series at different scales and places. The present study analyses the trend in rainfall pattern for Kharif season in three districts of Haryana viz. Yamunanagar, Kurukshetra and Panchkula from the year 1997 to 2014 along with wet spells and dry spells analyses.

1.1 Study Area and Data

Haryana is located in the northern part of India. It has a very fertile land and is called the green land of India. Geographically, it is situated at 29.0588°N latitude and 76.0856°E longitude. The area falls under monsoon climate region and annual temperature differs from 7.7 to 40.5 °C. Maximum amount of rainfall occurs during monsoon period, i.e. June to October. Pre-monsoon as the month of April, May and winter constituted the month of November, December and January. The economy of study area is comprised of agriculture and more than 80% population is practicing agriculture here. The climate of the region is characterised by arid to semi-arid. The annual precipitation of the region is about 589 mm. Rainfall of three districts viz. Yamunanagar, Kurukshetra and Panchkula shown in Fig. 1 is chosen for analysis. The daily rainfall data for the period (1997–2014) of these districts were taken from 'https://power.larc.nasa.gov'.

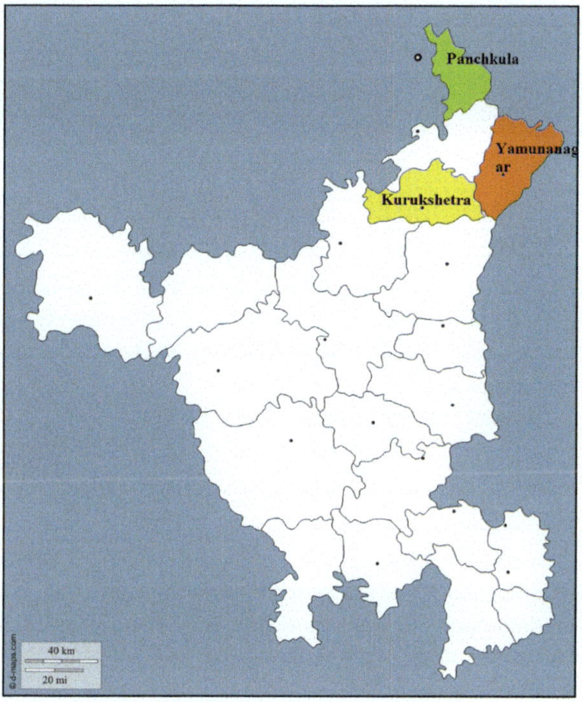

Fig. 1 Study area of Haryana

2 Methodology

Both parametric (regression analysis) and non-parametric can be used to test the trends in the metrological data (rainfall and rainy days) but non-parametric tests are more suitable for non-normally distributed data as compared to statistical tests Lanzante (1996), Jana et al. (2015). Also non-parametric tests can eliminate the problem aroused by data skew (Smith 2000).

In the present study trend, determination in rainfall has been done by Mann–Kendall non-parametric test and determination of slope magnitude by non-parametric Sen's slope estimator in MAKESENS software. Mann–Kendall test is the most frequently used non-parametric statistical tool for identifying the trends in hydro-metrological time series variables such as water quality, stream flow, temperature and precipitation. This is a test which is generally used for studying the temporal trends and spatial variations of hydro-metrological climate series. Also, Mann–Kendall test is preferred when various stations are being tested in a single study (Hirsch et al. 1991). Mann (1945) originally formulated this test as non-parametric test for detection of trend and then test statistic distribution had been given by Kendall (1975). Mostly, previous trend detection has been performed using this technique (Hirsch and Slack 1984; Gan 1988).

2.1 Mann–Kendall Test

Mann–Kendall test is applicable when the data values x_i of a time series can be assumed to obey the model

$$x_i = f(t_i) + \varepsilon_i \tag{1}$$

where $f(t)$ is a continuous monotonic increasing or decreasing function of time and the residuals ε_i can be assumed to be from the same distribution with zero mean. To test the null hypothesis of no trend, H_o (the observations x_i are randomly ordered in time, against the alternative hypothesis, H_1, where there is increasing or decreasing monotonic trends.

The test statistic is defined as follows:

$$S = \sum_{i=0}^{n-1} \sum_{j=i+1}^{n} \operatorname{sgn}(x_j - x_i) \tag{2}$$

where x_j and x_i are the annual values in years, j and i, j > i, respectively, and each of the data points x_i is taken as reference point which is compared with the rest of the data points x_j as

$$\operatorname{Sgn}(x_j - x_i) = \begin{cases} 0, & = (x_j - x_i) \\ +1, & > (x_j - x_i) \\ -1, & < (x_j - x_i) \end{cases} \tag{3}$$

If $n < 10$, the absolute value of S is compared directly to the theoretical distribution of S derived by Mann and Kendall and for $n \geq 10$, the normal approximation test is used (Gilbert 1987). The two-tailed test is used in MAKESENS for four significance levels α: 0.1, 0.05, 0.01 and 0.001. At certain probability level, H_o is rejected in favour of H_1 if the value of S equals or exceeds a specified value $S_{\alpha/2,}$ where $S_{\alpha/2}$ is the smallest S which has the probability less than $\alpha/2$ to appear in case of no trend. A positive value of S indicates an upward trend and a negative value indicates downward trend.

Minimum values of n with which these four significance levels can be reached are derived from the probability table for S as follows:

Significance level (α)	0.1	0.05	0.01	0.001
n required	≥4	≥5	≥6	≥7

The significance level 0.1 means that there is a 10% probability that we make a mistake when rejecting H_o of no trend. Thus, the significance level 0.001 means that the existence of a monotonic trend is very probable.

If n $\geq$ 10, the normal approximation test is used. First, the variance of S is calculated by the following equation which takes into account that ties (equal values) may be present, otherwise, it may reduce the validity of the normal approximation when the number of data values is close to 10.

$$Var(S) = \frac{n(n-1)(2n+5) - \sum_{i=1}^{m} t(i)(i-1)(2i+5)}{18} \quad (4)$$

Here, m is the number of tied groups and t_i is the number of data values in the ith group.

The test statistic Z is as follows:

$$Z_c = \begin{cases} \frac{S-1}{\sqrt{r(Var(S)}}, & S > 0 \\ 0, & S = 0 \\ \frac{S=1}{\sqrt{Var(S)}}, & S < 0 \end{cases} \quad (5)$$

A positive value of Z indicates an upward trend and negative value indicates downward trend. The static Z has a normal distribution. For testing an upward or downward monotonic trend (a two-tailed test) at α level of significance, H_0 is rejected if the value of $Z > Z_{1-\alpha/2}$, where $Z_{1-\alpha/2}$ is obtained from the standard normal cumulative distribution tables. In MAKESENS, the tested significance levels are 0.001, 0.01, 0.05 and 0.1.

2.2 Sen's Method

To estimate the magnitude of existing trend the Sen's non-parametric method is used. This method can be used where the trend assumed to be linear as

$$f(t) = Qt + B \quad (6)$$

Here, the slope of all the data pairs is computed as (Sen 1968)

$$T_i = \frac{(x_j - x_k)}{j-k} \quad \text{for } i = 1, 2, \ldots, n \quad (7)$$

where j > k and x_j and x_k are considered as data values at time j and k, respectively. The median of these N values of T_i is represented as Sen's estimator of slope as

$$Q_i = \begin{cases} T_{(N+1/2)} & N \text{ is odd} \\ 1/2(T_{(N/2)} + T_{(N+\frac{2}{2})}) & N \text{ is even} \end{cases} \quad (8)$$

At the end, a 100(1 − α)% two-sided confidence interval about the slope estimate is obtained by the non-parametric techniques based on the normal distribution. This method is valid for n as small as 10 unless there are many ties. MAKESENS software has been used for analyses.

2.3 Wet and Dry Spells

In this study, the analysis of wet and dry spells has been carried out using daily rainfall data for selected cities in Haryana. The daily rainfall data were examined according to the criteria of wet and dry spells. Wet spell is defined as the 7-day spell where the total amount of rainfall in 7 days equals to 25 mm or more and the condition that 3 out of these 7 days must be rainy with rainfall of more than 2.5 mm of each day. Dry spell like the drought means from agronomic point of view, a dry spell is not a drought of climatic magnitude but a period of a few days or weeks. A dry spell is interval between two wet spells of 7 days magnitude with at least 25 mm of rain. Dry spell at critical crop growth stages affects the growth and the ultimate production of crop.

3 Results and Discussion

Trend analysis of selected cities of Haryana has been done in the present study with 18 years of daily rainfall data from 1997 to 2014. Mann–Kendall and Sen's slope estimators have been used for the determination of the trend. Figures 2, 3 and 4 represent the trend analysis of monthly maximum rainfall for 18 years of Yamunanagar, Panchkula and Kurukshtera districts of Haryana for the months of July, August and September, respectively.

Trend of Yamunanagar for the months of July and October is decreasing with the values of slope −0.64 and −0.45, respectively, and for the months of August

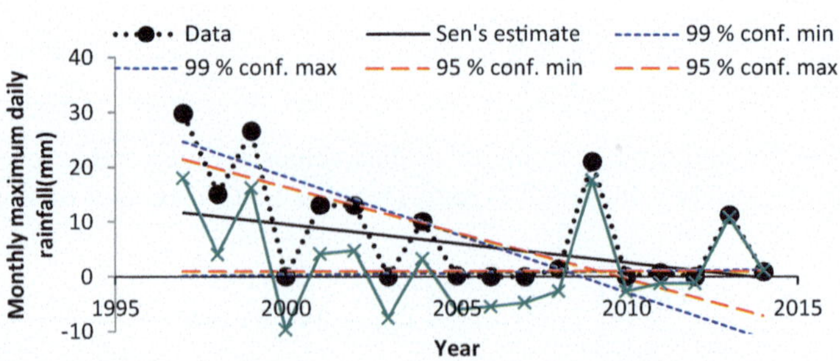

Fig. 2 Variation of monthly maximum daily rainfall for Yamunanagar (July)

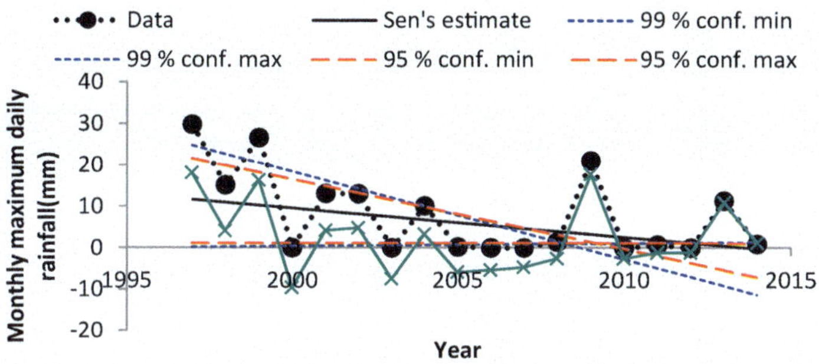

Fig. 3 Variation of monthly maximum daily rainfall for Kurukshetra (August)

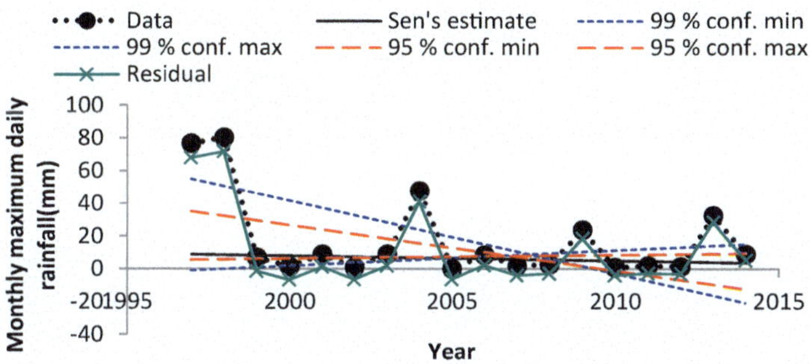

Fig. 4 Variation of monthly maximum daily rainfall for Panchkula (September)

and September is increasing with the slope of 0.15 and 1.67, respectively. Trend of Kurukshetra for the months of July and October is decreasing with the values of slope −1.10 and −1.74, respectively, and for the months of August and September is increasing with the values of 1.82 and 0.38, respectively. Trend of Panchkula for the month of October is decreasing with the value of slope −0.49 and for the months of July, August and September is increasing with the values of 0.61, 1.55 and 1.21, respectively.

From Table 1, probabilities of wet spells of 10 days or more in Yamunanagar in the month of July, August and September are 88%, 100% and 83%, respectively. From Table 2, probabilities of wet spells in Kurukshetra in the month of July, August and September are 77%, 61% and 50%, respectively. From Table 3, probabilities of dry spells in Panchkula in the month of July, August and September are 44%, 50% and 29%, respectively. From these analyses, it is seen that out of three districts Kurukshetra and Panchkula districts require irrigation and are more critical from irrigation point of view so that the crops could be sustained.

From Table 4, probabilities of dry spells of 7 days or more in Yamunanagar in the month of July, August and September are 17%, 39% and 77%, respectively. From Table 5, probabilities of dry spells in Kurukshetra in the month of July, August and September are 77%, 83% and 100%, respectively. From Table 6, probabilities

Table 1 Wet spell numbers and duration in days for Yamunanagar

Year	1997	'98	'99	'00	'01	'02	'03	'04	'05	'06	'07	'08	'09	'10	'11	'12	'13	'14
July	27	31	7, 17	26	28	7, 7	10, 21	11, 12	31	30	13, 18	31	26	31	30	4, 13	31	7, 7
Aug	27	30	7, 12	14	19	7, 14	31	27	9, 12	30	31	28	7, 12	28	31	30	31	19
Sep	18	16	7, 19	8	0	21	7, 10	13	25	9	11, 13	11, 11	11	17	15	7, 12	7, 7	15
Oct	7, 11	7	0	0	0	0	0	0	0	0	0	0	7	0	0	0	0	0

Table 2 Wet spell numbers and duration in days for Kurukshetra

Year	1997	'98	'99	'00	'01	'02	'03	'04	'05	'06	'07	'08	'09	'10	'11	'12	'13	'14
July	27	28	14	20	7, 11	0	8, 11	8	31	8	8, 9	13	10	9, 16	28	9, 11	23	19
Aug	7, 11	24	8	7	27	18	0	25	7	10	7, 7	16	7, 8	7, 7, 13	26	14, 14	26	11
Sep	23	12	13	0	0	0	9	13	9	0	7	11	8	16	10	12	17	13
Oct	15	0	0	0	0	0	0	11	0	0	0	0	0	0	0	0	0	0

Table 3 Wet spell numbers and duration in days for Panchkula

Year	1997	’98	’99	’00	’01	’02	’03	’04	’05	’06	’07	’08	’09	’10	’11	’12	’13	’14
July	7	8	14	8	10	0	8, 10	0	31	7	10, 13	8, 9	15	7, 15	7, 7	8	7, 15	8
Aug	7, 10	6, 12	8	13	8	11	7, 10	7, 15	0	0	7	9	7	9	8, 17	7, 13	12	7, 8
Sep	8	14	9	0	0	7	0	0	7	7	7	8	7	15	9, 30	7, 8	8	11
Oct	11	0	0	0	0	0	0	7	0	0	0	0	0	0	0	0	7	0

Table 4 Dry spell numbers and duration in days for Yamunanagar

Year	1997	'98	'99	'00	'01	'02	'03	'04	'05	'06	'07	'08	'09	'10	'11	'12	'13	'14
July	5	0	7	5	3	8, 9	0	8	0	0	0	0	5	0	0	4	0	3
Aug	4	0	3, 9	7	12	10	0	4	10	0	0	3	3, 8	3	0	0	0	12
Sep	3, 9	6, 8	22	22	31	9	6	8, 9	2, 3	21	5	7	4, 15	13	15	2, 8	5, 5	15
Oct	11	11, 13	31	31	31	31	31	31	31	31	31	31	24	31	31	31	31	31

Table 5 Dry spell numbers and duration in days for Kurukshetra

Year	1997	'98	'99	'00	'01	'02	'03	'04	'05	'06	'07	'08	'09	'10	'11	'12	'13	'14
July	0	0	17	11	7	31	7	11, 12	0	22	11	7	7, 15	7	0	10	9	12
Aug	13	7	20	7, 17	31	13	31	7	10, 13	10, 12	16	15	8, 10	0	7	0	4	20
Sep	8	7, 11	16	30	30	30	10, 11	14	17	30	22	8, 12	22	14	20	7, 13	18	16
Oct	16	31	31	31	31	31	31	14	31	31	31	31	31	31	31	31	31	31

Table 6 Dry spell numbers and duration in days for Panchkula

Year	1997	'98	'99	'00	'01	'02	'03	'04	'05	'06	'07	'08	'09	'10	'11	'12	'13	'14
July	7, 17	10, 13	17	11, 12	18	31	7	31	0	8, 15	9	8	8, 8	8	12	23	0	11, 12
Aug	12	8	20	8, 10	21	20	10	7	31	31	24	22	9, 15	8, 15	6	7	8, 11	17
Sep	7, 15	6, 10	17	30	30	23	30	30	7, 16	23	23	9, 13	23	16	8	4	22	18
Oct	9, 11	31	31	31	31	31	31	24	31	31	31	31	31	31	31	31	8, 15	31

of dry spells in Panchkula in the month of July, August and September are 88%, 94% and 94%, respectively. From these analyses, it is seen that out of three districts Kurukshetra and Panchkula districts irrigation requirements are more critical for crops sustenance.

4 Conclusions

The present study analysed the rainfall data for 18 years from 1997 to 2014 of selected cities of Haryana (Yamunanagar, Kurukshetra and Panchkula) for the determination of trend in maximum monthly daily rainfall during Kharif season using Mann–Kendall test along with Sen's slope estimator for the magnitude of trend. It can be concluded that rainfall of July in Yamunanagar and Kurukshetra is decreasing with slope >−0.6. This will have implication on Kharif crop sowing. Yamunanagar is having the maximum number of wet spells. Kurukshetra and Yamunanagar are having long periods of dry spells.

References

Chandler RE, Wheater HS (2002) Analysis of rainfall variability using generalized linear models: a case study from the west of Ireland. Water Resour Res 38(10)

CWC (Central water commision) (2005) Water data book. CWC, New Delhi. http://cwc.gov.in/main/downloads/Water_Data_Complete_Book.pdf

CWC (Central water commision) (2008–2009). Water data book. CWC, New Delhi. http://cwc.gov.in/main/downloads/AR_08_09.pdf

Goyal MK (2014) Statistical analysis of long term trends of rainfall during 1901–2002 at Assam, India. Water Resour Manag 28(6):1501–1515

Haylock M, Collins D, Trewin B, Rahimzadeh F, Tagipour A (2006) Global observed changes in daily climate extremes of temperature and precipitation. J Geophys Res Atmos 111(D5)

Jain SK, Kumar V, Saharia M (2013) Analysis of rainfall and temperature trends in northeast India. Int J Climatol 33(4):968–978

Jana C, Sharma GC, Alam NM, Mishra PK, Dubey SK, Kumar R (2015) Trend analysis of rainfall and rainy days of Agra in northern India. Int J Agric Stat Sci 12(1):263–270

Kruger AC (2006) Observed trends in daily precipitation indices in South Africa: 1910–2004. Int J Climatol 26(15):2275–2285

Kumar V, Jain SK, Singh Y (2010) Analysis of long-term rainfall trends in India. Hydrol Sci J J des Sci Hydrol 55(4):484–496

Luis MD, Raventós J, González-Hidalgo JC, Sánchez JR, Cortina J (2000) Spatial analysis of rainfall trends in the region of Valencia (East Spain). Int J Climatol 20(12):1451–1469

Mondal A, Kundu S, Mukhopadhyay A (2012) Rainfall trend analysis by Mann-Kendall test: a case study of north-eastern part of Cuttack district, Orissa. Int J Geol Earth Environ Sci 2(1):70–78

New M, Hewitson B, Stephenson DB, Tsiga A, Kruger A, Manhique A, Gomez B, Coelho CA, Masisi DN, Kululanga E, Mbambalala E (2006) Evidence of trends in daily climate extremes over southern and west Africa. J Geophys Res Atmos 111(D14)

Rajeevan M, Bhate J, Kale JD, Lal B (2006) High resolution daily gridded rainfall data for the Indian region: analysis of break and active monsoon spells. Curr Sci 296–306

Sridhar SI, Raviraj A (2017) Statistical trend analysis of rainfall in Amaravathi river basin using Mann-Kendall test. Curr World Environ 12(1):89–96

Sivakumar MVK (1992) Empirical analysis of dry spells for agricultural applications in West Africa. J Clim 5(5):532–539

Infiltration—A Critical Review

Satyam Garg and Arun Goel

Abstract Infiltration is an important process of hydrological cycle and needs to be studied for the proper and efficient management of water resources. Due to scarcity of water, especially in India, the study of various aspects of infiltration is of paramount importance. The rate and amount of infiltration are must to estimate the water requirements for an efficient irrigation system. The infiltration capacity is dependent mainly on soil texture, soil structure, vegetative cover, soil compaction, rainfall time, existing soil moisture content, etc. Infiltration characteristics can be estimated either by cylindrical infiltrometer method or by using mathematical models. The paper presents a review of research undertaken in the area of estimation of infiltration rate/capacity by using experimental methods and mathematical models, highlighting the major findings of the past studies. It is expected that outcome of this article may be utilized for selecting an appropriate technique for analyzing the problem of infiltration rate/capacity, which is urgently required for the management of available water resources in India.

Keywords Infiltration · Models · Soil compaction · Modeling · Infiltrometer

1 Introduction

Infiltration is a downward movement of water from soil surface to the subsurface layer. Infiltration is defined as the process by which a fluid passes through another substance traveling through pores and interstices (Simpson and Weiner 1989). For surface irrigation, water is pounded on the soil surface and the infiltration rate describes the flux into the soil profile. First, the water wets soil grains and then the extra water moves down due to resulting gravitational force. This downward movement of water with respect to time is known as infiltration rate and its unit is mm/h or cm/h.

S. Garg (✉) · A. Goel
Department of Civil Engineering, NIT Kurukshetra, Kurukshetra 136119, Haryana, India
e-mail: satyam.adg@gmail.com

A. K. Agnihotri et al. (eds.), *Sustainable Engineering*, Lecture Notes
in Civil Engineering 30, https://doi.org/10.1007/978-981-13-6717-5_11

The constant infiltration rates of different soils under different soil conditions were calculated by double ring infiltrometer method and compared with values calculated from Kostiakov (1932), which modifies Kostiakov model, Horton's model, Swartzendruber–Clague model, and Green–Ampt infiltration model. The assessment of the usability of various theories for evaluation of infiltration rate of particular soil under particular soil condition was carried out with correlation coefficients and standard errors. In light of its basic importance in ground water, the infiltration has a lot of consideration into geotechnical and hydraulic researchers. Certain variables, e.g., temperature of air, rain and soil surface, rain power, introductory soil dampness, and starting and remaining precipitation affect the infiltration capacity.

Tragically that the expressions "infiltration-capacity" and "infiltration rates" have not now and then been properly understood. Infiltration can happen at any rate from zero up to the capacity rate. On the other side, when the rain power is not as much as the capacity, at that point the infiltration rate is the same as capacity. If infiltration is occurring at capacity rate, it is preferred to as capacity, not simply as a rate. At the end of the day, there might be vast values of rates but there is just a single capacity at a specific time for a specific soil. Few important research works undertaken in the past for estimating the infiltration rates are highlighted in Table 1.

Table 1 Comparison of infiltration rate with different models and instruments

Sr. no	Name of author(s)	Year	Models\instruments	Objectives of study	Results obtained	Conclusions
1	E. G. Youngs	1991	Ring infiltrometer rainfall simulator and tension infiltrometer	Unsaturated Soils infiltration rate	20–200 mm/h of infiltration rate	Highlights 3-d flow nonuniformity in infiltration rate of soils
2	Basant Maheshwari	1996	Modified double ring infiltrometer	Use of water level sensors and software for data recording	Infiltration rate is 50–250 mm/h	Precise results up to a minute level accuracy of infiltration data are obtained
3	Carol P. Harden, P. Delmas Scruggs	2003	Single ring infiltrometer	Relation between rainfall, runoff and infiltration at three different mountainous regions	Infiltration rate for three regions vary from 17 to 34, 16 to 113, 2 to 99 mm/h	Infiltration is higher than rainfall, surface runoff is rare

(continued)

Table 1 (continued)

Sr. no	Name of author(s)	Year	Models\instruments	Objectives of study	Results obtained	Conclusions
4	S. Mohan and Kumari Sangeeta	2005	Green–Ampt, modified Kostiakov and Horton model	Estimation of recharge in four Neyveli regions	12–20% of rainfall is converted into recharge	Horton's model was found best fitted for recharge estimation
5	Fantaw Yimer and A. Abdelkadir	2008	Double ring infiltrometer with 30 cm inner and 50 cm outer diameter	Different land use type and slopes of Ethiopia	For cultivated and grazed 0.15 cm/min; 0.50 cm/h for forest land	Infiltration rate and cumulative infiltration are less cultivated and grazed than forestland
6	Chandrakantj jejurkar and Milind Rajurkar	2014	Double ring infiltrometer and Kostiakov and modified Kostiakov models	Infiltration rate and cumulative infiltration for varying soil physical properties	0.15–2.24 cm/h of infiltration rate for clay soil	Moisture content, electrical conductivity and soil texture affect infiltration rate: Kostiakov model is best fit
7	Debjit Roy	2015	Conel sprinkler infiltrometer	Infiltration rate of drained and undrained soils.	0.041–0.023 mm/min for drained and undrained respectively	Subsurface drained soil has more porosity, water holding capacity and infiltration rate
8	Isabelle Braud	2017	Single and double ring infiltrometer; tension infiltrometer	Unsaturated hydraulic conductivity	Unsaturated hydraulic conductivity is 38–996.6 mm/h	It varies with respect to site geology, soil type and vegetative cover of soil

2 Factors Affecting Infiltration Rate

The infiltration rate is controlled by the communication of various physical and synthetic soil attributes. These soil properties shift starting with one area to the next and change after some time because of social practices (e.g., culturing and compaction), water administration and organic procedures. The details of some of the important studies related to the factors affecting infiltration rate are summarized in Table 2.

Table 2 Properties of soil affecting infiltration rate

Sr no.	Author(s)	Years	Properties	Objective of the study	Results/outcome
1	Van Es	1991	Soil texture	Clay and silt contents infiltration characteristics	Infiltration rate is positively correlated with clay and negatively with silt content
	M. Childs	1993		Sandy soil infiltration characteristics	Higher infiltration rate in sandy soils
	Brady and Weil	2002		Effect of coarse sand in fine clay soils	Reduction in infiltration rate if coarse sand is present in clay soils
2	Brown	1988	Soil erosion	Variation of infiltration along furrow length	Infiltration rate is 1.5–2 times more in upstream than downstream of furrow
	Antonio and Alvarez	2003		Relationship between inflow rate and infiltration rate	Flow depth and wetted perimeter of nonerosive discharge effect furrow infiltration
	Horst	2007		Flow regime effect on cross section of furrow	Surge inflow causes removal of material from side wall and deposit to bed of furrow
3	Trout and Mackey	1988	Soil structure and compaction	Effect on infiltration rate due to compaction	20% higher infiltration rate in uncompact furrows and more than a 50% reduction in alternate wheeled furrows

(continued)

Table 2 (continued)

Sr no.	Author(s)	Years	Properties	Objective of the study	Results/outcome
	Jaynes and Hunsaker	1989		Bulk density and infiltration volume relation	25% of the variation in infiltrated volumes is by the variance in surface bulk density
	Lal and Shukla	2004		Causes and effect of soil pore size on infiltration rate	Biological activity, shrinkage from temperature or moisture effects, changes soil pore size and further affect infiltration rate
4	Mailhol and Gonzalez	1993	Soil moisture content	Cracking component of soil moisture content in clay soils	Degree of soil cracking is directly proportional to initial infiltration rate
	Raine	1998		Spatially and temporally variation of moisture content and infiltration	Due to rainfall, uniformity of previous irrigations, evaporation and plant extraction moisture content changes
	Robertson	2004		Relationship between initial infiltration and moisture content	Inverse linear relationship between initial infiltration and moisture content

3 Measurement of Infiltration

Measurement of infiltration rate/capacity can be done either by using different mathematical models suggested by researchers or by using instruments like infiltrometers, etc.

3.1 *Infiltration Models*

Following are few important infiltration models suggested by different researchers. These models have some parameters those can be obtained from various curves and field data as explained below.

3.1.1 Kostiakov Model

The relation between infiltration and time is given by Kostiakov in the following form:

$$F = at^{x} \quad (1)$$

$$\text{Further modified as} \quad F = at^{x} + f_1 \quad (2)$$

In this equation, cumulative infiltration is represented by F (mm), the slope between time and cumulative infiltration curve is represented by x, cumulative infiltration after unit time is denoted by a, with unit (mmha), time is represented by t (hours), and final infiltration is f_1 in (mm/h).

3.1.2 Horton Model

Horton (1941) describes relationship between initial infiltration rate, final infiltration rate, and cumulative infiltration as given below:

$$F = f_c + (f_o - f_c)e^{(-kt)} \quad (3)$$

Cumulative infiltration is represented by F (mm), final rate of infiltration is denoted by f_c in (mm/h), and initial rate of infiltration is denoted by f_o in (mm/h). In Fig. 1, graph showing variation of infiltration rate with respect to time duration is plotted using Horton's equation.

3.1.3 Green–Ampt Model

Green and Ampt (1911) gives two equations, one for infiltration rate and another for cumulative infiltration, which are as follows:

$$F(t) = k_1 t + a \ln (1 + F(t)/a) \quad (4)$$

$$f(t) = k_1 + k_2/F(t) \quad (5)$$

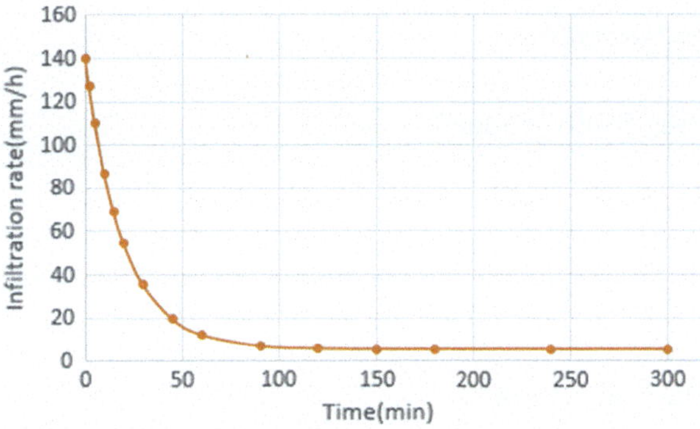

Fig. 1 Infiltration rate (mm/h) versus time (h) graph by Horton equation

Soil saturated hydraulic conductivity is k_1 at time t (mm/h), a is constant, cumulative infiltration is represented by F(t) (mm), and rate of infiltration is denoted by f in (mm/h).

3.1.4 Swartzendruber–Clague Model

This is a semi in empirical model, according to which infiltration capacity is given by following relation:

$$f_t = K_s + [\left(Se^{-as\sqrt{t}}\right) \div \left(2\sqrt{t}\right)] \quad (6)$$

Cumulative infiltration capacity is as follows

$$F_t = K_s t + \left[S\left(1 - e^{(-as\sqrt{t})}\right)\right]/a_s \quad (7)$$

a_s is coefficient of Swartzendruber–Clague model, S is sorptivity of soil.

The standard values of model parameters of mathematical model are given in Table 3.

Table 3 Values of different models parameters under different soil conditions

Horton's model $f = f_c + (f_o - f_c)e^{-kt}$	Green–Ampt model $f = m + n/f$		Kostiakov model $f = at^b$		Modified Kostiakov model $f = at^b + c$		
k	m	n	A	b	a	b	c
1.4–6.36	0.98–5.7	1.823–3.65	0.13–2.5	0.5–0.7	0.1–3.6	0.3–0.8	0.13–3.1

3.2 Field Measurement

3.2.1 Single Ring Infiltrometer

Ring-type infiltrometer, being maybe the most widely recognized sort of infiltrometer utilized as it is modest to build and work, requires moderately little water, and just a single individual can set up and run a few tests at the same time.

3.2.2 Double Ring Infiltrometer

Two concentric rings of stainless steel are commonly employed. The larger ring forms a buffer compartment around the inner to account for lateral flow. The rings are jacked or hammered into the ground 5–10 cm. Care is taken to minimize disturbance of the soil surface and the soil structure during installation. A specific and constant head of water (less than 5 cm of depth) is maintained in both rings, while the rate of water usage from the inner ring is measured. The period of time required to accomplish enduring infiltration rate ranges from 2 to 6 h contingent on soil type, surface, and soil moisture conditions.

A series of tests done on similar soils by Bedwany and Schumacher (1979) found that infiltration rates for the double ring were up to 10 times greater than that of the basin. The difference is due to lateral seepage and soil cracking. Chow et al. (1988) reported data in which the double ring infiltrometer gave rates eight times larger than a rainfall infiltrometer for overgrazed pasture and two times larger for lightly grazed. Duos of both sides unwrapped stainless steel rings are used whose diameters are 15–30, 15–45, 30–45, 30–60 cm, and height is 60 cm. One side of the ring will have a cutting edge so that it will get-up-and-go inside the soil surface easily without any damage to the soil or to the cylinder. A serious limitation to the use of ring infiltrometers is the method of placement. Hammering or jacking the rings into the shattered soils, more commonly caused by hammering, can disturb the interface between the soil and the metal ring resulting in leakage and abnormally high and variable infiltration rates (Fig. 2).

In one place for constantly 5–6 days, infiltration readings were taken so that the amount of infiltration rate can be acknowledged. At a time, interval of 5 min or 10 min was taken and the fall in the water level was measured. Comparison of different mathematical infiltration models using correlation coefficient and standard error with field infiltration rate could be done for finding which model is best fitted, as done by Dagadu and Nimbalkar (2012). The metal plate will shield the soil surface of the inner cylinder so that it can reduce the force of water poured inside the cylinder. In this method, water logging area can be taken into consideration. For places which are waterlogged, double ring infiltrometer can be used from where the amount of water which will directly percolate toward the groundwater and recharge it without any loss or wastage of water can be calculated.

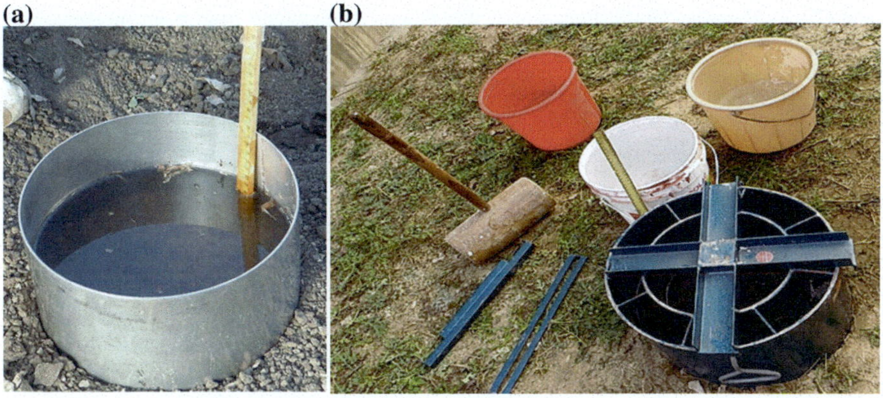

Fig. 2 a Single ring infiltrometer; b double ring infiltrometer

4 Conclusion

Infiltration capacity/rate depends upon soil properties like moisture content, electrical conductivity, soil surface, etc. Other factors which affect infiltration are site geology, soil composition, and vegetative covers of soil. Many past studies have concluded that infiltration rate and cumulative infiltration are comparatively less in the cultivated and grazed land than the forestland. The infiltration rate can be obtained either from field measurement by using infiltrometers or by applying the mathematical models on the observed data as mentioned by many researchers.

Out of the several available mathematical models, the Horton's model was best suited for recharge estimation, whereas Kostiakov's model is giving better fit for infiltration rate estimation. Many studies have also concluded that the double ring infiltrometer has been performing better as compared to the single ring infiltrometer, when used for infiltration studies in the field. It is expected that outcome of this paper will be useful for understanding the use and application of ring infiltrometer and mathematical models for estimation of infiltration rate/capacity of given soil under specific field conditions successfully.

References

Abdelkadir A, Yimer F (2011) Soil water property variations in three adjacent land use types in the Rift Valley area of Ethiopia. J Arid Environ 75(11):1067–1071

Bedwany AL, Schumacher AE (1979) A comprehensive study of two methods of infiltration rate measurements on some soils in Lesotho. Montreal Engineering Company Limited, Calgary, Alberta. 11pp

Brady NC, Weil RR (2002) The nature and properties of soils, 13th. Prentice Hall, Upper Saddle River, N.J.

Braud (2017) Mapping topsoil field-saturated hydraulic conductivity from point measurements using different methods. J Hydrol Hydromech 65
Brown MJ et al (1988) Sediment, erosion and water-intake in furrows. Irrig Sci 9(1):45–55
Childs JL et al (1993) Spatial and seasonal variation of furrow infiltration. J Irrig Drain Eng 119(1):74–90
Chow VT, Maidment DR, Mays LW (1988) Applied hydrology. McGraw-Hill Inc., Singapore, pp 1–572
Dagadu JS, Nimbalkar PT (2012) Infiltration studies of different soil under different soil conditions and comparison of infiltration models with field data. Int J Adv Eng Technol 154–157
Green WH, Ampt GA (1911) Studies of soil physics—part I: the flow of air and water through soils. J Agric Sci 4(1):1–24
Harden CP, Delmas Scruggs P (2003) Infiltration on mountain slopes: a comparison of three environment. Geomorphology 55:5–34
Horton RE (1941) An approach towards a physical interpretation of infiltration capacity. Sci Soc Am J 5:399–417
Kostiakov AN (1932) On the dynamics of the coefficient of water percolation in soils and on the necessity of studying it from a dynamic point of view for the purposes of amelioration. Part A, Moscow
Maheshwari BL (1996) Development of an automated double-ring infiltrometers. Aust J Soil Res 34:709–714
Mohan S, Sangeeta K (2005) Recharge estimation using infiltration models, ISH. J Hydraul Eng 11:1–10
Raine SR et al (1998) The effect of variable infiltration on design and management guidelines for surface irrigation. ASSCI national soils conference
Simpson JA, Weiner ESC (1989) The Oxford English dictionary, 2nd edn. Clarendon Press, Oxford University Press, Oxford, New York
Singh RV, Bhakar SR (2004) Comparison of different infiltration equations on different land covers
Van Es HM et al (1991) Infiltration variability and correlations with surface soil properties for an eroded Hapludult. Soil Sci Soc Am J 55(2):486–492
Yimer F, Abdelkadir A (2008) Effects of different land use types on infiltration capacity in a catchment in the highlands of Ethiopia, The Authors. Journal compilation @ 2008 British Society of Soil Science

Economy of Design of Rigid Pavement with Tied Concrete Shoulders

Pooja Singh and S. N. Sachdeva

Abstract Roads play a vital role in the development of a country. A good network of highways is important as it helps in the development of trade, commerce and other activities that characterise a vibrant and forward-looking nation. As from the recent past years, India is shifting toward the construction of rigid pavements, and it is high time that all the factors that affect the economy of its construction should be taken into consideration. One of the main factors is the type of shoulders provided. Even a small decrease in the thickness of pavement leads to an appreciable saving of the material used in its construction, thereby affecting the economy of the project. Therefore, an attempt has been made to design a two-lane two-way National Highway proposed to be laid in Haryana by considering various types of shoulders, viz., tied concrete shoulders and earthen shoulders. It is observed that by providing tied concrete shoulder, there is an appreciable reduction in the slab thickness appreciably which affects the economy of project, thereby attaining the goals of sustainable development in an effective manner.

1 Introduction

One of the factors affecting the thickness of cement concrete pavement is the construction of shoulders. Well-designed and maintained shoulders play an important role in the design of cement concrete pavement. Tied concrete shoulders provide a lateral support to the main slab of pavement and also results in additional stiffness. Tied concrete shoulders reduce the pavement stresses and edge deflection and increase the load distribution capability significantly. Moreover, the wide shoulders can also serve as an extra lane, thereby maintaining the desired level of service and can be used as temporary parking area in populated urban regions. Also, if rough texture is provided to it, it will bring in an additional safety of vehicle especially during night hours. This will also cut the economy of the future project as this concrete shoulder itself can be extended to make a new lane.

P. Singh (✉) · S. N. Sachdeva
Department of Civil Engineering, NIT Kurukshetra, Kurukshetra, Haryana, India
e-mail: pooja.nitkkr@gmail.com

A. K. Agnihotri et al. (eds.), *Sustainable Engineering*, Lecture Notes in Civil Engineering 30, https://doi.org/10.1007/978-981-13-6717-5_12

The design of rigid pavement is done as per IRC: 58-2015: Guidelines for design of plain jointed rigid pavements for highways (4th revision), which is based upon the fatigue damage analysis for both bottom-up cracking and top-down cracking. If the sum of cumulative fatigue damage caused by different axle loads is less than one for bottom-up and top-down cracking, the thickness of the pavement is safe.

In this paper, the design of two-lane two-way National highway is done which is proposed to be constructed in Haryana. The design has been done with dry lean concrete as subbase of thickness 150 mm and GSB drainage layer of 150 mm thickness beneath M40 grade PQC slab.

AASHTO introduced the concept of tied concrete shoulders in the year 1993 in which a procedure is provided for design of rigid pavements with tied concrete shoulders.

In the final report of "Guide for Mechanistic-Empirical design of new and rehabilitated pavement structures", Part 3. Design Analysis, Chapter 2: Shoulders, the advantages and requirements of paved shoulders are mentioned and also it is mentioned that the concrete shoulders should be tied to the mainline with properly spaced and sized tie bars. Tied concrete shoulders that maintain the tight joint and good load transfer will reduce pavement stresses and edge deflections. Tie bars are needed between the mainline pavement and concrete shoulders to keep the longitudinal joint tight to provide the necessary load transfer and avoid opening of the joint and a safety problem.

Colley et al. (1977) (Portland Cement Association) in their research paper on "Evaluation of Concrete Pavements With Tied Shoulders or Widened Lanes" have taken field and laboratory pavements which were instrumented and load tested to evaluate the effect of widened lanes, concrete shoulders and slab thickness on measured strains and deflections, and it is reported that concrete shoulders were effective in reducing the magnitude of measured strains and deflections.

2 Methodology

A rigid pavement for two-lane divided state highway into state of Haryana is designed. The axle load data is given in Table 1.

The design life for the study is taken as 30 years. The growth of traffic over the years is taken as 7.5%. The variation of thickness of slabs for pavement with and without tied concrete shoulders to be observed is for the design traffic value of 2500 commercial vehicle per day in each direction. The percentages of front single axle, rear single axle, rear tandem axle and rear tridem axle are taken as 45, 15, 25 and 15%, respectively. Average number of axles per commercial vehicle is assumed to be 2.35. The percentage of commercial vehicles with spacing between the front axle and the first rear axle less than 4.5 m is taken as 55%, and it is assumed that 50% of the vehicles travel during the night hours. Design flexural strength of concrete is taken as 4.95 MPa with a unit weight of concrete as 24 kN/m^3 and elastic modulus as 30,000 MPa. The analysis is done assuming both tied concrete shoulders and

Table 1 Axle load distribution

Single axle load (kN)	Frequency (%)	Tandem axle load (kN)	Frequency (%)	Tridem axle load (kN)	Frequency (%)
190	18.15	390	14.5	545	5.23
180	17.43	370	10.5	515	4.85
170	18.27	350	3.63	485	3.44
160	12.98	330	2.5	455	7.12
150	2.98	310	2.69	425	10.11
140	1.62	290	1.26	395	12.01
130	2.62	270	3.9	365	15.57
120	2.65	250	5.19	335	13.28
110	2.65	230	6.3	305	4.55
100	3.25	210	6.4	275	3.16
90	3.25	190	8.9	245	3.10
80	14.15	179	34.23	215	17.58
	100		100		100

earthen shoulders. Dowel bars are also assumed to be provided. As the design of pavement is based upon fatigue damage analysis, therefore, for the design to be safe, the minimum thickness for which the cumulative fatigue damage comes out to be less than or equal to one is provided in this case.

3 Results and Discussion

When dry lean concrete of 150 mm thickness is provided as subbase with tied concrete shoulders and untied concrete shoulders and dowel bars across the transverse joints, the most economical pavement thickness for subgrade CBR values ranging from 2 to 10% are shown in Fig. 1.

It can be seen that the safe thickness of pavement for 2% CBR of subgrade soil reduces to 29 cm from 32.5 cm when the pavement is designed with tied concrete shoulders. For 4, 6, 8 and 10% CBR of subgrade soil, the safe thickness with tied concrete shoulders remains constant and comes out to be 28.5 cm

From Fig. 2, it can be seen that there is a saving of 13.63% in thickness of pavement for 6, 8 and 10% CBR of subgrade soil, whereas for 2 and 4% CBR of subgrade soil, the % saving in thickness are 10.77 and 12.12%, respectively.

For a stretch of 1 km and for pavement thickness of 29 cm, when the tied concrete shoulders is provided, number of tie bars to be provided comes out to be around 4847 which is three times of it when the pavement is constructed with earthen shoulders. But this increase in number of tie bars does not affect the economy of project adversely

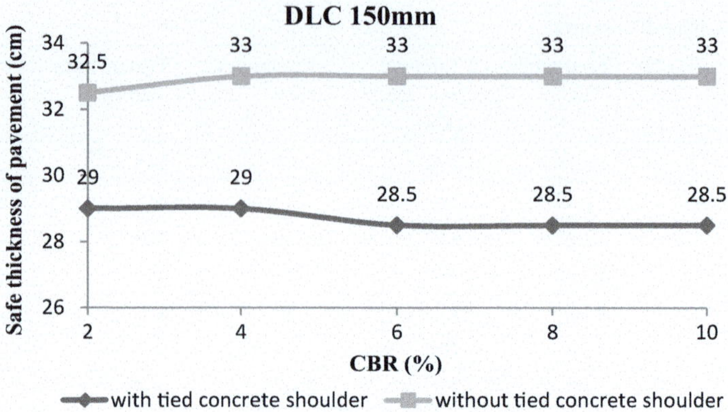

Fig. 1 Safe thickness of pavement at different subgrade strengths

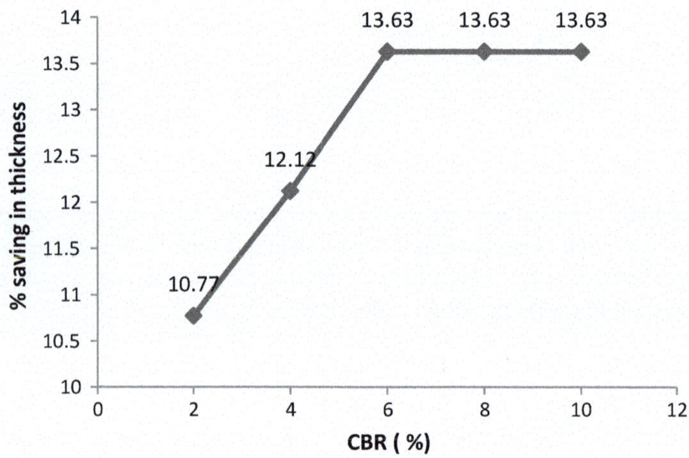

Fig. 2 Percentage saving in thickness of pavement versus CBR of subgrade soil

as it serves the purpose in long run. The design of dowel bars for tied concrete shoulders is done in the same way as that of the main slab. As per IRC: 58-2015, for a single axle load of 200 kN and a positive temperature differential of 17 °C, this axle load causes higher flexural stresses in thicker slabs for higher k-values. For a thickness in the region of 280 mm, there is practically no effect of modulus of subgrade reaction on flexural stresses. Increasing the subgrade modulus to high values does not help in thickness design due to high curling stresses caused by a stiff support.

4 Conclusions

1. In the analysis of the pavement with earthen shoulders, subgrade strength has insignificant effect on the slab thickness.
2. Analysis with tied concrete shoulders shows somewhat different results, but still after a subgrade strength, thickness remains constant.
3. In the analysis of pavement with tied concrete shoulder and earthen shoulders, there is a saving in thickness of pavement of around 4.5 cm when we provide tied concrete shoulders.
4. Although the number of tie bars and dowel bars to be used in tied concrete shoulders case comes out to be more, as this paved concrete shoulder can itself be extended as a new lane for future purpose so ultimately, it results in economising the future project on that road.

References

AASHTO (1993) Guide for design of pavement structures

Colley BE, Ball CG, Arriyavat P (1977) Evaluation of concrete pavements with lane widening, tied concrete shoulders, and thickened pavement. Minnesota Department of Transportation

Indian Road Congress, IRC: 58-2015, Guidelines for the design of plain jointed rigid pavement for highways

Singh S, Sachdeva SN (2015) Thickness requirement of a rigid pavement with varying conditions of subgrade, sub-base and shoulders. IOSR J Mech Civil Eng 52–57. e-ISSN:2278-1684, p-ISSN:2320-334X

Hallmark SL, McDonald TJ, Tian Y, Andersen DJ Safety benefits of paved shoulders. Report no.-InTrans Project 05-239, Center for Transportation Research and Education Iowa State University

Application of Aspergillus Niger for Biomachining of Aluminium Alloy 4004

Pallvi Verma, Amanpreet Kaur Sodhi and Neeraj Bhanot

Abstract Biomachining is one of the latest technologies of coming future in which metal removal takes place with very less amount of energy. To reduce the environmental concerning problems in other modes of machining, biomachining is now developed to remove the surface of workpiece by using metabolic activity of microorganisms with various advantages like low energy consumption and no heat generation during machining, etc. In this study, an attempt has been made to check the feasibility of biomachining on aluminium alloy 4004 using Aspergillus niger. The trials have been carried out at room temperature for regular interval of 24 hours till first 7 days based on which specific metal removal rate has been analysed to confirm the scope of its application.

Keywords Biomachining · Aspergillus niger · Aluminium alloy 4004 · Specific metal removal rate

1 Introduction

Machining is a technique that is used to remove the surface of material and make it in desired shape and dimensions. Conventional machining is commonly used for removal of metal from workpiece. Conventional machining is fast and efficient process for metal removal but has various negative impacts on environment as well as human health (Kovoor et al. 2012). The problem related to conventional machining is that it results in metallic wastes like metal chips, metal curling, metal cutting fluids, etc. that creates major environmental as well as health problems like skin problems, respiratory diseases and even cancer. To overcome these types of issues, an alternate method of machining is adopted, i.e. Biomachining.

Biomachining is a technique in which removal of metal surface from a workpiece is done using microorganisms. Biological techniques are nowadays most promising technique used for material processing. Biotechnology results in widely used

P. Verma (✉) · A. K. Sodhi · N. Bhanot
Guru Nanak Dev Engineering College, Ludhiana, Punjab, India
e-mail: pallviverma100@gmail.com

A. K. Agnihotri et al. (eds.), *Sustainable Engineering*, Lecture Notes in Civil Engineering 30, https://doi.org/10.1007/978-981-13-6717-5_13

of microorganisms for processing of materials (Hocheng et al. 2012). The use of microorganisms for the metal removal is the most environmental friendly process than other means (Zhang et al. 2014). Jadhav and Hocheng (2014) also suggested the feasibility of Aspergillus niger for biomachining of various metals at large as well as nanoscale. The potential of organic acid producing fungi for the biomachining of aluminium alloy 4004 has not been explored yet. Considering this, the present study indicates biomachining of aluminium alloy 4004 by using Aspergillus niger.

Aluminium alloy 4004 has a low melting point and narrow melt range, which makes it suitable for brazing aluminium alloys. In addition, the increased silicon content compared to other aluminium filler metals provides increased fluidity as well as reduced shrinkage. The use of AL 4004 also significantly reduces hot cracking during the brazing process. The increased magnesium content aids in preventing oxidation of the braze surface during heating (Material data AL 4004). Thus, aluminium alloy is widely used nowadays for construction purposes. The problems related to waste generation by conventional machining during metal removal is overcome by this study. This study focuses to check the feasibility of Aspergillus niger for biomachining of aluminium alloy 4004.

2 Literature Review

Uno et al. (1996) discussed the basic fundamental study on biomachining. Biomachining is a biological process using microorganisms such as acidithiobacillus ferrooxidans and acidithiobacillus thiooxidans, which are also used in the mineral refinement process of bioleaching. So in recent days, this process will come widely under consideration for machining of metal surfaces and various researches have been also carried out on the different materials including steel, copper, aluminium and various alloys. Ting et al. (2000) discussed the potential of microorganism for the biomachining process. Thiobacillus ferrooxidans are used for biomachining as an alternate method to conventional machining. Chang et al. (2008) checked and analysed specific metal removal rate from 1 cm^2 area in hour by a solution that contains 1 x 10^8 cells/ml. The results conclude that specific metal removal rates for copper, aluminium and nickel are 0.5 mg/h, 0.06 mg/h and 0.93 mg/h, respectively. Istiyanto et al. (2014) checked the profile and surface of biomachining uni-crystal copper. These results in a u-shaped profile of a rectangular groove having width and depth that is proportional to the machining time. Jadhav and Hocheng (2014) showed the effect of Aspergillus niger on various metals like tin, copper and aluminium. Complete material removal was achieved for tin metal by 100 ml culture supernatant, at 150 rpm shaking speed and at 30 °C temperature, in 72 h. These results suggest that it is feasible to apply A. Niger 34770 culture supernatant for biomachining of various metals at large scale as well as at micro- and nanoscale. Muhammad et al. (2015a, b) first showed the effect of variation in parameter on specific metal removal rate and material removal rate. Second, Taguchi design of experiment is applied to find the optimal value of various parameters for metal removal rate. Then, optimum value is predicted and verified.

Various studies have been conducted for biomachining of different metal workpieces using different microorganisms at particular temperatures, time and shaking rates. However, the current study deals with the biomachining of aluminium alloy 4004 using Aspergillus niger, which has not been attempted earlier as per the best knowledge of the authors. In the present study, feasibility of biomachining on aluminium alloy 4004 using Aspergillus niger has been checked and analysed at a regular interval of 24 hours for 7 days at room temperature. Specific metal removal rate (SMRR) is then calculated at every interval of 24 hours, to identify the trend of SMRR over the period of time.

3 Experimental Methodology

The current study has been completed in four phases discussed as follows.

3.1 Procurement of Culture

Aspergillus niger was obtained from Institute of Microbial Technology, Chandigarh. The culture is delivered after 15 days of its procurement. The culture we get is in the powdered form and packed in glass tubes.

3.2 Growth of Aspergillus Niger

Potato dextrose agar medium is used for growth of Aspergillus niger. PDA medium contains the following substances: 200 g potatoes, 20 g agar, 20 g dextrose and 1L distilled water (Aryal 2015). These components are mixed properly under the extreme sterilized conditions. Sterilization has been done using autoclave at 121 °C and 15 psi pressure for 30 min. After sterilization, this medium is put into agar plates at normal room temperature at sterilized conditions. Then, inoculation of that particular culture is done in these agar medium plates using inoculation needle, and then these plates are placed in incubator for 4–7 days for the growth of Aspergillus niger.

3.3 Workpiece Preparation

Workpiece is aluminium alloy 4004, in trapezium shape having an area of 16 cm^2. This aluminium alloy 4004 is purchased from metal Shop **P. S. SURA AND SONS** located at Gill Chowk, Ludhiana. This particular aluminium alloy 4004 is widely used

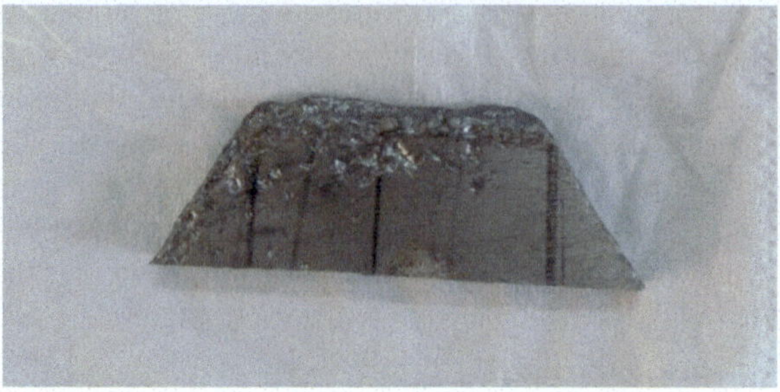

Fig. 1 Workpiece of aluminium alloy 4004

for manufacturing of pistons. Aluminium alloy 4004 has following composition of other metals: Si 9.0–10.5%, Fe 0.8%, Cu 0.25%, Mn 0.1%, Mg 1.0–2.0%, Zn 0.20% and Al 99.0%. Workpiece after cutting is washed with distilled water and ethanol and dried properly before its use as shown in Fig. 1.

3.4 *Specific Metal Removal Rate*

SMRR of that particular workpiece must be calculated. For calculating SMRR, workpiece must be placed in 250 ml conical flask having 20 ml of PDA broth with 1×10^8 cells of microorganism per ml. These cells must count using hemocytometer. The **hemocytometer** is a device originally designed and usually used for count of blood cells. This sample is then placed in shaker at room temperature 26 °C for 4–7 days. Initial weight of workpiece must be taken and after every 24-hour weight is measured for 4–7 days to analyse the specific metal removal rate (SMRR). Specific metal removal rate (SMRR) is calculated using the following formula described by Jadhav and Hocheng (2014):

$$\text{SMRR} = \frac{\text{Amount of metal removed (mg)}}{\text{Time(h) x Area}\,(\text{cm}^2)}$$

4 Experimental Results and Discussions

After every 24 hours for 7 days, weight of workpiece is taken and amount of surface metal removal rate (SMRR) is calculated. Initial weight of workpiece was 5.931 g. Table 1 presents various different values of SMRR at the regular intervals of 24 hours for 7 days. SMRR is calculated using the above-mentioned formula, which gives us the best value at which the metal removal rate is maximum.

From the above table, we conclude that best value of SMRR is at 48 h where it is maximum as compared to others.

Table 1 SMRR results

Time (h)	Weight of workpiece (gm)	Amount of metal removed with respect to initial weight (gm)	Surface metal removal rate (SMRR) (mg/h.cm^2)
24	5.929	0.002	0.0052
48	5.910	0.021	0.027
72	5.900	0.031	0.0269
96	5.891	0.040	0.026
120	5.882	0.049	0.0255
144	5.875	0.056	0.0243
168	5.870	0.061	0.0226

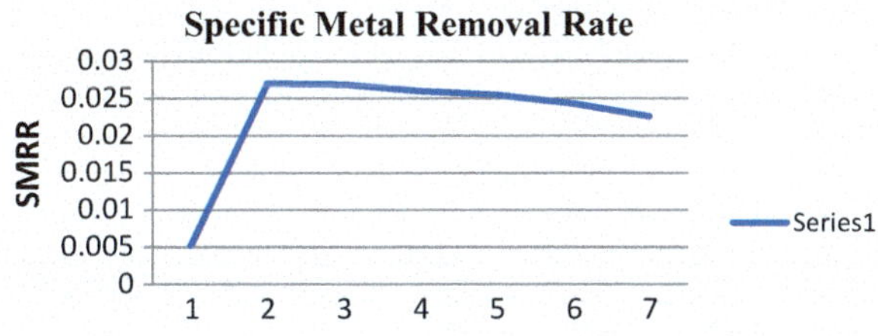

Fig. 2 Graphical representation of SMRR and DAYS

Figure 2 represents the graphical representation of specific metal removal rate along with the interval of number of days after every 24 hours.

This graph shows that at day 2, i.e. at 48 h, SMRR is maximum and then with the increase in number of days, the values of SMRR are continuously decreasing due to decrease in metabolism of Aspergillus niger on that particular workpiece, i.e. aluminium alloy 4004 with the passage of time.

5 Conclusions

This whole study concludes that specific metal removal rate is best at 48 h. Although the value of SMRR is less, in future we can make it efficient by changing the cell concentration. The metabolism of Aspergillus niger is good on aluminium alloy 4004, which gives the better results at very initial stage after 48 h, and then with the passage of time decrease in metabolic activity has been observed. The optimum value helps for our further study where we can vary various parameters such as concentration, shaking rate and temperature around this optimum value of 48 h. These results show that the proposed topic **is feasible**. **So**, biomachining for workpiece of aluminium alloy 4004 is possible by using **Aspergillus niger**. These results suggest that it is feasible to apply Aspergillus niger culture supernatant for biomachining of this particular metal at large scale as well as at micro- and nanoscale. Based on concurrent results, efforts will be made to explore the possibilities of effective biomachining for enhanced productivity along with analysing the surface characterization of the samples for improved cutting quality.

References

Aryal S (2015) Potato Dextrose Agar (PDA)—Principle, uses, composition, procedure and colony characteristics. https://microbiologyinfo.com/potato-dextrose-agar-pda-principle-uses-composition-procedure-and-colony-characteristics/

Averam O, Stroud I, Xirouchakis P (2011) A multi-criteria decision method for sustainability assessment of the use phase of machine tool systems. Int J Manuf Technol 53(5):811–828

Byrne G, Scholta E (1993) Environmentally clean machining processes—a strategic approach. Annals CIRP 42:471–474

Chang HJ, Hocheng H, Chang YH, Shih A (2008) Metal removal rate of Thiobacillus thiooxidans without pre-secreted metabolite. J Mater Process Technol 20(1):560–564

Hocheng H, Jadhav HH, Chang HJ (2012) Biomachining rates of various metals by Acidithiobacillus thiooxidans. Int J Surf Sci Eng 6:101–111

Istiyanto J, Taufiqurrakhman M, Kiswanto G, Santoso I, Ko JT (2014) Inclination angle effect on surface of copper in biomachining. Appl Mech Mater 660(8):23–27

Jadhav U, Hocheng H (2014) Use of Aspergillus Niger 34770 culture supernatant for tin metal removal. Corros Sci 82(9):248–254

Johnson D, Warner R, Shih A (2007) Surface roughness and material removal rate in machining using microorganisms. J Manufact Sci Eng 129(7):223–227

Kovoor PP, Idris RM, Hassan HM, Yahya TFT (2012) A study conducted on the impact of effluent waste from machining process on the environment by water analysis. Int J Energy Environ Eng 1–12

Muhammad I, Ullah SMS, Han SD, Ko JT (2015a) Selection of optimum process parameters of biomachining for maximum metal removal rate. Int J Precis Eng Manuf-Green Technol 2(4):307–313

Muhammad I, Jeong YJ, Zhen Y, ko JT (2015) Development of an empirical model for improving surface roughness of biomachined copper 8(2):3015–3016

Mat Web Material property Data 4004 Aluminium Composition Spec. http://www.matweb.com/search/datasheettext.aspx?matguid=6f46be39c79743e6829f77009eb8afce

Tan XC, Wang YY, Gu B, Mu Z, Yang C (2011) Improved methods for production manufacturing process in environmentally benign manufacturing. Energy 4:1391–1409

The Prince and Izant Company, AL 4004 Technical data. https://princeizant.com/uploads/Technical_Data_Sheets_2016/Aluminum/AL_4004_BAlSi-7_TDS.pdf

Ting YP, Kumar AS, Rahman M, Chia BK (2000) Innovative use of Thiobacillus ferrooxidans for the biological machining of metals. Acta Biotechnol 20(2):87–96

Uno Y, Kaneeda T, Yokomizo S (1996) Fundamental study on biomachining: machining of metals by Thiobacillus ferrooxidans. Jpn Soc Mech Eng Int J Ser C Dyn Control Robot Des Manuf 39(4):837–842

Wang S, To S, Chan YC, Cheung FC (2010) A study of the cutting-induced heating effecton the machined surface in ultra-precision raster milling of 6061 Al alloy. Int J Adv Manuf Technol 51(6):69–78

TiO_2-Assisted Photocatalytic Degradation of Herbicide 4-Chlorophenoxyacetic Acid: Slurry and Fixed-Bed Approach

Anoop Verma, Amrit Pal Toor, Palak Bansal, Vikas Sangal and Amit Sobti

Abstract The photocatalytic degradation of a recalcitrant herbicide 4-chlorophenoxyacetic acid (4-CPA) was investigated using suspended and supported TiO_2. The photocatalytic process was intensified with optimization of various operating parameters in slurry mode-like dose of TiO_2, H_2O_2, A/V ratio of reactor, pH, and UV intensity. An almost 93% degradation of 4-CPA was achieved after 2 h of UV irradiations while it enhanced to 99% under natural solar irradiations within the same treatment time. For fixed-bed studies, novel TiO_2 immobilized spherical clay beads were used which yielded approximately 97% degradation of 4-CPA after 5 h of treatment. Excellent stability and durability of the immobilized catalyst were confirmed through SEM/EDS analysis. TiO_2-coated clay beads were successfully recycled for more than 30 cycles without any significant reduction in the degradation efficiency of 4-CPA. The mineralization studies of 4-CPA were carried out by monitoring the generation of chloride ions along with reduction in COD. The intermediates formed during the degradation of 4-CPA were identified through GC-MS analysis. The results obtained in this study especially in fixed-mode using highly durable TiO_2 immobilized clay beads can open up new channels in this particular field for the removal of recalcitrant pollutants present in aquatic environment.

Keywords Photocatalysis · 4-chlorophenoxyacetic acid · Clay beads · Degradation · Mineralization

---

The original version of this chapter was revised: The affiliation of authors Amit Sobti and Amrit Pal Toor, has been updated. The erratum to the chapter is available at https://doi.org/10.1007/978-981-13-6717-5_39

A. Verma (✉) · P. Bansal · V. Sangal
School of Energy and Environment, Thapar Institute of Engineering and Technology, Patiala, India
e-mail: anoop.kumar@thapar.edu

A. P. Toor · A. Sobti
Dr. SSB University Institute of Chemical Engineering and Technology, Panjab University, Chandigarh 160014, India

A. K. Agnihotri et al. (eds.), *Sustainable Engineering*, Lecture Notes in Civil Engineering 30, https://doi.org/10.1007/978-981-13-6717-5_14

1 Introduction

The frequent detection of pesticides or herbicides in these water resources has prompted the scientists to look for the advanced water treatment methods to completely remove such pollutants from aquatic environment. Photodegradations using oxides of semiconductor such as TiO_2 have proven to be a promising approach among advanced treatment technologies in context to the environmental remediation of persistent organic pollutants (Chen and Mao 2007). Semiconductor-mediated photocatalysis has received immense attention during the last few years for the degradation and mineralization of bio-recalcitrant organic pollutants like herbicides or pesticides (Affam and Chaudhuri 2013). The mechanism of TiO_2 photocatalysis is now well established where it shows the generation of highly reactive species, i.e., hydroxyl radicals ($^{\bullet}OH$) that have the capacity to oxidize and mineralize a large variety of pollutants yielding H_2O, CO_2, and inorganic ions (Sadik et al. 2007).

Although slurry photocatalysis has acquired more prominence among advanced oxidation processes (AOPs), separation of catalyst from the aqueous solution post-treatment always presents a technical hindrance toward its commercial applications. Thus, applications of photocatalysis are always preferred in fixed form. Numerous of studies are reported in literature in context to fixed-bed photocatalysis using different support materials (Benhabiles et al. 2017; Chekir et al. 2017; Li et al. 2015; Manassero et al. 2017).

The use of such TiO_2 immobilized supports in photocatalytic process has nullified the expensive catalyst separation step post-treatment (Gaya and Abdullah 2008). But, the prolonged use of supported catalyst with retained activity and stability for longer period always remained a big concern. None of the reported studies has ever covered this issue regarding the extended durability of immobilized catalyst. The present study covers most of these concerns by using highly durable TiO_2 immobilized novel clay beads to study the detailed degradation of 4-chlorophenoxyacetic acid (4-CPA).

4-CPA comes under the category chlorophenoxy herbicides and is employed as plant growth regulator. These chlorinated compounds are chemically resistant, bio-recalcitrant, and potentially toxic for humans as well as animals (Lang and Luňák 2002; Torun and Şolpan 2016). Though photocatalysis-mediated degradation of 4-CPA has been investigated before (Singh et al. 2007), no efforts have been devoted toward the practical feasibility of degradation process along with its degradation mechanism.

In this view, a detailed study on the photodegradation of 4-CPA has been undertaken using suspended and supported TiO_2 under artificial UV as well as natural solar irradiations. The process was intensified with optimization of various operating parameters like dose of TiO_2, H_2O_2, A/V ratio of reactor, pH, and UV intensity in case of slurry photocatalysis. For fixed-bed studies, spherical clay beads were used as TiO_2 immobilization supports. To the best of our knowledge, this is the first reported study related to the photocatalytic degradation and mineralization of 4-CPA especially in fixed-mode along with its degradation mechanism.

2 Methodology

2.1 *Chemicals*

4-CPA (>98%) was purchased from TCI chemicals (India). TiO_2 P-25 was acquired from Evonik Industries, India. Hydrogen peroxide, H_2O_2 (30% w/v), was bought from Ranbaxy, India. Solutions of NaOH (0.1 N) and HCl (0.1 N) were used for setting the pH, and their concentrated solutions were obtained from Merck. Clay was obtained from local market, Patiala (India). All solutions were prepared using double-distilled water.

2.2 *TiO_2 Immobilization*

Around 400 clay beads of different sizes were prepared manually using a mixture of clay. From these beads, almost 300 beads of uniform size (~10 mm diameter) were selected and used for TiO_2 coating. TiO_2 was coated on clay beads (Fig. 1) using dip-coating method as described in our previous studies (Verma et al. 2014). Even the previously used TiO_2-coated beads were also used for checking their efficacy in degrading the 4-CPA.

2.3 *Experimental Setup and Procedure*

For performing the photocatalytic experiments, rectangular wooden UV chamber (1.37 × 0.9 × 1.0 m) comprising of seven UV tubes (36 W) having wavelength of 365 nm was used (Verma et al. 2014). The batch reactor was made up of borosilicate glass (Height = 5.2 cm, Diameter = 17.4 cm with a capacity of 1200 mL). The batch reactor was placed over laboratory jack under artificial UV irradiations. The required UV intensity could be achieved by varying the distance of reactor from the UV tubes

Fig. 1 Actual photograph of freshly coated TiO_2 clay beads

(Toor et al. 2006). The UV intensity was altered from 10 to 25 W m^{-2} (measured through radiometer).

The photocatalytic experiments were performed with 200 mL of 4-CPA solution (100 mg L^{-1}) in a batch reactor with desired amount of TiO_2 under UV radiation as well as sunlight with constant stirring. For fixed-bed studies, same batch reactor was used containing TiO_2 immobilized clay beads. Desired amount of aeration was supplied by means of air spargers. Approximately, 3–4 mL sample was collected at regular time intervals from the vessel using a syringe and was filtered in case of slurry experiments through millipore filter (0.45 μm) and was subsequently analyzed.

All experiments were carried out in triplicate for confirming the results. The standard deviation for all the reactions was in the range of 2–6%.

2.4 Analytical Determination

UV–Vis spectrophotometer (LABINDIA, model no. T60 U) was used for photodegradation analysis of 4-cpa at $\lambda_{max} = 275$ nm. Standard APHA methods (1989) were followed for monitoring the chloride ion generation and COD determination. The intermediates formed during the degradation process were identified through GC-MS analysis. Gas chromatography (45X-GC, Bruker) is integrated with the mass detector (MS-SCION-45P, Bruker) and with MS wax column with diameter and length of 0.25 and 30 m. Helium is the carrier gas with a flow rate of 1.0 mL min^{-1}. 80–410 was used as the range for scan. 280 °C was the injection temperature. The temperature of GC oven was set from 40 to 260 °C with 10° increase in temperature, 1:10 GC split ratio was employed, and 1.0 mL sample was injected automatically. The morphology and elemental composition of freshly coated clay beads and recycled beads was studied using SEM (JSM-6510LV, JEOL, Japan) and EDS (INCAX-act, Oxford Instruments, United Kingdom).

3 Results and Discussion

3.1 Degradation Studies of 4-CPA Using Slurry Photocatalysis

3.1.1 Preliminary Studies

Preliminary reactions were conducted to study the effect of individual parameters on the degradation rate of 4-CPA. A negligible amount of degradation (7.8%) was observed when aqueous solution of 4-CPA was irradiated under UV light in a batch reactor. Adsorption experiments in the presence of only TiO_2 yielded insignificant amount of reduction (15%) in the concentration of 4-CPA. Likewise, when H_2O_2

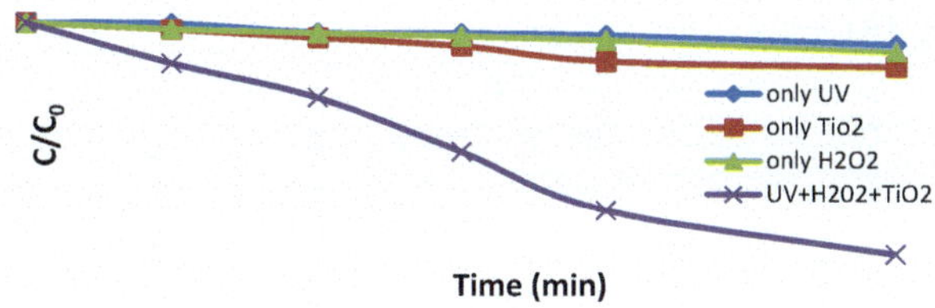

Fig. 2 Degradation of 4-CPA in the presence and absence of TiO_2, H_2O_2, and UV

alone was used, only 10% degradation was attained which might be due to the production of very less OH radicals. On the other hand, the photocatalytic treatment ($UV + TiO_2 + H_2O_2$) of 4-CPA showed significant degradation, i.e., 93% after 90 min (Fig. 2) due to the substantial increase in the generation of hydroxyl radicals. Thus, it leads to an assumption that adsorption and desorption of model compound, i.e., 4-CPA and formation of reaction intermediates, are slower than the generation of electron/hole pairs.

3.1.2 Effect of TiO_2 Concentration

The high concentration of catalyst is not recommended as it becomes difficult to separate the catalyst from slurry at higher concentration. Moreover, it makes the process more expensive. In our study, concentration of TiO_2 was varied in the range 0.25–0.875 g L^{-1}. The photocatalytic experiments for the degradation of 4-CPA were carried out under UV irradiations at natural pH of 4-CPA solution, i.e., 3.5. Figure 3a demonstrates that the degradation of 4-CPA initially increased by increasing the TiO_2 concentration up to 0.875 g L^{-1}. Thereafter, it decreased with further increase in photocatalyst concentration. The degradation rate constant (k) increased from 0.004 to 0.0102 min^{-1} up to 0.875 g L^{-1}. The probable reasons for the decrease in degradation efficiency at higher dose of TiO_2 might be the insufficient amount of hydroxyl radical generation due to the reduction in light penetration, enhanced scattering of light, formation of aggregates, and settling of catalyst under high concentration (Bansal et al. 2016). The best suitable amount of catalyst selected was 0.625 g L^{-1} of TiO_2 at which approximately 85.53% degradation was achieved after irradiation of 2 h.

3.1.3 Effect of H_2O_2 Addition

The role of oxidant (H_2O_2) in increasing the rate of photodegradation is now evident and well established in literature (Malato et al. 2000; Poulios et al. 2003). Actually, the presence of H_2O_2 generates $^{\cdot}OH$ radical when the photon excited electron from the conduction band is accepted. It leads to the generation of $^{\cdot}OH$ radicals according to reaction (1) (Kositzi et al. 2004).

$$H_2O_2 + {}^{\cdot}O_2^- \rightarrow {}^{\cdot}OH + OH^- + O_2 \tag{1}$$

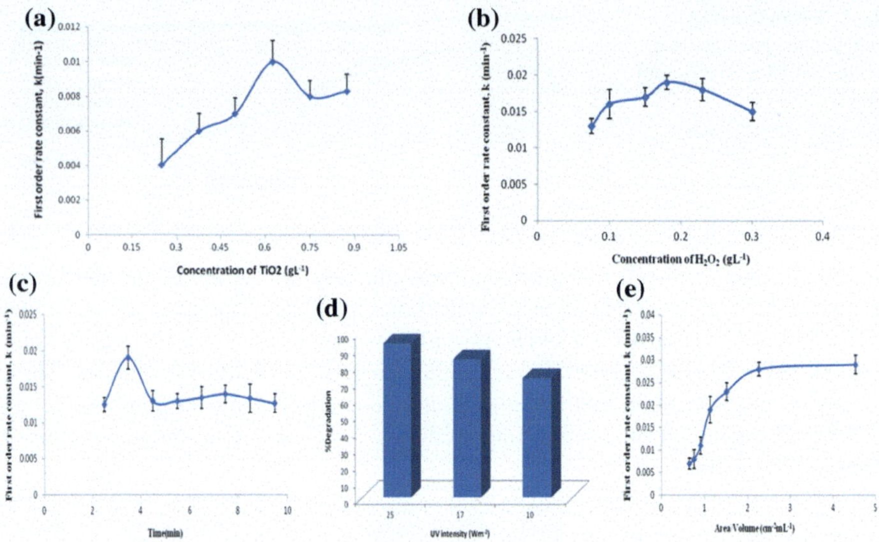

Fig. 3 Plot of first-order rate constant, k versus variation in **a** TiO_2 **b** H_2O_2 **c** pH, **d** UV intensity, and **e** A/V ratio during photocatalytic degradation of 4-CPA

In order to calculate the optimum concentration of H_2O_2, the experiments were carried out by varying the initial concentration of H_2O_2 from 0.075 to 0.3 g L^{-1} with optimum TiO_2 concentration of 0.625 g L^{-1} and at natural pH of the solution, i.e., 3.5. Photocatalytic degradation rate constant (k) varied from 0.013 to 0.019 min^{-1} by varying the amount of H_2O_2 from 0.075 to 0.18 g L^{-1}. The degradation increased to 93% with the addition of 0.18 g L^{-1} of H_2O_2 (Fig. 3b). The degradation rate constant (k) started decreasing after 0.18 g L^{-1} of H_2O_2. The possible reason may be the scavenging of $^{\cdot}OH$ and hole which forms HO_2 as shown in reactions (2) and (3) (Behnajady et al. 2006) at high concentrations of H_2O_2.

$$^{\cdot}OH + H_2O_2 \rightarrow H_2O + HO_2^{\cdot} \quad (2)$$

$$H_2O_2 + h^+ \rightarrow H^+ + HO_2^{\cdot} \quad (3)$$

Therefore, the best suitable amount of oxidant selected was 0.18 g L^{-1} of H_2O_2 which resulted in 93% of degradation after 2 h of irradiation at 0.625 g L^{-1} of TiO_2 at its natural pH, i.e., 3.5.

3.1.4 Effect of PH

In order to observe the effect of initial pH on photocatalytic removal of 4-CPA, pH was altered in the range of 2.5–9.5 at optimized dose of TiO_2, i.e., 0.625 g L^{-1}. The initial pH of the reaction solution shows the adsorption of the compound and surface charge characteristics of catalyst. At acidic pH, the surface charge of TiO_2 is positive and becomes negative at basic pH (Toor et al. 2007). Thus, pH plays an important part in the adsorption/desorption characteristics of the catalyst (Hoffmann et al. 1995). The degradation of 4-CPA in the present study was maximum at its natural pH, i.e., 3.5 (acidic). Thereafter, a considerable decrease in the degradation efficiency was observed at alkaline pH as depicted in Fig. 3c. The value of rate constant, "k", was considerably higher (0.02 min^{-1}) in acidic pH as compared to k value (0.001 min^{-1}) in basic pH. All studies were conducted at solution's natural pH, i.e., 3.5 at which approximately 93% degradation was attained under optimized conditions.

3.1.5 Effect of UV Intensity

Utilization of solar radiations for degrading contaminants is the most important application of AOP. Throughout the year, the intensity of solar radiations is not constant (Dubey et al. 2009). Optimized concentration of catalyst and H_2O_2 was added into the 100 mg L^{-1} aqueous solution of 4-CPA and irradiated under UV light while keeping A/V constant at 1.13 cm^2 mL^{-1}. The degradation efficiency of 4-CPA was increased with an increment in the value of rate constant from 0.008 to 0.019 min^{-1} with increasing the intensity of UV from 10 to 25 W m^{-2} (Fig. 3d). The possible reason may be the increase in production of OH radicals with increasing UV intensity.

3.1.6 Effect of Area/Volume Ratio of Batch Reactor

For field-scale applications, one of the major limitations of AOP is the obstruction of depth of reactor for the penetration of solar radiations. Less depth and more exposed area are generally recommended for the effective penetration of solar light (Toor et al. 2006). This can be achieved by varying area of the reactor and keeping the volume constant or volume of solution to be treated can be varied while keeping the area constant. In the present study, the Area/Volume (A/V) of batch reactor was varied from 0.648 to 4.53 cm^2 mL^{-1} by varying the volume of solution and keeping the aperture constant. A noticeable increase in the degradation of 4-CPA was achieved by increasing the A/V ratio as "k" changed from 0.0072 to 0.0263 min^{-1} as shown in Fig. 3e. Depth of the solution decreased by increasing the A/V which leads to the effective penetration of photons of light which results in higher yield of OH radical, thereby improving the photocatalytic performance (Verma et al. 2014). At lower penetration of light, electron–hole recombination is more which decreases the yield of hydroxyl radicals, thereby reducing the degradation rate. However, A/V ratio of 1.1343 cm^2 mL^{-1} was optimized in the present study.

The photocatalytic experiments were carried out in sunlight also. The percentage degradation was increased to 99% from 93% when photocatalysis was carried out in natural sunlight. Hence, the technology is economically feasible and can present a viable solution to industries for treating recalcitrant compounds.

3.2 Degradation Studies of 4-CPA Using Fixed-Bed Photocatalysis

The major problem with slurry photocatalysis regarding field-scale applications is the separation of catalyst from the treated solution. In this context, efforts were made to study the degradation of 4-CPA using fixed-bed catalysis employing TiO_2 immobilized clay beads. The fixed-bed studies were carried out under sunlight at optimized dose of H_2O_2 (0.18 g L^{-1}) and pH (3.5). An almost 83% degradation of 4-CPA was attained after 5 h of solar irradiations (Fig. 4a).

In the present study, the number of catalyst-coated beads was also varied in terms of percentage of the area covered by them and correspondingly degradation of the 4-CPA was analyzed. It was observed that best degradation (83%) was achieved when entire bottom surface of the batch reactor was covered with beads (with no overlapping) as all catalyst binding sites were available to bind the pollutant leading to efficient degradation, while degradation was decreased by reducing the number of beads (i.e., almost 50% of the total bottom surface area was covered with beads) due to the reduction in available active sites of the catalyst. In case when beads were overlapped one over the other (increasing the number of beads), the percentage of degradation was again reduced. This might be due to the decrease in the availability of the effective surface area for the catalyst to bind with the pollutant and trapping of the 4-CPA solution in the void spaces.

To seek the practical applications of technique, efforts were made to check the durability of immobilized catalyst for the degradation of 4-CPA. The clay beads were recycled for 30 to 40 times without any significant decrease in the degradation of target compound. Slight decrease in degradation efficiency of clay beads was possibly

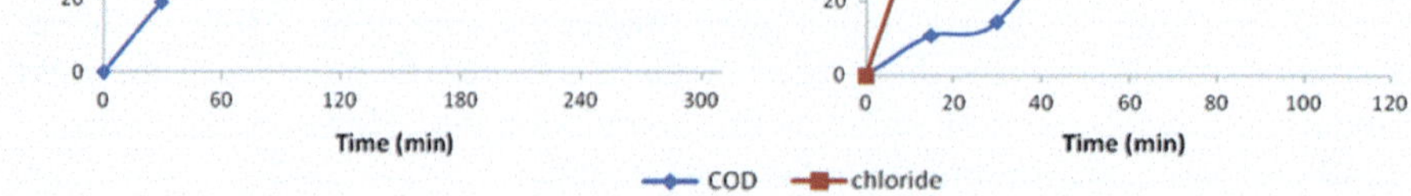

Fig. 4 **a** Plot of % degradation versus time, **b** COD and chloride removal during fixed-bed photocatalytic degradation of 4-CPA

due to the blocking of certain catalytic sites or loss of catalyst during subsequent reactions. The catalyst-coated clay beads were heated in oven at 100 °C after each cycle for the period of 60 min for reactivation. Stability of the catalyst on the surface of clay beads was confirmed by SEM-EDS analysis. The smooth morphology (SEM image) as shown in Fig. 5 along with the presence of "Ti" and "O" in EDS analysis (Table 1) even in case of recycled beads clearly proved the intactness of catalyst film after 30 cycles.

3.3 Mineralization Studies

Formation of inorganic anions and cations with the generation of CO_2 generally indicates the complete mineralization of compounds during photocatalytic treatment (Evgenidou et al. 2005). In this context, reduction of COD and generation of chloride ions were monitored for studying mineralization of 4-CPA under optimized conditions. 85.7% COD reduction was achieved after 2 h, while generation of chloride ions reached a maximum during the first hour and thereafter, it became constant after 1 h (Fig. 4b). The intermediates formed during the degradation process were identified through GC-MS analysis. Based on these intermediates, a tentative pathway for the degradation of 4-CPA was proposed as shown in Fig. 6. The decarboxylation

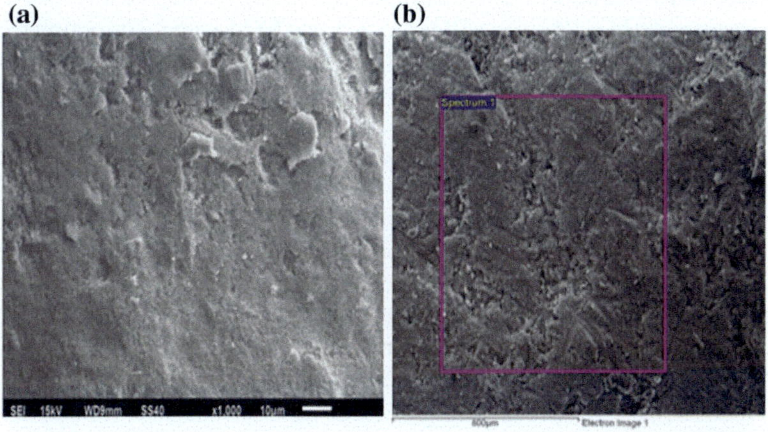

Fig. 5 SEM images **a** freshly coated clay bead, **b** used clay bead after 30 cycles

Table 1 EDS data for fresh and recycled clay beads

Element	C	O	Si	Ti
Fresh	11.99	71.8	1.7	14.6
Recycled	13.8	69.5	2.9	13.8

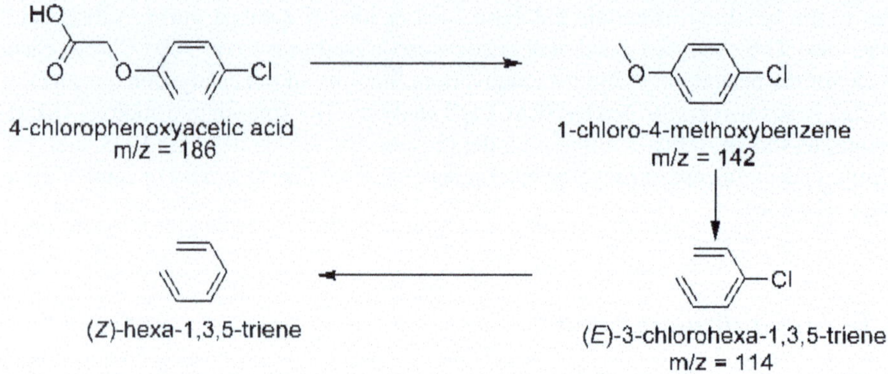

Fig. 6 Tentative pathway for the degradation of 4-CPA

of 4-CPA could lead to the production of 1-chloro-4-methoxybenzene which could further breakdown to yield further by-products through the cleavage of cyclic ring.

4 Conclusions

The present study was undertaken to show the photocatalytic degradation of 4-chloro phenoxyacetic acid (4-CPA) using TiO_2 in slurry and fixed-mode. The optimized conditions obtained for better degradation of the herbicide were TiO_2 0.625 g L^{-1}, H_2O_2 0.18 g L^{-1}, and C_0 100 mg L^{-1}. UV/H_2O_2/TiO_2 photocatalysis resulted in 93.05% degradation of 4-CPA in aqueous solution. For fixed-bed studies, TiO_2 immobilized clay beads were used resulting in 83% degradation of 4-CPA after 5 h of solar irradiations. The clay beads were durable enough even after 30 recycles as confirmed through SEM-EDS analysis. The mineralization of 4-CPA was confirmed by reduction in COD (85.7%) along with the generation of chloride ions (75%) after 2 h of irradiation. To the best of our knowledge, no study has been carried out so far for the degradation of 4-CPA through slurry and fixed-bed photocatalysis. Hence, best efforts were made to obtain a viable solution to treat wastewater containing these types of recalcitrant compounds using photocatalysis.

References

Affam AC, Chaudhuri M (2013) Degradation of pesticides chlorpyrifos, cypermethrin and chlorothalonil in aqueous solution by TiO_2 photocatalysis. J Environ Manage 130:160–165

American Public Health Association (APHA) (1989) Standard methods for the examination of water and wastewater, 19th edn, APHA. Washington, DC

Bansal P, Verma A, Aggarwal K, Singh A, Gupta S (2016) Investigations on the degradation of an antibiotic Cephalexin using suspended and supported TiO_2: mineralization and durability studies. Can J Chem Eng 94:1269–1276

Behnajady MA, Modirshahla N, Hamzavi R (2006) Kinetic study on photocatalytic degradation of C.I. Acid Yellow 23 by ZnO photocatalyst. J Hazard Mat B133:226–232

Benhabiles O, Chekir N, Tassalit D (2017) Solar photocatalytic degradation of methylene blue in a fixed bed reactor. Desalin Water Treat 60:285–290

Chekir N, Tassalit D, Benhabiles O, Merzouk NK, Ghenna M, Abdessemed A, Issaadi R (2017) A comparative study of Tartrazine degradation using UV and solar fixed bed reactors. Int J Hydrogen Energy 42:8948–8954

Chen X, Mao SS (2007) Titanium dioxide nanomaterials: synthesis, properties, modifications and applications. Chem Rev 107:2891–2959

Dubey SK, Kumar A, Srivastava P, Rajor A (2009) Solar photo-catalytic treatment of textile wastewater for biodegradability enhancement. Int J Environ Eng 1:152–164

Evgenidou E, Fytianos K, Poulios I (2005) Photocatlytic oxidation of dimethoate in aqueous solutions. J Photochem Photobiol A: Chem 175:29–38

Gaya UI, Abdullah AH (2008) Heterogeneous photocatalytic degradation of organic contaminants over titanium dioxide: a review of fundamentals, progress and problems. J Photochem Photobiol C: Photochem Rev 9:1–12

Hoffmann MR, Martin ST, Choi W, Bahnemann DW (1995) Environmental applications of semiconductor photocatalysis. Chem Rev 95:69–96

Kositzi M, Antoniadis A, Poulios I, Kiridis I, Malato S (2004) Solar photocatalytic treatment of simulated dyestuff effluents. Sol Energy 77:591–600

Lang K, Luňák S (2002) Photocatalytic degradation of 4-chlorophenoxyacetic acid in the presence of an iron complex and hydrogen peroxide. Photochem Photobiol Sci 1:588–591

Li D, Zhu Q, Han C, Yang Y, Jiang W, Zhang Z (2015) Photocatalytic degradation of recalcitrant organic pollutants in water using a novel cylindrical multi-column photoreactor packed with TiO_2-coated silica gel beads. J Hazard Mat 285:398–408

Malato S, Blanco J, Maldonado M, Fernandez-Ibanez P, Campos A (2000) Optimising solar photocatalytic mineralisation of pesticides by adding inorganic oxidising species; application to the recycling of pesticide containers. Appl Catal B: Environ 28:163–174

Manassero A, Satuf ML, Alfano OM (2017) Photocatalytic degradation of an emerging pollutant by TiO_2-coated glass rings: a kinetic study. Environ Sci Poll Res 24:6031–6039

Poulios I, Micropoulou E, Panou R, Kostopoulou E (2003) Photooxidation of eosin Y in the presence of semiconducting oxides. Appl Catal B: Environ 41:345–355

Sadik WA, Nashed AW, El-Demerdash A-GM (2007) Photodecolourization of ponceau 4R by heterogeneous photocatalysis. J Photochem Photobiol A: Chem189:135–140

Singh HK, Saquib M, Haque MM, Muneer M (2007) Heterogeneous photocatalysed degradation of 4-chlorophenoxyacetic acid in aqueous suspensions. J Hazard Mat 142:374–380

Toor AP, Verma A, Jotshi CK, Bajpai PK, Singh V (2006) Photocatalytic degradation of Direct Yellow 12 dye using UV/TiO_2 in a shallow pond slurry reactor. Dyes Pigm 68:53–60

Toor AP, Verma A, Singh V, Jotshi CK, Bajpai PK (2007) Treatment of bleaching effluent from the pulp and paper industry by photocatalytic oxidation. TAPPI J 6:9–13

Torun M, Şolpan D (2016) Analytical studies on degradation mechanism of herbicide 4-chlorophenoxyacetic acid in water by gamma/H_2O_2 and gamma/ozone processes. Environ Tech 37:2494–2507

Verma A, Prakash NT, Toor AP (2014) An efficient TiO_2 coated immobilized system for the degradation studies of herbicide isoproturon: durability studies. Chemosphere 109:7–13

Comparative Study on the Photocatalytic Degradation of Paraquat Using Tungsten-Doped TiO_2 Under UV and Sunlight

Manpreet Kaur, Anoop Verma, Hema Setia and Amrit Pal Toor

Abstract Photocatalytic degradation of paraquat dichloride a poisonous dipyridilium herbicide present as a pollutant in surface water has been studied in a batch photoreactor using P25 TiO_2 and tungsten (W)-doped TiO_2. Effect of varying parameters such as substrate concentration, catalyst loading, pH of the solution, light conditions, type of catalyst, and amount of dopant has also been studied. It was observed that synthesized catalyst with 2.5 wt% W-doped TiO_2 exhibited higher percentage degradation efficiency than bare TiO_2 under optimum conditions of catalyst loading, pH, and substrate concentration under solar irradiation. W-TiO_2 catalyst has been synthesized and characterized using UV–Vis DRS, XRD, SEM-EDS, and TEM techniques. Maximum degradation of 98% was observed for 25 ppm paraquat, at a pH of 6.5, catalyst loading of 1 g L^{-1}, and 2.5 wt% W-doped TiO_2 under sunlight.

Keywords Paraquat · Photocatalysis · Degradation · Doping · Tungsten-doped TiO_2 · Wastewater

1 Introduction

Herbicides are used to control, eliminate, or destroy pests in order to protect the food crop. However, their extensive and exhaustive use over the past decade has posed a threat to the environment. Due to their high toxicity and persistence, they have been found to have adverse ecological impact on aquatic life and human health. Therefore,

M. Kaur
Energy Research Centre, Panjab University, Chandigarh 160014, India

A. Verma
School of Energy and Environment, Thapar Institute of Engineering and Technology, Patiala, India

H. Setia (✉)
University Institute of Engineering and Technology, Panjab University, Chandigarh 160014, India
e-mail: hemasetia@pu.ac.in

A. P. Toor
Dr SSB University Institute of Chemical Engineering and Technology, Panjab University, Chandigarh 160014, India

A. K. Agnihotri et al. (eds.), *Sustainable Engineering*, Lecture Notes in Civil Engineering 30, https://doi.org/10.1007/978-981-13-6717-5_15

it becomes imperative to assess the fate of such herbicides and develop a mechanism for its removal and degradation in soil and waste bodies. Depending upon the environmental conditions, herbicides can persist in many environmental compartments for a long period of time. Starting from their point of application in field, these compounds undergo different chemical and physical processes like adsorption on to soil matrix, leaching into groundwater, and transport to surface runoff water. The different strategies that can reduce the pesticides concentration involve direct photolysis, wherein the degradation or transformation to other products takes place in the presence of sunlight. Advanced oxidation processes offer the advantages of destroying such pollutants by oxidizing them into mineral salts and nontoxic compounds under normal conditions of temperature and pressure.

Numerous research documents have demonstrated that heterogeneous photocatalysis, a representative of advanced oxidation process, is an effective process for the degradation of a variety of toxic substances, such as fungicides and herbicides, under different light irradiations (Shibin et al. 2015; Verma et al. 2014; Kaur et al. 2016; Grover et al. 2017). Recently, doping, co-doping, immobilization, impregnation, and supported catalyst systems have been investigated to enhance the degradation of such organic compounds (Cunff et al. 2015; Montañez et al. 2015; Maulidiyah et al. 2017; Kalantaria et al. 2017; Sakee and Wanchanthuek 2017; Kruanetr and Wanchanthuek 2017; Moctezuma et al. 1999; Cantavenera et al. 2007). Zahedi et al. (2015) demonstrated 84.39% degradation of 10 ppm paraquat in 5 h at pH 5.8 using N,S codoped TiO_2 thin films under visible irradiation. Eleburuike et al. (2016) achieved 80.798% degradation of paraquat in 4 h with CeO_2-modified TiO_2 nanotubes and catalyst loading of 0.38 g under UV irradiation. Limited data is available on the degradation of paraquat under visible illumination using surface-modified TiO_2. Surface modification of TiO_2 with metal ions can significantly enhance its photocatalytic activity under economical sunlight. Although photocatalytic degradation of paraquat using titania under UV light has been widely studied, the aim of the present study is to achieve higher degradation efficiencies using modified TiO_2 under naturally available solar illumination.

2 Methodology

2.1 Materials

All the chemicals procured were used as received without any further purification. Paraquat was obtained from Sigma Aldrich (>99% pure, technical). Titanium dioxide (P25) (purity 97%, average particle size-30 nm, 80% anatase) was procured from Evonik India. Tungsten trioxide (WO_3) was obtained from Central Drug House. Distilled water was used throughout the experiment for preparing solutions. Millipore membrane filters (0.22 μm) were used to filter the samples before analysis.

Table 1 Crystallite size calculated from Scherrer's equation

Sample	Crystallite size (nm)	Bandgap energy (eV)
P25 TiO_2	32.64	3.20
1.5 wt% W-TiO_2	30.22	3.16
2.0 wt% W-TiO_2	29.94	3.00
2.5 wt% W-TiO_2	27.58	2.80
3.0 wt% W-TiO_2	31.37	2.94

2.2 *Synthesis of W-Doped TiO_2 Nanoparticles*

Tungsten-doped titanium dioxide (W-TiO_2) was prepared by mechanical mixing through homogenization and magnetic stirring. Aeroxide TiO_2 P25 was used as the precursor. Precisely, calculated amount of TiO_2 was dispersed in 50 ml distilled water under continuous stirring. Tungsten trioxide was taken as the source for doping TiO_2. Appropriate concentrations of WO_3 were then dispersed in 100 ml of distilled water by continuous stirring for 30 min at room temperature. This solution was gradually added to TiO_2 solution under vigorous stirring to get a uniformly dispersed solution. The sediments thus obtained were dried in an oven and calcined at 400 °C for 4 h.

3 Results and Discussion

3.1 *Characterization of Catalyst*

The X-ray diffraction patterns of TiO_2 P 25 and 2.5 wt% W-doped TiO_2 are illustrated in Fig. 1. The anatase phase located at 2θ equal to 25.4°, 37.8°, 48.1°, and 55.2° corresponds to (101), (004), (112), (200) and (211) planes, whereas peaks located at 2θ equal to 27.5°, 36.2°, 39.3°, 41.3°, and 54.4° corresponds to the rutile phase which were also observed in W-TiO_2 (Zahedi et al. 2015). Since the XRD pattern of W-doped TiO_2 does not show any major shift or distortion to the main characteristic peak of TiO_2, we can say that doping of TiO_2 with W does not change the basic crystalline state of TiO_2. The results suggest that W was well dispersed on the surface of TiO_2 particles. A small percentage of rutile phases were also observed in W-TiO_2 as were observed in P25 TiO_2 suggesting that this method of doping with W does not favor the anatase to rutile or rutile-to-anatase phase transition.

The crystalline size of W-doped TiO_2 was calculated using the Scherrer's equation (Toor et al. 2013) and is listed in Table 1. The smallest crystallite size of the W-doped TiO_2 was 27.58 nm for 2.5 wt% W-doped TiO_2; hence, the TEM analysis of this sample was carried out.

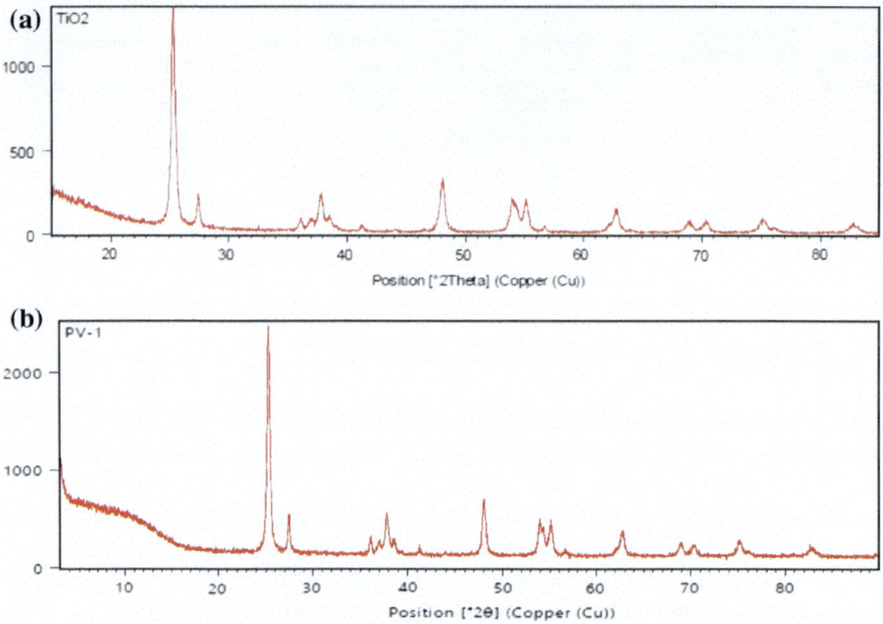

Fig. 1 **a** XRD pattern of P25 TiO_2, **b** 2.5wt% W-TiO_2

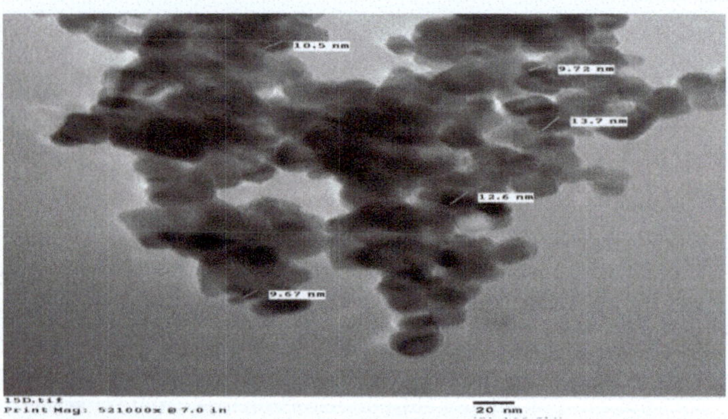

Fig. 2 TEM image of the 2.5 wt% W-TiO_2 nanoparticles

The morphology of the synthesized 2.5 wt% W-doped TiO_2 was characterized by TEM, and the result is shown in Fig. 2. The particles of W-TiO_2 are seen to be nearly cubic in shape.

The elemental composition of the synthesized W-TiO_2 was examined by energy-dispersive spectroscopy. Figure 3 illustrates the EDS for 2.5 wt% W-doped TiO_2. EDS results confirmed the presence of 2.45 wt% tungsten element in the W-TiO_2 sample, thus confirming the doping of tungsten over TiO_2.

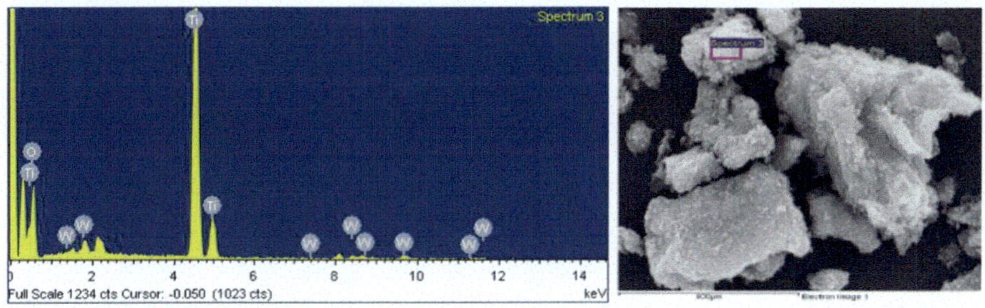

Fig. 3 SEM and EDS image of 2.5 wt% W-doped TiO_2

To study the optical response difference of synthesized W-TiO_2 catalyst, the absorption spectra and diffuse reflectance spectra (DRS) analysis were performed with a UV–Vis spectrophotometer (Fig. 4a). The UV–Vis absorption spectra were measured in the wavelength range of 200–700 nm taking $BaSO_4$ as reference compound. Optical bandgap was calculated using the relation $\alpha h\nu = A\,(h\nu - E_g)^n$, where α is the absorption coefficient and is calculated using Beer–Lambert relation: $\alpha = 2.303\ A/t$, A is the absorbance, t is the film thickness, h is the Plank's constant, and E_g is the optical bandgap energy (Baruwati et al. 2006). The value of exponent n depends on the type of transition and can take values of 1/2, 2, 3/2, and 3 for allowed direct, allowed indirect, forbidden direct, and forbidden indirect transitions, respectively (Manjunatha and Paul 2015; Dhaouadi and Adhoum 2009; Zahedi et al. 2015). The linear part of the plot of $(\alpha h\nu)^2$ versus $h\nu$ (Fig. 4b) was extrapolated to obtain the E_g values for the W-TiO_2 catalyst with varying wt% of tungsten in TiO_2. The values so obtained (Table 1) reveal that the bandgap energy for the doped particles is smaller than those for the pure TiO_2. Further, the bandgap energy was observed to be lowest at 2.8 eV for 2.5 wt% doped W-TiO_2. Low values of bandgap energy enable the absorption of light in the visible region.

3.2 *Degradation Studies*

The irradiation experiments were performed in a UV chamber consisting of UV fluorescent lamps (Philips, 20 W) having wavelength of 365 nm. The UV intensity was measured and kept constant at 25 Wm^{-2} for the entire experimentation. Herbicide solutions were prepared by dissolving appropriate amounts in distilled water. Hemispherical-shaped shallow pond slurry reactor was used to carry out the batch experiments. For each run, a 200 ml sample of the herbicide solution with varying amount of catalyst was irradiated under UV or sunlight for 180 min. The concentration of paraquat was determined using a UV–Vis spectrophotometer. The λ_{max} was observed at 257 nm (Moctezuma et al. 1999). Prior to irradiation, the solutions were kept in dark for 30 min under stirring conditions for adsorption of the herbi-

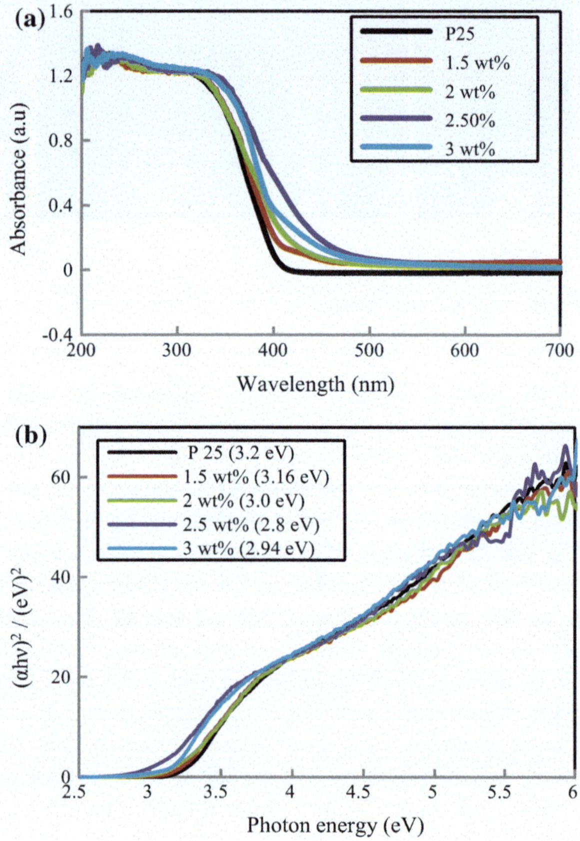

Fig. 4 **a** Diffuse reflectance spectra of TiO_2(P25) and synthesized W-TiO_2 nanoparticles, **b** plot of $(Ah\nu)^2$ versus $h\nu$ for the W-TiO_2 nanoparticles with varying wt% of tungsten

cide on the surface of the catalyst. No significant photolytic and photodegradation of paraquat was observed. Similar results have been observed by Florêncio et al. (2004) and Cantavenera et al. (2007).

3.3 *Effect of Catalyst Loading*

To determine the minimum catalyst loading required for the optimum degradation of paraquat, reactions were carried out by varying the TiO_2 loading from 0.5 to 1.5 g L^{-1}. The results demonstrate that photocatalytic activity increased linearly up to 96% for a catalyst loading of 1 g L^{-1} (Fig. 5). However, the degradation started decreasing with further increase in the amount of catalyst. This can be attributed to an increase in the solution turbidity that blocks the penetration of light into the solution (Wei and Wan 1991) also referred to as shielding effect. Similar studies on the complete degradation of paraquat have been reported. For same catalyst loading, Florêncio

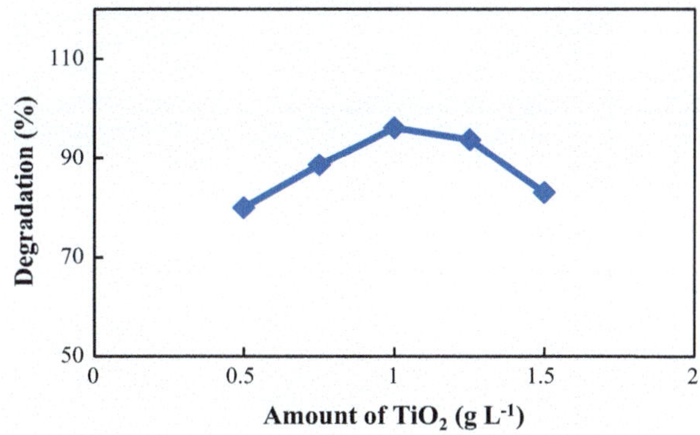

Fig. 5 Effect of amount of P25-TiO_2 on the photocatalytic degradation of paraquat (25 ppm, pH 6.5, UV irradiation for 3 h)

et al. (2004) have reported almost complete degradation for 7 ppm paraquat solution at an intensity of 126 klx, whereas in this study a degradation of 96% was achieved for 25 ppm paraquat at an intensity of 25 Wm^{-2} which is the intensity of UV available in solar light (i.e., 3–4% of the solar irradiation).

3.4 *Effect of Substrate Concentration*

The effect of substrate concentration is an important parameter for studying the degradation rate. The degradation of paraquat was studied for 25, 50, 75, and 100 ppm. It was observed that as the concentration of the substrate is increased from 25 to 100 ppm the percentage degradation decreased from 96 to 51%. Lower substrate concentrations were found to be favorable for photocatalytic degradation of paraquat. At higher concentrations, the OH radicals produced are not sufficient to degrade the substrate molecules which are absorbed on or near the catalyst surface (Fig. 6).

3.5 *Effect of pH*

Effect of pH is also an important parameter because it commands the surface charge properties of the catalyst and therefore the adsorption of the substrate. The pH studies in the range 4–11 were carried out for 25 ppm paraquat with catalyst amount of 1 g L^{-1} under UV irradiation. A significant variation in the degradation of paraquat was observed with change in the pH. The results depicted in Fig. 7 show that at pH 4 little or negligible degradation of paraquat takes place. Paraquat is positively charged, so low pH hinders the adsorption on the catalyst surface leading to the observed low rates of degradation. While basic pH favors the degradation of paraquat, the

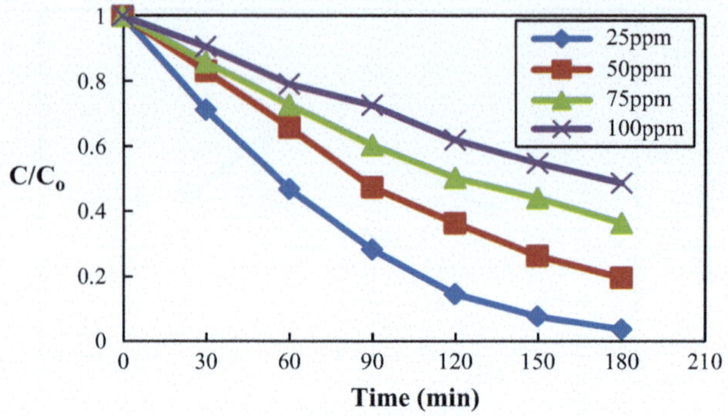

Fig. 6 Effect of substrate concentration on the photocatalytic degradation of paraquat (P25 TiO_2 $1gL^{-1}$, pH 6.5, UV irradiation)

Fig. 7 Effect of solution pH on the photocatalytic degradation of paraquat (25 ppm, TiO_2 $1gL^{-1}$, UV irradiation)

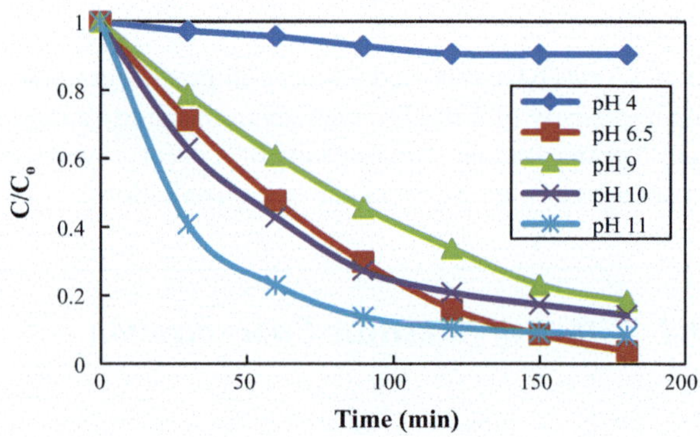

maximum degradation of up to 96% was observed at the natural pH of aqueous paraquat that is 6.5. Under alkaline conditions, the surface of the catalyst becomes negatively charged which explains the enhanced adsorption and degradation of the cationic paraquat molecule (Moctezuma et al. 1999; Shamsedini et al. 2017; Garcia et al. 2017). Similar results have been reported for the photocatalytic degradation of 3,4-dichlorophenol at different pH (Toor et al. 2005).

3.6 Activity Comparison of TiO_2 and Doped W-TiO_2 Under Solar Illumination

The activity of TiO_2 and the synthesized W-doped TiO_2 with varying weight percentages of W was studied under natural solar irradiation and optimized parameters of catalyst loading, substrate concentration, and solution pH. The experiments were carried out on sunny days in clear skies with average solar intensity of 430 klx. A comparison of the solar photocatalytic degradation of paraquat with TiO_2 and vary-

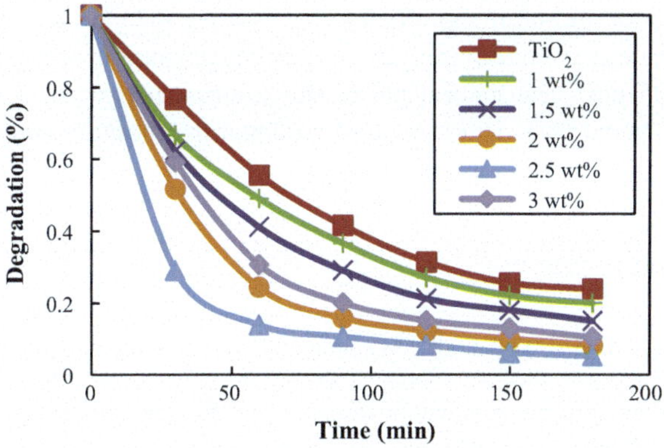

Fig. 8 Solar photocatalytic degradation of paraquat over TiO_2 (1 g L^{-1}) and different levels of W-doped TiO_2 (25 ppm, W-TiO_2 amount-1 g L^{-1}, pH 6.5)

ing weight percent of dopant in TiO_2 is depicted in Fig. 8. Results clearly display enhanced photocatalytic activity of W-doped TiO_2 in comparison to bare TiO_2. P25 TiO_2 shows 83% degradation, while 94% degradation was observed with 2.5 wt% tungsten-doped TiO_2 under solar illumination. Around 90% degradation occurs in the first 90 min of irradiation after which the rate of degradation is very low. Doping improves the photocatalytic efficiency by reducing the bandgap of TiO_2 (Bokare et al. 2013). The doped atoms either enter the lattice of TiO_2 or disperse on its surface, thereby affecting the electronic structure and narrow the bandgap of doped TiO_2 which absorbs under visible light illumination. The results imply that W-doped P25 (TiO_2) can significantly improve the photocatalytic degradation as compared to bare P25 (TiO_2) due to induced bandgap contraction leading to the enhancement of optical absorption of photocatalyst under visible illumination. Also, doping decreases the crystallite size which represents the specific surface area of the catalyst which in turn affects the photoactivity. The smaller the size and the greater the surface area, the more the light received per unit area and the higher the electronic excitations which results in an increase in the photo-efficiency of the catalyst. Hence, a low percentage degradation was observed for 3 wt% W-doped TiO_2 which has a higher crystallite size than 2.5 wt% W-doped TiO_2.

4 Conclusions

In summary, the present investigation demonstrates that W-TiO_2 (tungsten-doped P25 TiO_2) can effectively degrade paraquat herbicide under sunlight irradiation. The synthesized catalyst has been characterized in detail. Doping with tungsten narrowed the bandgap energy of the synthesized catalyst to 2.8 eV that relates to the enhancement of photocatalytic activity under visible light. The photocatalytic degradation of paraquat solution by W-doped TiO_2 photocatalysts was experimentally investigated via its absorption spectra. The results conclude that 2.5 wt% W-doped P25 (TiO_2)

is an optimum and efficient catalyst for the photocatalytic degradation of herbicide paraquat under sunlight. The catalyst amount of 1 g L^{-1}, initial concentration of 25 ppm and natural pH of the solution (pH = 6.5) are found to be optimum for around 94% degradation of paraquat in 3 h under sunlight.

References

Baruwati B, Kumar DK, Manorama SV (2006) Hydrothermal synthesis of highly crystalline ZnO nanoparticles: a competitive sensor for LPG and EtOH. Sensor Actuat B-Chem 119(2):676–682

Bokare A, Sanap A, Pai M et al (2013) Antibacterial activities of Nd doped and Ag coated TiO_2 nanoparticles under solar light irradiation. Colloids Surf B Biointerfaces 102:273–280

Cantavenera MJ, Catanzaro I, Loddo V et al (2007) Photocatalytic degradation of paraquat and genotoxicity of its intermediate products. J Photochem Photobiol A Chem 185:277–282

Cunff JL, Tomasic V, Wittine O (2015) Photocatalytic degradation of the herbicide terbuthylazine: preparation, characterization and photoactivity of the immobilized thin layer of TiO_2/chitosan. J Photochem Photobiol A Chem 309:22–29

Dhaouadi A, Adhoum N (2009) Degradation of paraquat herbicide by electrochemical advanced oxidation methods. J Electroanal Chem 637(1–2):33–42

Eleburuike NA, Bakar WAWA, Rusmidah A et al (2016) Photocatalytic degradation of paraquat dichloride over CeO_2-modified TiO_2 nanotubes and the optimization of parameters by response surface methodology. RCS Adv 6:104082–104093

Florêncio MH, Pires E, Castro AL et al (2004) Photodegradation of diquat and paraquat in aqueous solutions by titanium dioxide: evolution of degradation reactions and characterization of intermediates. Chemosphere 55:345–355

Garcia OA, Valencia JE, Romero R et al (2017) W and Mo doped TiO_2: synthesis, characterization and photocatalytic activity. Fuel 198:31–41

Grover IS, Prajapat RC, Singh S et al (2017) Highly photoactive Au-TiO_2 nanowires for improved photo-degradation of propiconazole fungicide under UV/sunlight irradiation. Sol Energy 144:612–618

Kalantaria K, Kalbasia M, Sohrabia M et al (2017) Enhancing the photocatalytic oxidation of dibenzothiophene using visible light responsive Fe and N co-doped TiO_2 nanoparticles. Ceram Int 43:973–981

Kaur T, Sraw A, Toor AP et al (2016) Utilization of solar energy for the degradation of carbendazim and propiconazole by Fe doped TiO_2. Sol Energy 125:65–76

Kruanetr S, Wanchanthuek R (2017) Studies on preparation and characterization of Fe/TiO_2 catalyst in photocatalysis applications. Mater Res Express 4(7):076507

Manjunatha KN, Paul S (2015) Investigation of optical properties of nickel oxide thin films deposited on different substrates. Appl Surf Sci 352:10–15

Maulidiyah TA, Nurwahidah AT, Wibowo D et al (2017) Photoelectrocatalyst of Fe co-doped N-TiO_2/Ti nanotubes: pesticide degradation of thiamethoxam under UV–visible lights. Environ Nanotechnol Monit Manag 8:103–111

Moctezuma E, Leyva E, Monreal E et al (1999) Photocatalytic degradation of the herbicide "Paraquat". Chemosphere 39:511–517

Montañez JP, Gómez S, Pierella LB et al (2015) Photodegradation of herbicide dicamba with TiO_2 immobilized on HZSM-11 zeolite. Int J Environ Res 9(4):1237–1244

Sakee U, Wanchanthuek R (2017) Catalytic activity of bimetallic Zn/TiO_2 catalyst for degradation of herbicide paraquat: synthesis and characterization. Mater Res Express 4(11):115504

Shamsedini N, Dehghani M, Nasseri S et al (2017) Photocatalytic degradation of atrazine herbicide with illuminated Fe^{+3}-TiO_2 nanoparticles. J Environ Health Sci 15:7

Shibin OM, Yesodharan S, Yesodharan EP (2015) Sunlight induced photocatalytic degradation of herbicide diquat in water in presence of ZnO. J Environ Chem Eng 3(2):1107–1116

Toor AP, Verma A, Jotshi CK et al (2005) Photocatalytic degradation of 3,4-dichlorophenol using TiO_2 in a shallow pond slurry reactor. Indian J Chem Techn 12:75–81

Toor AP, Yadav N, Wanchoo RK (2013) Enhancement in photocatalytic activity of nano-TiO_2 photocatalyst by carbon doping. Mater Sci Forum 757:271–284

Verma A, Prakash NT, Toor AP (2014) Photocatalytic degradation of herbicide isoproturon in TiO_2 Aqueous Suspensions: study of reaction intermediates and degradation pathways. Environ Prog Sustain Energy 33(2):402–409

Wei TY, Wan CC (1991) Heterogeneous photocatalytic oxidation of phenol with titanium dioxide powders. Ind Eng Chem Res 30(6):1293–1300

Zahedi F, Behpour M, Ghoreishi SM et al (2015) Photocatalytic degradation of paraquat herbicide in the presence TiO_2 nanostructure thin films under visible and sun light irradiation using continuous flow photoreactor. Sol Energy 120:287–295

Pervious Concrete—A Review on Its Properties and Applications

Rekha Singh and Sanjay Goel

Abstract This paper presents various past studies on pervious concrete where cement is partially replaced by using one or more mineral admixtures and the effect of mineral admixtures on the efficiency of pervious concrete. A systematic literature review is carried out on the properties of pervious concrete to categorize and evaluate the published literature on usage of pervious concrete and the results of experimental investigation. This review study includes types of supplementary cementitious materials (SCM) like fly ash, silica fume and other waste material, etc., which has been utilized in the making of pervious concrete. Following a brief outline of the important parameters for designing and analysis of pervious concrete using SCM as well as critical observation from the previous literature is discussed. Pervious Pavements are one of the most important components of a city's infrastructure covering significant portions of urban land.

Keywords Pervious concrete · Supplementary cementitious materials · Mechanical properties · Durability

1 Introduction

Nowadays, environmental issues have become major concern around the world hence pervious Concrete is increasingly used in variety of infrastructures, pavements, and overlays subjected to low to heavy traffic load. Due to this extended applications, greater strength and durability become concern of porous concrete. Permeable Concrete Pavements are different than conventional concrete pavements. Pervious Concrete is a special type of concrete characterized by an interconnected network of pore structures and high void content/porosity typically in the range of 15–35% by volume.

R. Singh (✉) · S. Goel
Department of Civil Engineering, I.K.G. Punjab Technical University, Kapurthala, India
e-mail: rekhasingh14@gmail.com

A. K. Agnihotri et al. (eds.), *Sustainable Engineering*, Lecture Notes in Civil Engineering 30, https://doi.org/10.1007/978-981-13-6717-5_16

Storm water or surface runoff can carry and deposit untreated pollutants such as sediments, nutrients and pesticides into natural water bodies. Pollutant in discharges remains a significant source of environmental impacts to water quality. Under the clean water act (CWA), Environment protection agency (EPA) has implemented pollution control program such as set water quality standards for all contaminates in surface water, setting waste water limitation from industry.

The potential of application of pervious concrete is staggeringly high. It can replace normal concrete in several usages. In cities where normal concrete pavements have almost finished ability of water to get soaked by the soil and the respiration system of urban trees, pervious concrete provides solution to both. Other than pavements, parking lots, pedestrian paths and medians of roads can use it for greater ecological benefits. The pervious concrete thus helps in reducing risk of floods and storm water runoff, help in removing pollutants from water before percolating into soil, and ground water recharge during rains. Since water is absorbed by porous structure of pervious concrete, it helps in increasing friction and prevents skidding of tire during rainy season. Due to low cement content, the pervious concrete are cheaper to normal concrete thus acting as suitable low-cost sustainable product; although its maintenance is high since its essential to prevent clogging of pores to maximize its benefits, still the overall low costs makes it a viable alternative. In this open literature several pervious concrete mixes with different sizes and types of aggregates, mix proportions and admixtures have been investigated.

2 Material Characteristics

Pervious concrete uses the same materials as normal concrete. The porosity in pervious concrete is created by the little or no fine aggregate and the use of coarse aggregate either single sized or in combination to allow relatively low particle packing. Typically, the volume of aggregates in pervious concrete is about 50–65% compared to conventional concrete, which is about 60–75%. In order to retain sufficient voids in the material, pervious concrete is usually made up of aggregates sizes in the range of 19–9.5 mm (ACI 522R-10); however, many researchers with the main aim to increase the strength properties have used coarse aggregates of size in between 9.5 and 2.36 mm. The void size is not purely governed by the size of coarse aggregate, but also depends upon their gradation, and the thickness of cement paste. A higher porosity of the material does not ensure a higher permeability, as the permeability is a function of the void surface area, void size, and their distribution. The porosity decreases with the increase in paste thickness. When the paste contents become more, the active voids size decrease and reduce the porosity of the sample. Torres et al. (2015) investigated that larger the paste thickness, the smaller the void space. The water-to-cement ratio being a very important variable is lower compared to those used in the conventional concrete mix, and has been historically varied over the range of 0.25–0.40 with the main intention to provide sufficient cement coating for the aggregates. Further, the aggregate-to-cement ratio has been varied in the range of 3:1 to as high as 6:1.

For pervious concrete, various mix designs and proportioning has been recommended in past research studies. The method given in ACI (522R-10 2010) was simple as it only consider physical properties of aggregates, dry-rodded volume of coarse aggregate (in a unit volume of concrete) and porosity (selected to fulfill the desired application) but did not take into account the effect of compaction. Similarly Zouaghi (2002) and Zheng et al. (2012) proposed mix design considering physical properties of aggregates but the relationship among the parameters is applied only for locally available gravel. All these require correction step through experimentation in order to adjust the composition in all the proposed mix designs.

Nguyen et al. (2014) proposed a modified method for mix design of pervious concrete which was based on the assumption that the cementitious paste act only as a coating and does not fulfill the void among the grains of gravel. This method of mixture proportioning is divided into three steps: the **determination of aggregate volume**, **cement paste volume,** and **water–cement ratio**. The typical mix proportions used in the past studies are tabulated in Table 1.

Many previous studies have reported the properties of pervious concrete by varying the water-to-cement ratio, aggregate-to-cement ratio, aggregate size and binder material type. The range of water-to-cementitious ratio varies in the range of 0.20–0.42 and aggregate-to-cementitious ratio 3:1–8:1. The aggregate density content varies in the range of 1400–1800 kg/m^3 for various mix designs of pervious concrete; however, numbers of such studies are very limited.

3 Literature Review

The sustainability aspect in pervious concrete has been inspected upon with partial substitution of cement with supplementary cementitious materials like silica fume, fly ash, rice husk ash, and blast furnace slag. Utilization of recycled aggregates in pervious concrete enhances its ecological advantages. Use of metakaolin is not investigated enough in pervious concrete as in conventional concrete. The literature study reported use of variety of supplementary cementitious materials as listed in Table 2.

The literature review also represents numerous compaction techniques used in making pervious concrete where hand rodded method was found inappropriate and standard proctor method was used often (As reported by many researchers, i.e., Ghafoori and Dutta 1995; Putman and Neptune 2011).

Concrete curing is also important as it is required to maintain sufficient moisture to allow cement hydration and concrete microstructure development. Curing has been shown to effect concrete durability as well as concrete strength. The current method of curing pervious concrete involves covering the fresh concrete with plastic sheets for the required curing period.

Table 1 Summary of mix proportions of pervious concrete from literature

References	Year	Cementitious material (kg/m^3)	Aggregates (kg/m^3)	W/CM	Aggregate/CM ratio	Water (kg/m^3)
Khankhaje et al.	2016	340	1460	0.32	–	109
Chandrappa and Biligiri	2016	321–487	1373–1692	0.25–0.35	3.0–5.0	84–161
Yahia and Daddy Kabagire	2014	195–535	1500–1700	0.30	–	–
Nguyen et al.	2013	309	1525	0.30	4.9	93
Lim et al.	2013	242–495	1560	0.20–0.30	3.15–6.44	73.4–148.5
Bassuoni and Sonebi	2010	315–415	1200–1400	0.28–0.40	4–6	125–154
Huang et al.	2010	320–353	1440–1587	0.35	4.5	112.1–123.4
Neithalath et al.	2010	309–312	1544–1568	0.33	5	101–104
Kevern et al.	2009a, b	180–380	1510–1820	0.24–0.30	4.0–10	50–100
Crouch et al.	2007	287–345	1542–1620	0.30	4.5–5.6	87–105
Ghafoori and Dutta	1995	300–413	1651–1800	0.37–0.42	4–6	125–154

In another study carried out by Kevern et al. (2009a, b), combination of four different pervious concrete mixtures was used to study the effect of cured samples using six common curing methods. The samples cured under plastic sheets had the best abrasion resistance and highest flexure strength although there was no effect of duration of curing days (Kevern et al. 2009a) (Table 3).

Table 2 Pervious concrete with supplementary cementitious material

References	Admixture	Tests	Observation
Gaedicke et al. (2016)	GGBFS	Permeability test, compressive strength	Moderation of permeability and compressive strength by 20% and 17% respectively for similar parameters
Suozzo and Dewoolkar (2014)	Class F fly ash	Compressive strength, permeability coefficient	Elastomeric rubber pad and sulfur capping methods were found to provide comparable compressive strength measurements
Crouch et al. (2007)	Class C fly ash	Compressive strength	The compressive strength and modulus of elasticity appeared to be higher at similar void contents
Hesami et al. (2014)	Rice husk ash	Compressive, tensile, flexural strengths, permeability coefficient	The mechanical properties initially increase up to its peak value and then decrease gradually The permeability decreases with an increase in RHA value till its peak, 12% of RHA can be more suitable for concrete pavements
Yang and Jiang (2003)	Silica fumes	Compressive, tensile, flexural strengths, permeability coefficient	By using smaller aggregate, SF, and SP, pervious concrete of both high strength and good water penetration could be produced
Lian et al. (2011)	Sand and Silica fumes	Compressive, tensile, flexural strengths, permeability coefficient	With the optimum w/c ratio of 0.32, compressive strength of 40 MPa and permeability above 2 mm/s can be achieved with 28 days testing
Chen et al. (2013)	Class C fly ash and Silica fumes	Compressive strength tests, flexural strength, fracture toughness and fatigue life	Polymer-modified PC displays much higher fracture toughness and far longer fatigue life than SPC at any stress level

Table 3 Recent studies reported using one or more mineral admixture

References	Study performed	Significant observation	Comments
Wu et al. (2016)	The effects of replacing sandstone with incineration bottom ash and different water-to-cement ratios with different aggregate sizes in pervious concrete bricks preparation	Amongst the various mix proportions, the specimen using bottom ash aggregate of size #4 with w/c ratio of 0.55 demonstrated the maximum compressive strength The strength of incineration bottom ash permeable bricks was only 20 MPa and recommended only for sidewalks, bicycle ways and general landscaping	A useful review has been presented by considering experimental validation but need to work on a mix proportion with available aggregate type
Zhong and Wille (2016a, b)	Research has been carried out linking matrix strength, total porosity, aggregate size and mean pore size of pervious concrete to its compressive strength. The linear path function is tailored for pervious concrete and validated by an Image analysis method	The proposed equation using matrix strength, total porosity, aggregate size and mean pore size predict well the compressive strength of pervious concrete	Proposed model for the stress–strain relationship of pervious concrete was found to agree well with the experimental results of 27 pervious concrete series
Zhong and Wille (2016a, b)	Pervious concrete was investigated for its compressive strength and porosity. The mixture designs include silica fumes with silica powder in three different matrices, three different sizes of quartz aggregates and three different aggregate-to-binder ratios	A decrease in aggregate size increases the strength of pervious concrete. The influence becomes more pronounced with an increase of matrix strength The aggregate-to-binder ratio affects the strength of pervious concrete only through altering the total porosity. Based on the analyzed correlation of matrix strength, porosity, aggregate size, and aggregate to binder ratio, a semi-empirical model is proposed predicting the compressive strength of pervious concrete using matrix strengths	A useful study for explaining the effect of various parameters on pervious concrete

(continued)

Table 3 (continued)

References	Study performed	Significant observation	Comments
Zhong and Wille (2015)	Research emphasizes the development and characterization of high performance pervious concrete aiming at improved mechanical resistance and advanced durability properties. Freeze–thaw tests have been carried out to compare the durability performance of conventional pervious concrete with high performance pervious concrete. Silica fume with a very low carbon content and supplemental material size in between SF and cement was used	Use of optimized ultra-high performance matrix (UHPM) along with balanced aggregate to binder ratio (A/B) and tailored aggregate size could potentially advance and broaden the application of pervious concrete without sacrificing its hydraulic conductivity	A useful study for improved mechanical performance and enhanced durability properties of HPPC, further research is needed to address concerns such as potential shrinkage, cracking and higher material cost prior to the application of this material
Giustozzi (2016)	Studied four types of polymers for polymer-modified concrete with two different w/c ratios on cement grouts and mortars. Mechanical and durability properties on prepared samples have been carried out	Results showed a delayed curing of polymer-modified mixes as well as an increased mechanical resistance and durability to raveling and freeze–thaw cycles The addition of small percentage of fine sand to the pervious matrix demonstrated to be efficient in increasing mechanical performance, particularly, and often being better that polymer modification alone	Results highlighted how some of the polymers were useful to improve the durability of the material Polyvinyl acetate polymer was observed to be a good option
Chang et al. (2016)	Studied the mechanical properties of pervious concrete made with electric arc furnace slag (EAFS) and alkali activated slag cement (AASC)	Results @ 28 Days: (a) Compressive Strength-35 MPa (b) Water permeability coefficient-0.49 cm/s (c) BPN-79 (d) Sound absorption coefficient 0.94@125 Hz	A pervious concrete with high compressive strength could be achieved by using EAFS and AASC

4 Conclusion

Pervious concrete is a sustainable pavement material in low volume on and off road applications and has been engaging designers and contractors for long. This paper has reviewed past studies reported by researches on properties of pervious concrete such as mechanical, hydrological and durability properties.

- Most of the studies carried out on pervious concrete are limited to laboratory conditions. There is a gap in the laboratory and field practices in terms of unit weight, compaction efforts, porosity, curing technique, fatigue, etc.
- Standardization of physical, mechanical, durability, hydrological, and other important properties is required for wide range applications of pervious concrete.
- Standardization of mix designs and testing methods for the application of pervious concrete as studies reported are discrete in nature. Different mix design and proportion have been reported in past studies.
- The maintenance of pervious concrete is highly challenging as its porous structure gets clogged and requires regular maintenance. The initial and maintenance costs of pervious concrete for long periods need to be collected for proper implementation.
- Overall; The Pervious Concrete is promising material in pavements for storm water management, low volume roads such as pedestrian, walk ways and driveway.

References

ACI 522R-10 (Reapproved 2011) Report on pervious concrete. American Concrete Institute, Farmington Hills, MI, USA

Bassuoni MT, Sonebi M (2010) Pervious concrete: a sustainable drainage solution. Concr Concr Soc 44:14–16

Chandrappa AK, Biligiri KP (2016) Comprehensive investigation of permeability characteristics of pervious concrete. A hydrodynamic approach. Constr Build Mater 123:627–637

Chang JJ, Yeih W, Chung TJ, Huang R (2016) Properties of pervious concrete made with electric arc furnace slag and alkali-activated slag cement. Constr Build Mater 109:34–40

Chen Y, Wang K, Wang X, Zhou W (2013) Strength, fracture and fatigue of pervious concrete. Constr Build Mater 42:97–104

Crouch LK, Pitt J, Hewitt R (2007) Aggregate effects on pervious Portland cement concrete static modulus of elasticity. J Mater Civ Eng 19:561–568

Gaedicke C, Torres A, Huynh KCT, Marines A (2016) A method to correlate splitting tensile strength and compressive strength of pervious concrete cylinders and cores. Constr Build Mater 125:271–278

Ghafoori N, Dutta S (1995) Laboratory investigation of compacted no-fines concrete for paving materials. J Mater Civ Eng 7(3):183–191

Giustozzi F (2016) Polymer modified pervious concrete for durable and sustainable transportation infrastructures. Constr Build Mater 111:502–512

Hesami S, Ahmadi S, Nematzadeh M (2014) Effects of rice husk ash and fiber on mechanical properties of pervious concrete pavement. Constr Build Mater 53:680–691

Huang B, Wu H, Shu X, Burdette EG (2010) Laboratory evaluation of permeability and strength of polymer-modified pervious concrete. Constr Build Mater 24:818–823

Kevern JT, Schaefer VR, Wang K (2009a) The effect of curing regime on pervious concrete abrasion resistance. J Test Eval 37(4)

Kevern JT, Schaefer VR, Wang K (2009b) Evaluation of pervious concrete workability using gyratory compaction. J Mater Civil Eng 21:764–770

Khankhaje E, Salim MR, Mirza J, Hussin MW, Rafieizonooz M (2016) Properties of sustainable lightweight pervious concrete containing oil palm kernel shell as coarse aggregate. Constr Build Mater 126:1054–1065

Lian C, Zhuge Y, Beecham S (2011) The relationship between porosity and strength for porous concrete. Constr Build Mater 25:4294–4298

Lim E, Twan KH, Fwa TF (2013) Effect of mix proportion on strength and permeability of pervious concrete for use in pavement. Proc Eastr Asia Soc Trans Studvol 9

Neithalath N, Sumanasooriya MS, Deo O (2010) Characterizing pore volume, sizes, and connectivity in pervious concretes for permeability prediction. Mater Charact 61:802–813 (Elsevier)

Nguyen DH, Boutouil M, Sebaibi N, Leleyter L, Baraud F (2013) Valorization of seashell by products in pervious concrete pavers. Constr Build Mater 46:151–160

Nguyen DH, Sebaibi N, Boutouil M, Leleyter l, Baraud F (2014) A modified method for the design of pervious concrete mix. Constr Build Mat 73:271–282

Putman BJ, Neptune AI (2011) Comparison of test specimen preparation techniques for pervious concrete pavements. Constr Build Mater 25:3480–3485

Suozzo M, Dewoolkar MM (2014) Evaluation of strength and hydraulic testing methods of pervious concrete. ACI Mater J 111(1)

Technical guidance on implementing the stormwater runoff Environment protection agency. EPA 2017

Torres A, Jiong H, Ramos A (2015) The effect of the cementitious paste thickness on the performance of pervious concrete. Constr Build Mater 95:850–859

Wu MH, Lin CL, Huang WC, Chen JW (2016) Characteristics of pervious concrete using incineration bottom ash in place of sandstone graded material. Constr Build Mater 111:618–624

Yahia A, Daddy Kabagire K (2014) New approach to proportion pervious concrete. Constr Build Mater 62:38–46

Yang J, Jiang G (2003) Experimental study on properties of pervious concrete pavement materials. Cem Concr Resea 33:381–386

Zheng M, Chen S, Wang B (2012) Mix design method for permeable base of porous concrete. Int J Pavement Res Technol 5:102–107

Zhong R, Wille K (2015) Material design and characterization of high performance pervious concrete. Constr Build Mater 98:51–60

Zhong R, Wille K (2016a) Compression response of normal and high strength pervious concrete. Constr Build Mater 109:177–187

Zhong R, Wille K (2016b) Linking pore system characteristics to the compressive behavior of pervious concrete. Cem Concr Comp 70:130–138

Zouaghi A (2002) Technological problems of multi-performance porous concrete. In: Proceedings of the 1st fib congress, pp 233–242

Reclamation of Waterlogged Saline Lands in Haryana Through Subsurface Drainage Technology—A Review

Hemant Raheja, Arun Goel and D. S. Bundela

Abstract Irrigation induced waterlogging and soil salinization is one of the serious problems in irrigated agriculture in arid and semi-arid regions. At present, 2.96 Mha area in India is affected by soil salinity. About 50,000 ha area in Haryana state is waterlogged saline area having shallow water table (depth < 1.5 m) and soil salinity (ECe > 4 dS/m) as natural drainage system is not sufficient in affected areas for controlling the problem. Therefore, subsurface drainage (SSD) technology, an artificial drainage system, is needed to reclaim waterlogged saline lands in the state for enhancing crop productivity. A SSD system is designed to control soil salinity and waterlogging in an affected area. This technology improves the crop intensity significantly by 50–110% and yield by 45–120% and doubles the farmer's income. About 10,584 ha waterlogged saline area in nine districts of the state (Rohtak, Jhajjar, Palwal, Bhiwani, Sonipat, Kaithal, Jind, Fatehabad, and Sirsa) have been reclaimed by SSD technology by Haryana Operational Pilot Project (HOPP), Department of Agriculture and Farmers' Welfare, Govt. of Haryana in technical collaboration with ICAR-CSSRI Karnal. In this paper, the current status of waterlogged saline lands in Haryana, need of subsurface drainage technology, design parameters and performance of SSD systems in Haryana are discussed. These results and outcomes will be helpful for effective and large scale reclamation of irrigated saline lands in the country.

Keywords Waterlogging · Waterlogged saline soils · Salinity · Subsurface drainage (SSD) technology · Crop yield

1 Introduction

Salinity and waterlogging is one of the serious problems in irrigated lands of the arid and semi-arid region. In the World 230 Mha land (approximate 18%) out of the 1500 Mha area is cultivated. Heuperman et al. (2002) has reported that in the world

H. Raheja (✉) · A. Goel · D. S. Bundela
Department of Civil Engineering, NIT Kurukshetra, Kurukshetra 136119, Haryana, India
e-mail: raheja.hemant788@gmail.com

A. K. Agnihotri et al. (eds.), *Sustainable Engineering*, Lecture Notes in Civil Engineering 30, https://doi.org/10.1007/978-981-13-6717-5_17

more than one-third irrigated land is affected by soil salinity and waterlogging. In India 6.74 Mha cultivated land are affected by waterlogging and salinity problems (CSSRI 1996). Earlier, Bhattacharya (1992) reported in India nearly 3.3 million ha are affected by the problem of waterlogging. In India soil salinity and waterlogging problem has been developed in nineteenth century due to over irrigation or poor irrigation practices. When canal irrigation was introduced in the country are these problems found in canal commanded area. Waterlogging and salinization are two different processes which may occur singly or simultaneously (Smedema 1990). Where the groundwater quality for irrigation is not good salts accumulate on soil surface directly or by the capillary action as result creates the serious problems of waterlogging in area (Houk et al. 2006). Saline soils having pH < 8.2, ESP < 15, ECe > 4. Nearly 2.96 Mha area in the country and 16 states comes under the saline land, i.e., Punjab, Haryana, Rajasthan, Gujarat, Karnataka, Andhra Pardesh, Telengana, Tamil Nadu, West Bangal, Bihar, Odisha, Karla, Uttar Pradesh, Maharashtra, Madhya Pardesh. In Haryana more than 50,000 ha area is having shallow water table having a depth 1.5 m and almost 3,80,000 ha area having water table depth of 1.5–3.0 m. ICAR-CSSRI reported that 5.66 million tonnes of crop production and 8,000 crores monetary losses due to the salinity problem.

2 Methodology

Drainage System: A drainage system controls the salinity and waterlogging problems of irrigated area and it reduces the water table to a safe depth and salt balance in the rootzone (Konukcu et al. 2006). A drainage system can be divided as surface or subsurface based on its purpose and it can be natural or artificial (Abdel-Dayem et al. 2007). The surface drainage system is easy to construct and maintain and is for disposal of the surface runoff. But it takes the land and that cannot be cultivated. The other disadvantage of the surface drainage system is that it does not remove water from or below the rootzone. Because of these limitations with surface drainage, subsurface drainage is widely used for controlling the problem of salinity and waterlogging. Subsurface drainage system controls the salinity problem in arid and semi-arid regions (Rao and Visvanatha 1998).

2.1 Need of Subsurface Drainage System

Drainage reduces waterlogging and salinity in rootzone in arid and semi-arid regions. In humid and sub-humid regions, drainage removes excess water from the area and protects the crop from failure and gives the better option for the farmer for high yield of crop varieties (Ritzema 2007). Worldwide about 10–16% of the irrigated lands affected by the problem of salinity and waterlogging. The land loss due to waterlogging and salinity is about 0.5 Mha per year. To overcome the problem of waterlogging

Table 1 Depth to water table in different districts of Haryana (Haryana KisanAyog 2013)

Sr. no.	District	Geographical area (ha)	Water table depth (m)	
			0–1.5	1.5–3.0
1	Rohtak	166,777	16,516	78,694
2	Bhwani	487,072	8601	53,949
3	Hisar	386,052	3019	27,119
4	Jhajjar	186,770	12,980	75,651
5	Sonepat	226,053	2892	56,097
6	Jind	273,600	789	16,711
7	Sirsa	427,600	2325	8747
8	Palwal	136,455	499	3691
9	Fatehabad	249,110	477	1968
	Total	4,421,200	54,130	379,762

Source Critical evaluation… by Karma and Sharma (2015)

and salinity in irrigated land it is estimated that existing drainage systems are not sufficient and have to be replaced by new artificial systems (Ritzema 2009). Over the last 50 years, to protect the irrigated lands from waterlogging and salinity problem, subsurface drainage systems have been or are being installed (Ritzema et al. 2006). These drainage systems effectively reduce salinity from the root zone and maintained the water table and also increase crop yield and farm income. These systems are very cost-effective as compared to other measures (Ritzema and Schultz 2011).

2.2 *Status of Waterlogged Saline Area in Haryana*

According to Natural Resource Management Report of Haryana KisanAyog (2013), more than 50,000 ha area in Haryana is having shallow water table having a depth 1.5 m and almost 3,80,000 ha area having water table depth of 1.5–3.0 m. The district named as: Rohtak, Jhajjar, Bhiwani, Hisar, Sonepat, Jind, Fatehabad, Sirsa, and Palwal (Table 1).

2.3 *SSD System*

A SSD system consists a network of underground corrugated perforated pipes (i.e., laterals and collectors) installed at the designed drain depth and spacing. The outlet of the SSD system may be pumped outlet or gravity outlet. Drainage water is disposed into the open drain, pond, or rivers. The water enters in the lateral pipes through the perforations and flows towards the outlet under gravitational flow through the

Table 2 Design parameters of SSD for different conditions of India

Drainage coefficient (mm/day)			Drainage depth (mm)		Drainage spacing (m)	
Region	Range	Optimum	Type of outlet	Depth (m)	Soil texture	Spacing (m)
Arid	1–2	1	Gravity	0.9–1.2	Light	100–150
Semi-arid	1–3	2			Medium	50–100
Sub-humid	2–5	3	Pumped	1.2–1.8	Heavy	30–50

Source Guidelines and cost of… by Bundela et al. (2017)

collectors and main drains. The main components of a SSD system are described as follows:

1. Lateral drain: According to IS 9271:2004, lateral drains are the single wall corrugated (SWC) perforated pipes of unplasticized poly vinyl chloride (UPVC) having the outer diameters, 80 and 100 mm and length of 150–200 m respectively.
2. Collectors drain: Collector pipes may be single wall corrugated (SWC) or double wall corrugated (DWC) with the diameter of 100–315 mm. Collector pipe collects the water from the laterals and transports to the outlet, so nonperforated pipes are used. DWC non-perforated pipes are designed as per IS 16908 part 2:2013.
3. Manholes and sump well: Manholes and sump well are the prefabricated RCC pipes with the dia. 900 mm and 1200 mm and having a length of 3.75 m with three parts and 5 m with four parts (i.e., 1.25 m each), respectively.
4. Pump house and pump set: According to IS 1538:1986, where pumped outlet system is used, a pump house is constructed having dimensions of (3 × 3 × 2.4 m) and 6 HP diesel pump set is provided for the drainage.

2.4 Drainage Design Criteria

A SSD system installed at the designed spacing and depth for removal of excess salt and water from the affected area. The filter or envelope which is wrapped on the drain pipes is required based on soil properties. ASSD block or unit having area from 20 to 60 ha is designed to control soil salinity and waterlogging. The design parameters of SSD system for different conditions and guidelines for drain pipe size and slope and filter/envelope requirement are described in Tables 2 and 3.

Table 3 Guidelines for drain pipe size and slope and filter/envelope requirement

Pipe size (mm)	Grade (%)	Clay and SAR	Filter required
80–100 (lateral)	0.10	<30%	Yes
160–450 (collector)	0.09–0.05	30–40% and SAR > 8	Yes
		>40%	No

Source Guidelines and cost of… by Bundela et al. (2017)

2.5 *Drainage Design Equation*

The transient state Glover-Dumm equation is used in drainage design for irrigated lands under arid and semi-arid monsoon climate. However, the steady state Hooghoudt equation has been frequently used due to its simplicity and given as below

$$q = \left(8kdh + 4kh^2\right)/S^2 \quad (1)$$

where q = discharge rate per unit surface area ($m^3/d - m^2$), k = hydraulic conductivity (m/d), S = drain spacing (m), h = hydraulic head (m) and d = thickness of equivalent layer (m).

3 Discussion

3.1 *Implementation of Subsurface Drainage System*

During 1980s, ICAR-CSSRI installed a number of pilot SSD projects and were manually installed, the large project mechanically installed in states of Haryana, Punjab, Rajasthan, Andhra Pradesh, Gujarat, Karnataka, and Maharashtra. Government funding under schemes like CADA, RKVY, etc., for land reclamation. In India about 66,500 ha area is reclaimed by SSD technology. In Haryana, about 10,584 ha waterlogged saline area in nine districts of the state (Rohtak, Jhajjar, Palwal, Bhiwani, Sonipat, Kaithal, Jind, Fatehabad, and Sirsa) have been reclaimed through this technology by Haryana Operational Pilot Project (HOPP), Department of Agriculture and Farmers' Welfare, Govt. of Haryana in technical collaboration with CSSRI (Table 4). Out of 10,584 ha area provided with SSD over last two decades, 7453 ha area has been installed in past 10 years with funding under Ministry of Rural Development (MORD, GOI) and RKVY schemes. The cost for installation of SSD is Rs. 70,000 per hectare (excluding staff salaries) (Fig. 1).

Table 4 SSD sites under various schemes/projects in Haryana (HOPP, JULY 2012)

Sr. no.	Project/Funding and period	Project/Districts	Villages	No. of drainage block	Area (ha)	No. of benefi-ciaries
1	Indo-Dutch/MMM project (1999–2002)	Gohana (Sonipat)	Moi, Ball, Katwal and Rewara	23	1253	872
		Kalayat-I (Kaithal)	Kherllamba, Kalayat and Kurar		746	425
		Kalayat-II (Kaithal)	Kalayat, Kherilamba and Kolekhan		407	300
2	Ministry of rural development (MORD), GOI (2003–2009)	Beri (Jhajjar)	Beri, Dopana, Beri-Khas, Dhadlan, Gochhi and Bakra	14	805	873
		Charkhi Dadri	Rawaldhi, Khatiwas, Loharwas, Jhinjar and Achina	16	904	963
		Dhanana (Sonipat)	Rindhana, Dhanana, Dhanana-II and Jagsi	23	1137	836
		Rori (Sirsa)	Surtia, Darbra-Kalan, Manak-Diwan and NathusriChaupta	12	479	180
3	RKVY (2009–2015, contd.)	Banmandori (Fatehabad)	Banmandori, Gorakhpur and Kumarian		600	152
		Wazirpur (Jhajjar)	Wazirpur, Baghpur and Beri	23	1034	798
		Kanoli (Palwal)	KherilJeeta, Kanoll, and Madkola	10	351	233
		Mokharakheri (Rohtak)	MokharaKheri, Madina	18	100	546
		Siwana Mal (Jind)	Siwana Mal, Gangana	8	600	215
		Gharwal (Sonipat)	Gharwal		1000	169

Source Critical evaluation… by Karma and Sharma (2015)

Fig. 1 Soil condition before and after installation of SSD. *Source* CSSRI Annual Report

Table 5 Mean water table depth and ground water quality parameters of MokhraKheri blocks

Year	Water table depth (m)		EC (ds/m)		Ph	
	NP	P	NP	P	NP	P
2014	1.09	1.23	8.83	1.98	7.75	7.57
2015	1.04	1.28	8.75	1.79	7.69	7.49

Where *NP* Non-pumping blocks, *P* Partial or full pumping blocks
Source CSSRI Annual Report 2015–16

3.2 Critical Evaluation of Existing SSD Projects

CSSRI has been involved in monitoring and evaluation of SSD projects executed by HOPP in Haryana and detailed progress reports are regularly submitted for each project. During 2015–2016, the impact of SSD systems was evaluated by CSSRI for Siwana Mal (Jind), MokhraKheri (Rohtak) project sites.

In MokhraKheri, for partial or full pumping blocks soil salinity reduces throughout 0–90 cm soil profile in comparison to non-pumping blocks and shallow water table (depth < 1.5 m) conditions existed in all blocks due to irregular pumping in different months (Table 5).

In Siwanamal, the improvement in soil salinity in 38 ha area of block 4 is presented in Table 6. EM38 salinity survey conducted during May 2012 (before) and May 2015 (after) installation of SSD system. It can be seen that 98.7% and 97.1% area having (>4 ds/m) salinity in 0–15 cm and 15–30 cm layer during 2012 reduces respectively to 79.5% and 82.4% area during 2015 (Table 6) and notable increase in crop yields (30–110%) in rice; (20–120%) in wheat over pre-drainage levels were observed in SSD blocks with adequate pumping efforts of HOPP and individual farmers (Table 7).

4 Conclusions

Subsurface drainage technology is a proven technology for reclaiming waterlogged saline lands in the country. These design parameters and guidelines will be helpful for installation of SSD system. About 50,000 ha area in Haryana state is waterlogged

Table 6 Improvement in percent area under different soil salinity levels in Block 4 at Siwana Mal SSD site

Year	Depth (cm)	ECe (<4 ds/m)	ECe (4–8 ds/m)	ECe (8–16 ds/m)	ECe (>16 ds/m)
		Area %	Area %	Area %	Area %
2012	0–15	1.3	21.8	66.2	10.7
2015		20.5	54.3	23.2	2
2012	0–30	2.9	25.4	63.9	7.7
2015		17.6	37.2	35.6	9.6
2012	0–60	6.6	34.6	56.4	2.4
2015		18.7	46.0	30	5.3

Source CSSRI Annual Report 2015–16

Table 7 Soil and groundwater conditions in proposed SSD sites (2014–2015) based on HOPP-CSSRI visits

District	Villages	Area covered (ha)	Soil sample ECe (ds/m)	Water table depth (m)	Water EC	Drain availability
Sirsa 02-06-2014	Lohgarh	200	1.76–12.52	<1.0	0.27–25.50	
	Jottawali		9.33–9.54		1.41–4.16	No Drain
Jhajjar 30-06-2014	Palda	300	12.71–16.85	<1.0		Link drain
	Mangawas	400	3.53–4.79	<1.0	5.78	Main drain
	Bisan	500	1.06–31.38	1.39–2.11	10–40	Main drain
	Jahajgarh		6.33–12.3	<1.0		Link drain
	KheriKhumbar		10.84–37.31	<1.0	39.70	Main drain
Sonipat 30-06-2014	Ahulana	200	7.84–10.68	1.0–1.3	0.63–5.87	Main drain
Fatehabad Aug. 2014	Pili Mandori	600		<2	8.7	Link drain
	Thuian				7.5	Link drain
	Mehuwala				4.6	Link drain
Rohtak 15-09-2014	Kahni	500	20.2–39.6	<1.0	3–11.4	Main drain
	Shanghi		6.0–26.9		5.1–5.6	Link drain
	RithalNarwal		2.4–18		2–3.4	Main drain
Rohtak 08-12-2014	Baniyani	1000	6.0–56	7 1.4	3.8–19.5	Link drain
Sirsa 16-02-2015	GudiaKheda	100	2.68–11.36	<1.5	0.3–07.5	Main drain
	ShakkarManduri	1000	4.36–39.5	<1.5	4.2–11.5	No drain

saline area having shallow water table (depth < 1.5 m) and soil salinity (ECe > 4 dS/m) urgently needs subsurface drainage, a major part in Rohtak, Jhajjar and Sonepat districts, followed by Bhiwani, Hisar, Jind, Fatehabad, Sirsa, and Palwal. In MokhraKheri, soil salinity reduces and shallow water table (depth < 1.5 m) conditions existed in all blocks due to irregular pumping and in Siwana Mal soil salinity has been reduced and notable increase in crop yields (30–110%) in rice; (20–120%) in wheat due to the installation of subsurface drainage.

References

Abdel-Dayem S, Abdel-Gawad S, Fahmy H (2007) Drainage in Egypt: a story of determination, continuity, and success. Irrig Drain: J Int Comm Irrig Drain 56(S1):S101–S111

Bhattacharya AK (1992) Status paper on waterlogged/drainage affected areas. In: Proceedings on drainage congestion in Northern Bihar, vols. I and II. Water and Land Management Institute, Patna

Bundela DS, Kaledhonkar MJ, Gupta SK, Mohan Lal, Kamra SK, Sharma DK, Sharma PC, Chaudhari SK (2017) Guidelines and costs of sub-surface drainage technology for large scale reclamation of waterlogged saline soils. In: ICAR-CSSRI Karnal

Heuperman AF, Kapoor AS, Denecke HW (2002) Biodrainage: principles, experiences and applications (No. 6). Food and Agriculture Organization

Houk E, Frasier M, Schuck E (2006) The agricultural impacts of irrigation induced waterlogging and soil salinity in the Arkansas Basin. Agric Water Manag 85(1–2):175–183

Kamra SK, Sharma DK (2015) Critical evaluation of performance and organizational framework of subsurface drainage projects in Haryana and Maharashtra. Water Energy Int 72(4):64–72

Konukcu F, Gowing JW, Rose DA (2006) Dry drainage: a sustainable solution to waterlogging and salinity problems in irrigation areas? Agric Water Manag 83(1–2):1–12

Rao KVGK, Visvanatha NA (1998) Salinity control through subsurface drainage. In: Gupta SK, Sharma SK, Tyagi NK (eds) Proceedings of workshop on salinity management in agriculture, 2–5 Dec 1998. CSSRI, Karnal, India, pp 109–115

Ritzema HP (2007) Performance assessment of subsurface drainage systems: case studies from Egypt and Pakistan. Alterra-ILRI

Ritzema HP (2009) Drain for gain: making water management worth its salt: subsurface drainage practices in irrigated agriculture in semi-arid and arid regions

Ritzema HP, Schultz B (2011) Optimizing subsurface drainage practices in irrigated agriculture in the semiarid and arid regions: experiences from egypt, India and Pakistan. Irrig Drain 60:3–13

Ritzema HP, Nijland HJ, Croon FW (2006) Subsurface drainage practices: from manual installation to large-scale implementation. Agric Water Manag 86:60–71

Smedema LK (1990) Comparative study of the East Khairpur, Mardan and Drainage IV pipe drainage project. Project: PAK

Effect of Modulus of Elasticity and Water Table on Uplift Capacity of Granular Anchor Pile: An FEM Approach

Abhishek and Ravi Kumar Sharma

Abstract Expansive soils are considered as one of the most challenging soils due to their enormous volume variations. These volumetric alterations pose problem of stability of structures particularly the lightly encumbered structures founded on them. The bearing capacity of these soils is very low and uplift load resistance is almost nil. Development of cracks takes place on the structures originated on expansive soils. Granular Anchor Pile (GAP) is a new foundation practice to increase pullout behavior and bearing capacity of foundations founded on expansive soil. In the present paper, a numerical study was carried out to predict the behavior of GAP in expansive soil using three-dimensional finite element PLAXIS software. The effect of modulus of elasticity of soil and position of water table on the uplift capacity at different prescribed displacements was examined using load-displacement curves. With the increase in value of elasticity modulus of soil there was an increase in uplift resistance of the GAP system. With the rise in water table to the ground surface, the uplift capacity of GAP system decreased.

Keywords Expansive soil · Modulus of elasticity · Uplift capacity · Upward movement · Granular anchor pile

1 Introduction

Expansive soils are prone to alternate swelling and shrinkage on imbibition of water (Subba Rao 1999) which leads to formation of cracks on the structures founded over them. Due to these cracks, there is danger of instability of structures due to differential settlement. Various foundation techniques which are being used for treating with swelling and shrinkage problems of poor soils resulting alterations in moisture content may be characterized as physical modification techniques, addition of chemi-

Abhishek (✉) · R. K. Sharma
NIT, Hamirpur, Himachal Pradesh, India
e-mail: abhishek1@nith.ac.in

R. K. Sharma
e-mail: rksnithp61@gmail.com

A. K. Agnihotri et al. (eds.), *Sustainable Engineering*, Lecture Notes in Civil Engineering 30, https://doi.org/10.1007/978-981-13-6717-5_18

cals to poor soils and pullout resistant foundation techniques. The physical alteration practice comprises sand cushions (Satyanarayna 1966) and cohesive non-swelling (CNS) layer method (Katti 1979). The chemical alteration technique includes mixing of numerous chemicals such as lime, Portland cement and $CaCl_2$ as illustrated by Rao 1984, Hunter 1988, Sanker 1989, Petry and Little 1992, Kumar 1996, Rollings and Rollings 1996, Desai and Oza 1997, Rollings et al. 1999, Phanikumar 2000, Cokca 2001, Phanikumar and Reddayya 2001, Phanikumar and Sharma 2004, Acosta et al. 2013, Hoyos et al. 2004. The tension-resistant techniques includes under-reamed piles (Setty et al. 1966); Sharma et al. 1978; BabuShanker et al. 1980; Rao 1986; Dey et al. 1996), belled piers, and drilled piers (Chen 1988; Nelson and Miller 1992), pile foundation (Poulos and Davis 1980; Dinesh 1991). Granular pile is a new technique for coping with the problems of loose sand deposits and poor clayey soils.

Considering the effect of uplift forces, granular pile alone is not capable resist the pullout forces which is triggered on the footing by bulging of poor soils. Granular pile if at the bottom end is anchored to an anchor plate with the help of a rod then it becomes a tension-resistant GAP foundation system and is capable of resisting the tensile uplift forces and heave reduction (Phanikumar et al.1994; Phanikumar 1997; Sharma et al. 2004; Hari 2006). The design of GAP foundation should be done in such a way that it does not fail during pullout and side by side also resists the upward force in the foundation. This resistance towards the uplift force depends upon many factors such as swelling pressure, the relative density of the granular pile material the surface area of the GAP, the depth of embedment of the granular pile, angle of inclination, mode of loading, length and diameter of the GAP, etc. (Kranthikumar et al. 2016a, b; Phanikumar 2016; Abhishek and Sharma 2018).

1.1 Concept of Gap System

The GAP structure is a cost-effective and operational technique developed for refining the shear-strength assets of the soil and enhancing the uplift capacity soils. In the construction of GAP, a borehole is drilled in the expansive soil (Fig. 1) and then a casing pipe is inserted to that borehole to prevent lateral movement of surrounding soil. After inserting the casing pipe, an anchored rod of suitable thickness which is further connected to an anchor plate at one end, is then inserted into the borehole. After this, the borehole is filled with required quantity of backfill by compacting it in layers of same thickness so that required density can be ensured. The other end of anchor rod is connected to footing plate which is placed at top of granular pile. This arrangement is helpful in resisting the uplift forces (Fig. 2).

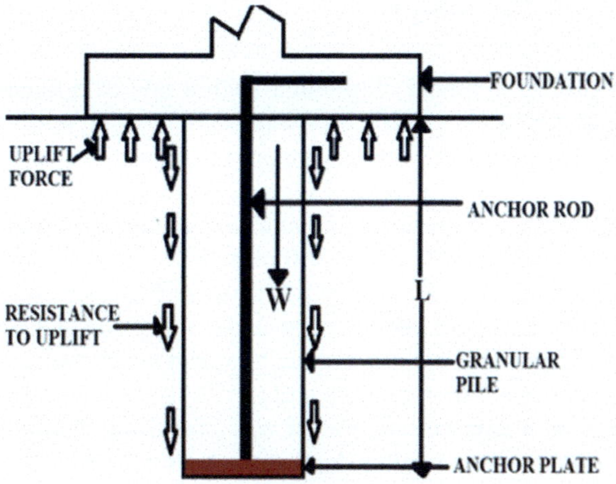

Fig. 1 Concept of Granular anchor pile and forces acting on GAP

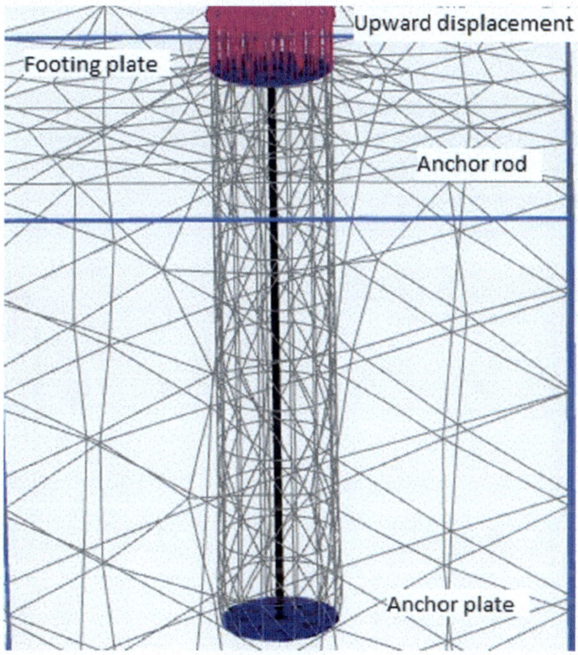

Fig. 2 Modeling of single GAP

2 Methodology

A prototype model of granular anchor pile of different lengths and diameters was modeled in PLAXIS 3D software for analyzing uplift capacity. The soil around the periphery of GAP was considered as expansive soil which extends to a depth of 600 mm below the ground surface. A soil layer was modeled using borehole option in PLAXIS 3D by taking suitable plan dimensions. The plan dimensions of soil layer were taken as 300 mm × 300 mm and depth was kept as 600 mm. Single granular pile was modeled in PLAXIS 3D software by using poly-curve and extrudes options. The modeling of anchor plate and footing was done with the plate elements. The node to node element was used to model the anchor rod of the GAP system to tie the anchor plate, footing plate and GAP together. For plate elements and node to node anchor, a linear elastic model was considered. Mohr−Coulomb criterion was used for defining the relationship between granular pile and clay.

The modeling of volume elements was done using ten-node tetrahedral elements and the plate elements were modeled using six-noded triangular elements. Generation of mesh plays a very important role in the calculation of accurate values in case of finite element method and therefore, coarse global finite element mesh was used for the analysis of modeling. For the single pile, a circular footing equal to that of the diameter of anchor plate was considered at the top (Fig. 3). The modulus of elasticity of pile was kept constant throughout the analysis. The modulus of elasticity of soil was varied from 4 to 10 MPa.

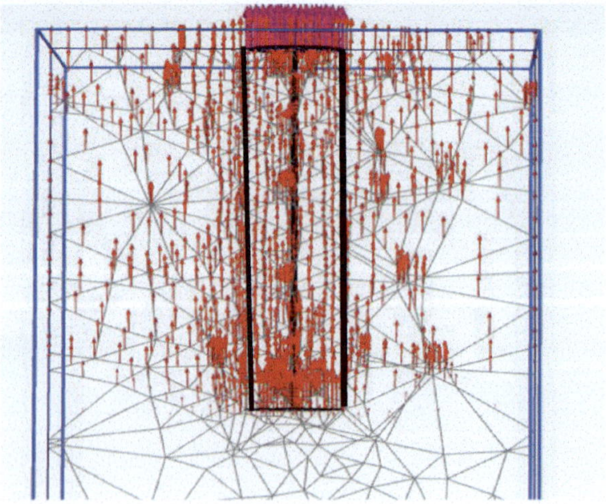

Fig. 3 Arrow diagram of deformation observed

Table 1 Properties of soil, GAP and structural element

Material and its properties	Clay	GAP	Anchor plate	Footing plate
γ_{unsat}, kN/m^3	17	22	–	–
γ_{sat}, kN/m^3	19	24	–	–
Cohesion (c), kN/m^2	25	0	–	–
Angle of shearing resistance (∅)	0	36	–	–
Poisson ratio (μ)	0.40	0.30	0.15	0.15
Modulus of elasticity (E), MPa	4, 6, 8, 10	11	2×10^5	2×10^5

Table 2 Different parameters used in numerical modeling

Length of GAP (L) (mm)	Diameter of GAP & anchor plate (D) (mm)	Thickness of anchor/footing plate (mm)	L/D ratio
100	10, 15, 20	5	10, 6.67, 5
200	20, 30, 40	5	10, 6.67, 5
300	30, 45, 60	5	10, 6.67, 5

2.1 *Material Used*

The material properties of soil, GAP and structural elements and parameters of GAP, footing used in the numerical study has been tabulated in Tables 1 and 2, respectively, as below:

3 Results and Discussion

3.1 *Effect of Modulus of Elasticity*

Figure 4 shows the variation of uplift forces with the upward movement of a GAP system for different values of modulus of elasticity of soil for pile length of 100 mm and diameter 10 mm. Figure 4 indicates an increase in the uplift capacity with an increase in the soil modulus. The percentage increase in the uplift capacity with increase in the elastic modulus of soil decreases. The percentage increase in value of uplift capacity was 14.70% when soil modulus was increased from 4 to 6 MPa, 10.20% for soil modulus increased from 6 to 8 MPa and 6.97% for soil modulus increased from 8 to 10 MPa for pile length of 100 mm and diameter 10 mm. Similar trend was obtained for other pile lengths and diameter values as illustrated in Table 3 and Figs. 5 and 6.

Fig. 4 Upward movement versus uplift capacity for L = 100 mm and D = 10 mm

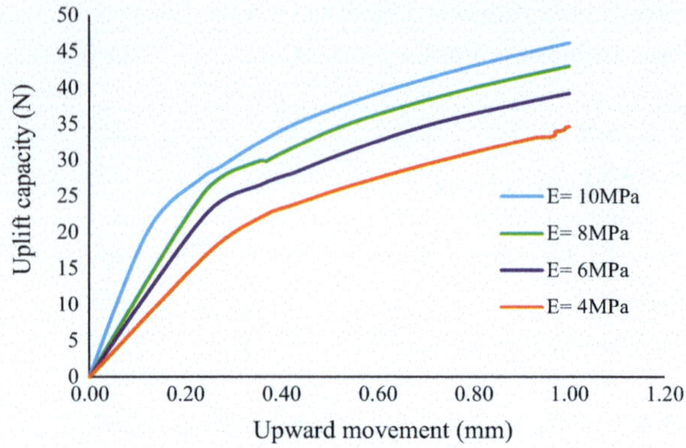

Table 3 Uplift capacity of single pile for different configurations

Length (mm)	Diameter (mm)	L/D ratio	Uplift capacity (N) at E = 4 MPa	Uplift capacity (N) at E = 6 MPa	Uplift capacity (N) at E = 8 MPa	Uplift capacity (N) at E = 10 MPa
100	10	10	34	39	43	46
100	15	6.67	56	63	69	73
100	20	5	92	104	110	114
200	20	10	97	111	120	127
200	30	6.67	186	209	222	229
200	40	10	284	293	298	302
300	30	10	196	213	232	237
300	45	6.67	399	415	427	429
300	60	10	466	470	472	475

Fig. 5 Upward movement versus uplift capacity for L = 200 mm and D = 20 mm

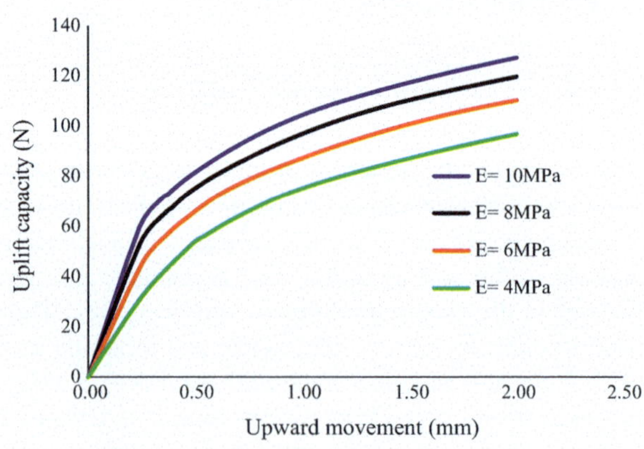

Fig. 6 Upward movement versus Modular ratio (Ep/E) for L = 100 mm and D = 10 mm

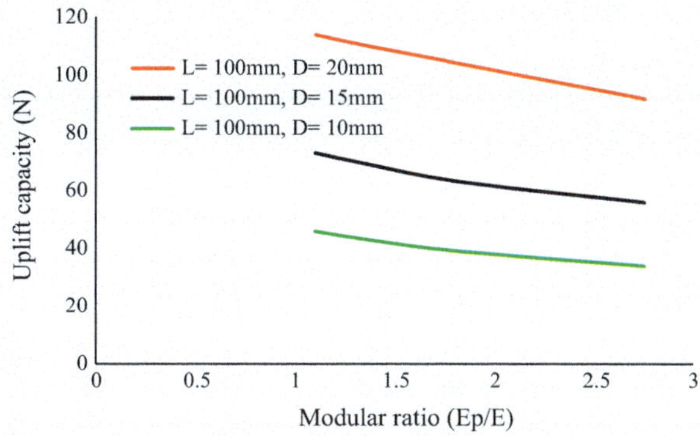

3.2 *Effect of Water Table*

Figure 7 describes the influence of water table with varying depth on the uplift capacity of GAP system for three different lengths having constant length-to-diameter ratio (L/D = 10). It is observed from Fig. 7 that with the increase in the depth of water table to the ground surface, there is decrease in uplift capacity of GAP system which is attributed due to reduction in the effective stress and shear strength of soil. Further, by lowering the depth of water table from 0D to ½D, the uplift capacity of GAP system increased by about 85% for pile having L = 100 mm. For the same length, when the depth of water table was lowered from 1/2D to D, the percentage increase was only limited to 30%. The difference is percentage increase may be caused due to the bulging of GAP system at the bottom. Figure 7 shows that there is a linear variation of uplift capacity with the depth of water table for all lengths.

Fig. 7 Effect of water table

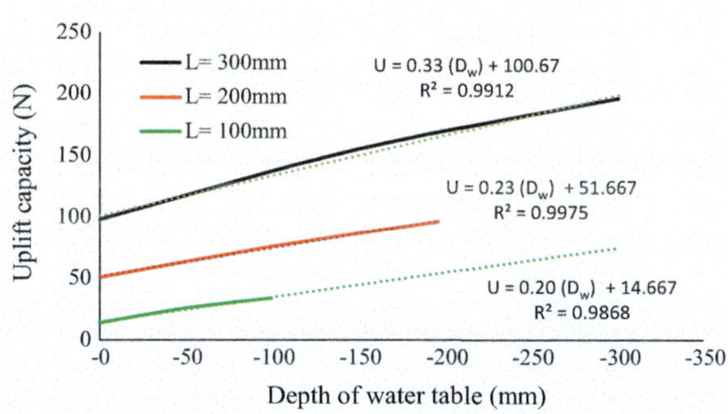

4 Conclusions

PLAXIS 3D, FEA was performed using to study the performance of the GAP system in a clayey soil. The effects of key parameters, such as elastic modulus of the surrounding soil and position of water table on the uplift capacity of a GAP were examined. The load-displacement response and ultimate pullout capacity of the GAP were also analyzed. Based on the present study, the following conclusions can be made:

1. There is an increase in the uplift capacity of granular anchor pile system on increasing the modulus of elasticity of soil. There is decrease in percentage increment on increasing soil modulus.
2. For higher modulus ratios of granular anchor pile system, the bulging failure is more predominant; whereas, for lower modular ratios the shaft failure occurred keeping constant modulus of elasticity of soil.
3. There is an increase in uplift capacity of granular anchor pile system on increasing the diameter of the GAP keeping soil modulus constant. The increase may be due to the increase in surface area on increasing pile diameter.
4. The uplift capacity of GAP system decreased linearly on increasing the water table depth towards ground surface.

References

Abhishek, Sharma RK (2018) A numerical study of granular pile anchors subjected to uplift forces in expansive soils using PLAXIS 3D. Indian Geotech J 1–10

Acosta H, Edil T, Benson C (2013) Soil stabilization and drying using fly ash. Geo Engineering Report No 03–03, Department of Civil and Environmental Engineering, University of Wisconsin-Madison, Madison, Wisconsin

BabuShanker N, Appayanna M, Shankaraiah B (1980) Design and testing of under – reamed piles. In: Proceedings of Indian geotechnical conference, Bombay vol 1, pp 203–206

Chen F (1988) Foundations on expansive soils. Elsevier, Amsterdam

Cokca E (2001) Use of class C fly ash for the stabilization of an expansive soil. J Geotech Geoenvironmental Eng 127(7):568–573

Desai I, Oza B (1997) Influence of anhydrous Calcium Chloride on shear strength of expansive soils. In: Proceedings of 1st international symposium on expansive soils, HBTI, Kanpur, India

Dey A, Minda R Gurung K (1996) A case study on under-reamed piles. In: Proceedings of 6th international conference and exhibition on piling and deep foundations, Bombay

Dinesh M (1991) Pile foundations. Oxford & IBH Publishing Co Pvt Ltd, Delhi

Hari Krishna P (2006) A study on the use of granular anchor piles to control heave of footings resting on expansive soil. PhD thesis, Kakatiya University Warangal

Hoyos L, Puppala A, Chainuwat P (2004) Dynamic properties of chemically stabilized sulphate rich clay. J Geotech Geoenvironmental Eng 130(2):153–162

Hunter D (1988) Lime-induced heave in sulfate - bearing clay soils. J Geotech Eng 114(2):150–167

Katti R (1979) Search for solutions for problems in black cotton soils. Indian Geotech J 9(1):1–80

Kranthikumar A, Sawant V, Kumar P, Shukla S (2016a) Numerical and experimental investigations of granular anchor piles in loose sandy soil subjected to uplift loading. Int J Geomech 17(2):1–10

Kranthikumar A, Sawant V, Shukla S (2016b) Numerical modeling of granular anchor pile system in loose sandy soil subjected to uplift loading. Int J Geosynth Ground Eng 15(2):1–7
Kumar V (1996) Fly ash utilization: a mission mode approach: ash ponds and ash disposal systems. Narosa Publishing House, New Delhi
Nelson D, Miller J (1992) Expansive soils: problems and practice in foundation and pavement engineering. Wiley, New York
Petry M, Little N (1992) Update on Sulfate-induced heave in treated clays: problematic Sulfate levels. Transportation Research Record. 1362, National Research Council, Washington D.C, pp 51–55
Phanikumar B (1997) A study of swelling characteristics of and granular pile anchor foundation system in expansive soils. PhD thesis, Jawaharlal Nehru Technological University Hyderabad, India
Phanikumar B (2000) Use of fly ash in treating problematic expansive soils. In: Proceeding of workshop on environmental geotechnology, Indian Geotechnical Society, Hyderabad, pp 36–39
PhaniKumar B (2016) Influence of geogrid reinforcement on pullout response of granular pile-anchors (GPAs) in expansive soils. Indian Geotech J 46(4):437–444
Phanikumar B, Reddayya S (2001) Swelling characteristics of fly ash-treated expansive soils. In: Proceeding of international symposium on suction, swelling, permeability, and structure of clays, IS Shizuoka, Japan, pp 121–123
Phanikumar B, Sharma R (2004) Effect of fly ash on engineering properties of expansive soils. J Geotech Geoenvironmental Eng 130(7):764–767
Phanikumar B, Prasada G, Srirama A (1994) Use of anchored granular column in minimising swell in expansive clays. In: Proceedings of Indian geotechnical conference, Warangal, India, pp 61–65
Poulos H, Davis E (1980) Pile foundation analysis and design. Wiley, New York
Rao A (1984) A study of swelling characteristics and behavior of expansive soils. PhD thesis, Kakatiya University, Warangal India
Rao B (1986) What techniques other than under-reamed piles have proven to be effective in minimising foundation problems in black cotton soils? - Response paper to the question. In: Indian geotechnical conference, New Delhi, vol 2, pp 145–149
Rollings M, Rollings R (1996) Geotechnical materials in construction. McGraw-Hill, New York
Rollings R, Burkes J, Rollings M (1999) Sulphate attack on cement-stabilized sand. Aus J Geodesy Photogram Surv 125(5):364–372
Sankar N (1989) Use of lime-soil piles for in-situ stabilization of black cotton soils. In: Proceedings of Indian geotechnical conference, vol 1, pp 149–153
Satyanarayana B (1966) Swelling pressure and related mechanical properties of black cotton soils. PhD thesis, Indian Institute of Science, Bangalore India
Setty K, Ravi Shankar A, Manohar C (1966) Behavior of under-reamed piles in Shedi soil. In: Proceedings of Indian geotechnical conference, Madras, India, vol 1, pp 286–289
Sharma D, Jain M, Prakash C (1978) Handbook on under reamed and bored compaction pile foundations. Central Building Research Institute, Roorkee India
Sharma R, Phanikumar B, Nagendra G (2004) Compressive load response of granular piles reinforced with geogrids. Can Geotech J 41(1):187–192
Subba Rao K (1999) Swell-shrink behavior of expansive soils – geotechnical challenges. Indian Geotech J 30(1):1–68

Aeration Through Weirs—A Critical Review

Akash Jaiswal and Arun Goel

Abstract Aeration is a process, which helps in the enhancement of concentration of dissolved oxygen in water. This concentration of dissolved oxygen (DO) is required for sustaining aquatic life and is also considered as an indicator of the quality of water. The DO can be enhanced by using either mechanical aerator or by hydraulic structures. Hydraulic structures accelerate the water–air–mass transfer by entraining a large amount of air in the form of bubble, thus increasing surface contact area for full oxygen transfer. Commonly hydraulic structures used for aeration are weirs, spillways, cascades, water jets and close conduit flow arrangements, etc. When the performance of weirs is compared with other aerating agents like hydraulic jump, labyrinth weirs, chute etc., it was found that weirs give maximum oxygen transfer efficiency. The researchers have used rectangular-, triangular-, trapezoidal- and semicircular-shaped weirs for DO studies in the past. The present paper looks into the literature review of weirs having free surface when employed for aeration purpose for wastewater treatment. It was found that the triangular-shaped weir has created maximum oxygen transfer efficiency out of all other shapes of weir. It is expected that findings of this study will help in the selection of particular weir configuration and shape of weir for creating maximum oxygen transfer efficiency in the rivers and streams for the purpose of treatment of water.

Keywords Aeration · Triangular weir · Dissolved oxygen

1 Introduction

Keeping water requirements and its scarcity in mind, much more emphasis is being given to the water quality. One of the most widely cited water quality parameters is the concentration of dissolved oxygen. It is very essential to maintain a minimum level of DO for rivers, streams, lakes for proper and healthy aquatic life. Reduction to DO concentration in the rivers and streams can create a great stress to the natural

A. Jaiswal (✉) · A. Goel
NIT Kurukshetra, Kurukshetra 1369119, Haryana, India
e-mail: akaskjaiswal@gmail.com

A. K. Agnihotri et al. (eds.), *Sustainable Engineering*, Lecture Notes in Civil Engineering 30, https://doi.org/10.1007/978-981-13-6717-5_19

aquatic cycle. For warm water aquatic life, water must contain 4 mg/l of DO and for cold water aquatic life, there must be 5 mg/l of DO concentration. The oxygen absorbed from the atmosphere replenishes the oxygen deficit, and this process is called as aeration.

Along with many possible ways to enhance DO concentration such as self-aeration, chemicals and hydraulic jump, presently, hydraulics structures are now being successfully used in maintaining the DO in the water stream (weirs, spillways). Hydraulic structures try to accelerate oxygen concentration by entraining the air in the stream in the form of large amount of air bubble, thus increasing the surface area of contact with air. Hydraulic heads occur naturally in the river stream so incurs no additional operating cost. This helps in making weir aeration economically competitive with other artificial oxygen accelerators including mechanical aerators.

2 Background

The material balance equation comparing the instantaneous rate of change in concentration of dissolved oxygen dC/dt and the rate of air–mass transfer is given by modified Rindels and Gulliver (1991) equation

$$dC/dt = K_l A/V (C_s - C) \quad (1)$$

where, K_l = liquid mass transfer coefficient; A = air/water interfacial surface area with the volume V; C_s = saturation concentration of dissolved oxygen; C = DO concentration. The term A/V is often termed as specific area per unit volume. Liquid mass transfer coefficient changes with the turbulence in the flow across the control volume.

For a predictive relationship, it is assumed that C_s is constant and is taken as the saturation concentration in the water, in equilibrium at local air pressure and temperature. Taking the above assumption, integration of Eq. (1) gives a relative equation between the upstream and downstream concentration of DO at a hydraulic structure and is given as below.

$$E = (C_d - C_u)/(C_s - C_u) = 1 - 1/r \quad (2)$$

where

C_d = concentration of DO at d/s; C_u = concentration of DO at u/s; C_s = saturation concentration of DO; r = deficit ratio, E = oxygen transfer efficiency.

In the limit of no air–mass transfer, r = 1, E = 0 and in the limit of complete air–mass transfer, r = ∞, E = 1.

Relation of Water temperature and saturation concentration of DO
Since saturation concentration of dissolved oxygen is a strong function of water temperature, thus the air–mass transfer efficiency is also temperature dependence. To provide a standard base for comparison of different systems, oxygen transfer efficiency is normalized to a 20 °C standard temperature. This is given by Eckenfelder and Ford (1970) equation

$$1 - E_{20} = (1 - E)^{1/f} \quad (3)$$

where E = air–mass transfer efficiency at measured tempt 't'; E_{20} = air–mass transfer efficiency at 20 °C tempt and f = term to adjust from 20 to T °C and is given by the following equation.

$$f = 1.0 + 0.02103(T - 20) + 8.261 * 10^{-5}(T - 20)^2 \quad (4)$$

where T = temperature.

Equation (4) is presently being used for all tempt adjustments.

3 Literature of Weirs and Aeration

Even though being in contact with water for a very short period of time, hydraulic structures increases the concentration of DO up to a large extent. The studies show that under the self-aeration process, the amount of oxygen transfer taking place over a length of several kilometres can itself be achieved just by using a single hydraulic structure. Aeration process is being excelled up to such extent primarily due to the addition of a large amount of oxygen in the form of bubble, which as a result increases the surface area of contact available of air–mass transfer.

Use of weir to accelerate the aeration process was introduced by (Gameson 1957) in his work entitled as weirs and aeration of rivers. Since then, this work has been evolved to include stepped cascade spillways (Tebbutt et al. 1977), side stream elevates pool aeration (Macaitis and Reclamation 1990), labyrinth weir and piano key weir (Lempérière and Ouamane 2003).

Most of the earlier works done have shown that triangular weir gives a better result for the purpose of aeration as compared to any other shapes of weir or even any other aerator geometry say cascades, labyrinth weir, etc.

4 Methodology

4.1 Weir Aeration

Gas transfer occurring at weirs is the most classical example of hydraulic structures aeration. The free over fall water in the form of jet from the weir plunging into downstream water causes entrainment of air bubbles, hence, accelerating the aeration process (Fig. 1).

After the introduction of concept of aeration by using weirs (Gamenson 1957), a lot of interest was developed on this topic and many important and notable works have been done till the date A number of researchers have worked to investigate factors affecting aeration, characteristics of weir, shapes of weir, mainly aiming to have best possible aeration efficiency. Investigations to evaluate the basic concept of aeration capacity and effect of drop height, configuration of weir, flow rate and temperature were considered by Van der karoon and Schram (1969), Gulliver and Rindels (1993), Gulliver et al. (1998), Watson et al. (1998). These researchers somehow reached to the same conclusion that triangular weir configuration has highest aeration efficiency potential as compared to other shaped weirs.

The cross sections of weirs used by these investigators are triangular, rectangular, trapezoidal and semicircular. These cross sections are shown in Fig. 2 as given below.

Since now it was known that triangular weir gives best results in terms of creating aeration potential, now, the researchers, namely Baylar et al. (2001), Emiroglu and Baylar (2003a, b) have worked on triangular shape weir. They made efforts to get the best configurations of triangular weir relating apex angle, No. of V notches etc.,

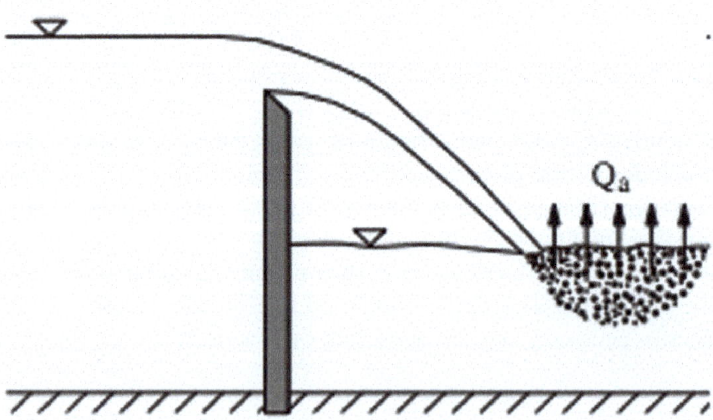

Fig. 1 Air entrainment at weir flow (Baylar et al. 2010)

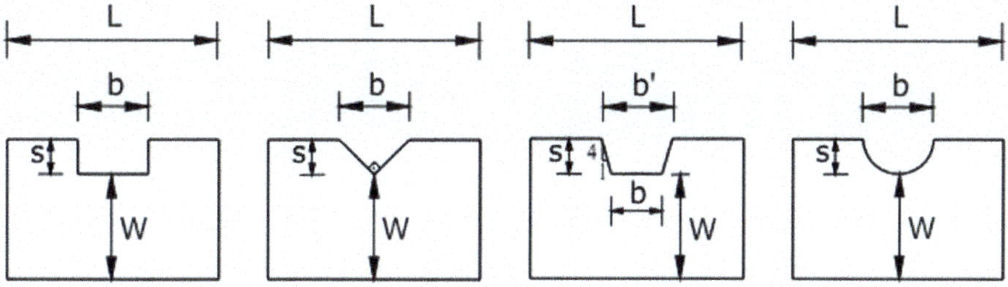

Fig. 2 Cross sections of weirs (Baylar and Bağatur 2000)

so that their aeration potential can be maximized. Apart from these, a few other researchers have also compared weirs with other aeration agents in terms of aeration potential. These research works have been summarized in Table 1 as listed below.

Along with the geometry and classification of weirs and factors affecting aeration, contemporarily some other researches have also been done in this field. A few of research works are listed in Table 2.

4.2 *Use of Weir in Field for Practical Application*

Working on the best configuration of weir so as to get maximum aeration potential (A_P). On parallel note, the researchers have also used weir to enhance aeration in the field situations in order to increase the quality of agricultural drains receiving domestic sewage by increasing the concentration of dissolved oxygen so that efficiency by oxygen transfer into wastewater is increased. A few notable research works and their results are listed in Table 3.

4.3 *Use of Modelling Techniques to Verify the Experimental Value Regarding Aeration Potential*

Along all these works and researches listed and mentioned above, one the other hand, a few have used modelling techniques to verify the experimental value regarding aeration potential, which is given below:

1. Baylar et al. (2008), had investigated the effect of free over fall jets from triangular sharp-crested weirs (featured three exchangeable weir elements: 45°, 90° and 135°) on their aeration efficiency. An expert system based on adaptive network-based fuzzy inference system (ANFIS) was obtained for predicting air entrainment rate and aeration efficiency of weirs.
2. The empirical equations, support vector machine (polynomial and radial basis function) models and linear regression techniques were applied by Goel (2015),

Table 1 Weirs and its geometries used in laboratory and their comparison

References			Weirs and its geometry	Specification of Weirs	Discharge range (Q)	Drop height range (H)	Results
Sr. no.	Author name(s)	Year					
1.	Van der Karoon and Schram	1969	Rectangular, semicircular and semicircular with flatten base	(a) Weir A–F No. of nappe split increases in rectangular weir	0–80 m^3/hr	0.2–0.9 m	a. At a fixed Q and H, aeration potential increases with No. of split nappe, up to a limit and then starts decreasing in rectangular weir
				(b) Weir G–I No. of nappe split increases in semicircular weir			b. At a fixed Q and H, aeration potential decreases with No. of split nappe in rectangular weir
				(c) Weir J semicircular weir with flatten base			c. Weir J shows increase in aeration potential
							d. For all weir, aeration potential increases with increase in H and decreases with increase in Q at fixed H

(continued)

Table 1 (continued)

References			Weirs and its geometry	Specification of Weirs	Discharge range (Q)	Drop height range (H)	Results
Sr. no.	Author name(s)	Year					
2.	Watson et al.	1998	Rectangular weir and chute	Smooth weir, cobble faced weir and a cobble faced chute	0–0.34 m^3/s	0.1–0.8 m	a. Aeration potential decreases in a sequence of cobble faced weir, smooth weir and cobble faced chute
3.	Baylar and Bagatur	2000	Rectangular, triangular, semicircular and trapezoidal	Weirs have same cross-sectional opening for comparison	0–4 l/s	0.15–1.05 m	Triangular weir is best out of these four weirs
4.	Kim and Walter	2001	Rectangular weir	Low drop	167, 334, and 502 m^3/hr per m	0.54, 1.04, and 1.36 m	Tail water depths are an important factor to predict aeration potential at low-drop weirs, but when designing for optimum tail water depth to maximize oxygen transfer it may not be that important
5.	Baylar et al.	2001	Triangular notch weir	Weir angle (α) as 45°, 90° and 135°	1.0–4.0 L/s	0.15–0.90 m	A 90° triangular notch weir is generally better than that of the other weirs

(continued)

Table 1 (continued)

References			Weirs and its geometry	Specification of Weirs	Discharge range (Q)	Drop height range (H)	Results
Sr. no.	Author name(s)	Year					
6.	Baylar and Emiroglu	2002	(a) Sharp-crested weirs	(a) Triangular weir angle (α) as 30°, 45°, 90°, and 135°	1.0–5.0 L/s	0.15–0.90 m	a. A 30° triangular sharp-crested weir is better than that of the other sharp-crested weirs
			(b) Rectangular, triangular, semicircular and trapezoidal	(b) 1,2 and 3 V notches over the width of the weir plate was studied for 30° triangular weir			b. A 30° triangular sharp-crested weir with two V notches have higher values than other variations in number of V notches
7.	Emiroglu and Baylar	2003a	(a) Broad-crested weirs	Triangular weir angle (α) as 30°, 45°, 90°, and 135°	1.0–5.0 L/s	0.15–0.90 m	a. A 30° triangular sharp-crested weir is better than that of the other sharp-crested weirs
			(b) Rectangular, triangular, semicircular and trapezoidal				b. Empirical relation for broad-crested weirs given was $Q_a = 2.160\cos\left(\frac{\alpha}{2}\right) Q^{1.030} H^{0.782}$ Where, Q_a = air entrainment rate, Q = weir discharge, H = drop height

(continued)

Table 1 (continued)

References			Weirs and its geometry	Specification of Weirs	Discharge range (Q)	Drop height range (H)	Results
Sr. no.	Author name(s)	Year					
8.	Emiroglu and Baylar	2003b	(a) Triangular weir (sharp-crested weir and Broad-crested weir)	(a) Triangular weir angle (α) as 90°	1.0–5.0 L/s	0.15–0.90 m	a. The broad-crested weirs generally give much better results than for the sharp-crested weir and the labyrinth weirs
			(b) Triangular labyrinth weirs	(b) Triangular labyrinth weirs angle (α) as 90°, sill length 0.30 m and sill slope (φ) as 22.5° and 45°			b. Aeration potential rate increases as the weir longitudinal slope of the broad-crested weirs and the weir sill slope of the labyrinth weirs became larger
9.	Raikar and Kamatagi	2015	Rectangular, triangular and hydraulic jump	–	0.5, 1, 1.5 and 2 (l/s)	15, 25, 35 and 45 m	a. Triangular weir have better aeration potential
							b. Weirs are more efficient than hydraulic jump

Table 2 Works on factor effecting aeration performance and relating equations

References			Weirs used	Works	Results
Sr. no.	Author name(s)	Year			
1.	Markofsky and Kobus	1978	Deep plunging pool	Presented a nomogram for constant values of $(r - \text{I})$ as a function of the nappe Froude and Reynolds number	a. Small-scale Froude model studies tend to underestimate reoxygenation in prototype. b. Laboratory experiments at prototype heights of fall tend to overestimate the reoxygenation if the flow rate is not considered
2.	Nakasone	1987	Weirs and Cascades	Proposed an equation that deals with fall height, discharge, and tail water depth	$\text{In}\, r_{20} = \text{A}(\text{D}\, 1.5\text{H}_c)^{\text{B}}\, \text{q}^{\text{C}}\, \text{H}^{\text{d}}$ where, A, B, C and D are constants depends on D, $\text{H}_{c,}$ H and q
3.	Gulliver and Rindels	1993	–	Worked on the relationship for a transfer efficiency from the mass transport equation, and adjustment of measurements for water temperature and unit discharge	a. The oxygen transfer a given structure is sensitive to head loss, temperature but little on discharge and tail water depth
					b. At a given structure, the head loss and tail water depth are relatively constant so not normally of concern
					c. Water temperature can vary greatly, and relationships exist to accurately index oxygen transfer to a common water temperature

(continued)

Table 2 (continued)

References			Weirs used	Works	Results
Sr. no.	Author name(s)	Year			
4.	Chanson	1995	Hydraulic jump	Results of free surface aeration and mathematical formula for hydraulic jump aeration are combined to predict the DO concentration at downstream of weir and spillways	Self-aeration might contribute to a large part of the oxygenation taking place at hydraulic structure for small discharge but for large discharges, the reduction or the disappearance of free surface aeration affects the substantially the aeration efficiency
5.	Gulliver et al.	1998	Weirs, spillways (gated and ungated), gated sills and gated conduit outlets	Field data of oxygen transfer at these four types of hydraulic structures are compared to the predictions by 12 equations in the literature	a. Avery and Novak's equation provides the best prediction of oxygen transfer for weirs
					b. Rindels and Gulliver's equation and Wilhelms's equation provide the best prediction for spillways (gated and ungated) and gated sills, respectively

Table 3 Works on use of weirs in field

References			Weir used	Purpose of use	Results
Sr. no	Author name(s)	Year			
1.	Heza et al.	2008	Conventional weir	To increase the quality of agricultural drains receiving domestic sewage	Conventional weirs can be efficiently used for the indirect aeration of water with high efficiencies
2.	Bagatur	2009	Weir and Stepped cascade	Alternative pre-aeration system in circular clarifiers	Weir and stepped cascade can be efficiently used as pre-aeration system
3.	Poommai et al.	2012	Rectangular weir	(1) To increase the efficiency by oxygen transfer into wastewater (2) Wastewater depths used in m are 0.03, 0.04, 0.05, 0.06 and 0.07	Circular clarifiers with weir and stepped cascade can be efficiently used as pre-aeration system

for prediction of air entrainment rate and aeration efficiency of a free over fall jets issuing from a triangular sharp-crested weir. It was pointed out that there exists a good relationship between the experimental values and the values obtained from empirical equations.

5 Conclusions

On the basis of literature review of research works on weirs used for aeration purpose, the followings conclusions are drawn.

1. Triangular weir gives the best A_P out of rectangular, triangular, semicircular and trapezoidal geometries of weirs.
2. Out of apex angles 30°, 45°, 90° and 135°, the triangular weir with 30° apex angle have maximum oxygen transfer efficiency.

3. In comparison between weirs, labyrinth weirs and hydraulic jumps, weirs give better aeration potential.
4. Weirs can be efficiently used as pre-aeration system for practical applications in the field.
5. Experimental results of weir aeration performance can be effectively checked and compared by using soft computing techniques (ANFIS, SVM, etc.).

References

Bagatur T (2009) Evaluation of pre-aeration with V- notch weir and cascades structures in clarifiers. J Environ Eng 135(3):176–184

Baylar A, Bağatur T (2000) Study of aeration efficiency at weirs. Turk J Eng Environ Sci 24(4):255–264

Baylar A, Bagatur T, Tuna A (2001) Aeration performance of triangular-notch weirs. Water Environ J 15(3):203–206

Baylar A, Emiroglu ME (2002) The effect of sharp-crested weir shape on air entrainment. Can J Civ Eng 29(3):375–383

Baylar A, Hanbay D, Ozpolat E (2008) An expert system for predicting aeration performance of weirs by using ANFIS. Expert Syst Appl 35(3):1214–1222

Baylar A, Unsal M, Ozkan F (2010) Hydraulic structures in water aeration processes. Water Air Soil Pollut 210(1–4):87–100

Chanson H (1995) Hydraulic design of stepped cascades, channels, weirs and spillways. Pergamon, Oxford, U.K

Eckenfelder WW Jr, Ford DL (1970) Water pollution control. Pemberton Press, Austin and New York

Emiroglu ME, Baylar A (2003a) The effect of broad-crested weir shape on air entrainment. J Hydraul Res 41(6):649–655

Emiroglu ME, Baylar A (2003b) Experimental study of the influence of different weir types on the rate of air entrainment. Water Qual Res J Can 38(4):769–783

Gameson ALH (1957) Weirs and aeration of rivers. J Inst Water Eng 11(5):477–490

Goel A (2015) Modeling aeration of sharp crested weirs by using support vector machines. WASET Int J Mech Aerosp Ind Mech Manuf Eng 7(12):2620–2625

Gulliver JS, Rindels AJ (1993) Measurement of air-water oxygen transfer at hydraulic structures. J Hydraulic Eng 119(3):327–349

Gulliver JS, Wilhelms SC, Parkhill KL (1998) Predictive capabilities in oxygen transfer at hydraulic structures. J Hydraulic Eng 124(7):664–671

Heza YBM, El-din M, Donia NS, Refaat AM (2008) Water quality improvement of drains polluted by wastewater using conventional weir. In: Welfth international water technology conference, IWTC12

Kim J, Walters RW (2001) Oxygen transfer at low drop weirs. J Environ Eng 127(7):604–610

Lempérière F, Ouamane A (2003) The Piano Keys weir: a new cost-effective solution for spillways. Int J Hydropower Dams 10(5):144–149

Macaitis B, Reclamation MW (1990) Side stream elevated pool aeration station design. MOO, p 199

Markofsky M, Kobus H (1978) Unified presentation of weir-aeration data. J Hydraulics Div 104:562–568

Nakasone H (1987) Study of aeration at weirs and cascades. J Environ Eng 113(1):64–81
Poommai S, Chunkao K, Bualerd S (2012) Variation of oxygen transfer along the rectangular weir crest distance of wastewater treatment pond. Procedia Environ Sci 13:498–512
Raikar RV, Kamatagi PB (2015) Use of hydraulic phenomena in enhancement of dissolved oxygen concentration. Int J Res Eng Technol 4(2)
Rindels AJ, Gulliver JS (1991) Oxygen transfer at spillways. Air-water mass transfer. In: Wilhelms S, Gulliver J (eds) Proceedings of 2nd international symposium, on gas transfer at water surfaces. ASCE, Reston, Va., pp 524–533
Tebbutt THY (1977) Reaeration performance of stepped cascades. J Inst of Water Engrs and Sci 31:285–297.
Van der karoon GT, Schram AH (1969) Weir aeration—part I: single free fall. Water 2(22):528–537
Watson CC, Walters RW, Hogan SA (1998) Aeration performance of low drop weirs. J Hydraulic Eng 124(1):65–71

Photocatalytic Activity of Bi-doped TiO_2 for Phenol Degradation Under UV and Sunlight Conditions

Rahul Singh, Vivek Kumar, Anoop Verma, Amit Sobti and Amrit Pal Toor

Abstract Bi-doped P25 TiO_2 nanoparticles were prepared by surface impregnation method using bismuth nitrate as bismuth source. The prepared catalyst was characterized by XRD, SEM-EDS, TEM and UV–Vis DRS techniques. The TEM images showed the agglomerated (particle–particle interaction) spherical-like morphology. The band gap energy of Bi-doped P25 TiO_2 nanoparticles was found to be less than undoped P25 TiO_2. The photocatalytic activity of the prepared catalyst was examined by the degradation of phenol under UV and sunlight conditions. Bi-doped P25 TiO_2 nanoparticles have shown higher photodegradation of phenol than undoped P25 TiO_2 under sunlight conditions. The optical response of TiO_2 has shifted from UV to visible region after Bi-doping. The parametric study was also carried out to obtain the maximum degradation of phenol. The maximum degradation of phenol was obtained at an optimum catalyst loading of 0.5 g/L, dopant concentration of 1 wt% and calcination temperature of 500 °C.

Keywords Phenol · Bismuth · TiO_2 · Photocatalytic degradation · Bi-doped TiO_2 · UV/sunlight

1 Introduction

Water contamination by organic pollutants, nowadays, has become a serious environmental issue worldwide. Some of the organic pollutants found in our water bodies are non-biodegradable, called 'Bio-Recalcitrant Organic Compound'. These contaminants become a key health concern, regardless of their low concentration, because of their exceptionally high endocrine disrupting potency and genotoxicity (Turki et al.

R. Singh · V. Kumar · A. Sobti (✉) · A. P. Toor
Dr. SSB University Institute of Chemical Engineering and Technology, Panjab University, Chandigarh 160014, India
e-mail: amitsobti2002@yahoo.com

A. Verma
School of Energy and Environment, Thapar Institute of Engineering and Technology (Deemed to be University), Patiala 147004, India

A. K. Agnihotri et al. (eds.), *Sustainable Engineering*, Lecture Notes in Civil Engineering 30, https://doi.org/10.1007/978-981-13-6717-5_20

2015). Phenols and their derivatives are well known for their bio-recalcitrant and severe toxicity (Ahmed et al. 2010), the removal of which is of high interest. Heterogeneous photocatalysis is a promising method which offers an immense prospective in environmental pollution control like the degradation of the organic pollutants in the wastewater (Chandraboss et al. 2013). Titanium dioxide (TiO_2) has been extensively used as photocatalyst for the degradation of organic pollutants in wastewater due to its low cost, non-toxicity, potential to degrade a wide variety of liquid pollutants (Teh and Mohamed 2011), high photostability, chemical robustness and high oxidizing capacity (Bagwasi et al. 2013). However, TiO_2 photocatalyst suffers from two major drawbacks, namely wide band gap and high recombination rate of photo-induced charge carriers. Due to its large band gap, TiO_2 shows its photocatalytic activity only under ultraviolet light, hence making it expensive because of the need of a UV light source. Therefore, the cost-effective way is to use TiO_2 under solar light. However, the large portion (45%) of solar light consists of visible light (low energy light), which cannot be used by TiO_2 because of its wide band gap (Bagwasi et al. 2013). Consequently, the development of new methods to extend its absorption edge towards visible light region with robust interfacial charge carrier transfer process is of great interest (Devi and Kavitha 2013). In order to utilize visible light of solar spectrum, metal and non-metal doping of TiO_2 has been considered as a proficient method. It reduces the band gap and improves the quantum efficiency by inhibiting the recombination of electron-hole pairs (Kaur et al. 2016; Teh and Mohamad 2011). A lot of work has been done for the degradation of organic pollutants using TiO_2 photocatalysts doped with Zn, Sn, Cu, Si, Ca, Al, Mn, Fe, Ni, V, Ag, rare-earth metal, etc. (Wang et al. 2008). However, the studies on degradation of phenol using Bismuth-doped TiO_2 are rather scarce. Therefore, the present study is focused on the preparation, characterization and application of the bismuth-doped TiO_2 photocatalyst for degradation of phenol.

2 Methodology

2.1 Materials and Methods

TiO_2 (P25) powder was procured from Evonik Industries, Germany having average particle size of 30 nm, purity was 97% and anatase-to-rutile ratio of 80:20. Phenol (used as a model pollutant) and nitric acid were purchased from Qualigens. Bismuth Nitrate Pentahydrate was purchased from Nice Laboratory reagents. Absolute ethanol was purchased from Changshu Yangyuan Chemical, China. All the solutions were prepared in distilled water. The phenol solution having 25 ppm as initial concentration was prepared, and 200 mL of this solution was taken separately in a batch reactor, which was placed under UV chamber for photocatalytic degradation studies. The detail of this experimental setup is given elsewhere (Sraw et al. 2014). The concentration of liquid samples and diffuse reflectance spectra of solid particles was analyzed using UV–Vis spectrophotometer (Shimadzu, model No. 2450). The crystalline structure of the doped and undoped TiO_2 was determined by a Pan analytical X'Pert Pro diffractometer (D/max rA) at 45 kV and 40 mA (Cu $K\alpha = 1.5406$Å). TEM analysis was used for the particle size and morphology determination (Hitachi Japan 100 kV, H7500). Elemental composition of the prepared dopants was determined by energy-dispersive X-ray spectroscopy (Bruker, X Flash 6130). BET surface area of the optimum dopant was determined using Quantachrome NovaWin—data acquisition and reduction version 11.04.

2.2 *Preparation of Bi-doped TiO_2 Photocatalyst*

Surface impregnation method was used for the preparation of bismuth-doped P25 TiO_2 nanoparticles. The required amount of bismuth nitrate pentahydrate was dissolved in 20 ml of distilled water and add 2–3 drops of nitric acid to make better solubility while keeping the total amount of bismuth unchanged. On the other hand, the required amount of P25 TiO_2 powder was dissolved in 25 ml of absolute ethanol. Both the solutions were kept under continuous stirring for 1 h. The bismuth nitrate solution was then added drop by drop into the P25 TiO_2 solution while stirring the solution continuously for 1 h. The solution was kept for continuous stirring for another 1 h, and then sonication was done for 90 min. The prepared mixture was kept undisturbed for 24 hours at room temperature followed by oven drying for 18–24 h at 80 °C. Finally, the obtained solid was grinded to make photocatalyst in powder form (fine particles), and then the activation of catalyst was carried out in muffle furnace by calcinating at different temperatures, i.e. 300, 400, 500 and 600 °C for 2 h.

3 Results and Discussion

3.1 *UV-Visible Diffuse Reflectance Spectra*

It is well known that the light absorption characteristics of TiO_2 are affected by metal ions doping (Zhou et al. 2006). Therefore, the samples of Bi-doped TiO_2 (0.5, 1, 2 and 3 wt% dopant concentration) and undoped TiO_2 were analyzed using diffuse reflectance spectrometer operated in the range of 200–700 nm. The UV-visible spectra of all the dopants and undoped TiO_2 is given in Fig. 1. It is observed from Fig. 1 that the optical absorbance edge of Bi-doped TiO_2 shifts towards the lower energy region compared to undoped TiO_2, therefore, the absorption spectra of Bi-doped TiO_2 shows stronger absorption in the visible region. Further, the band gap energies of doped and undoped TiO_2 were estimated using the following equation (Lopez and Gomez 2012):

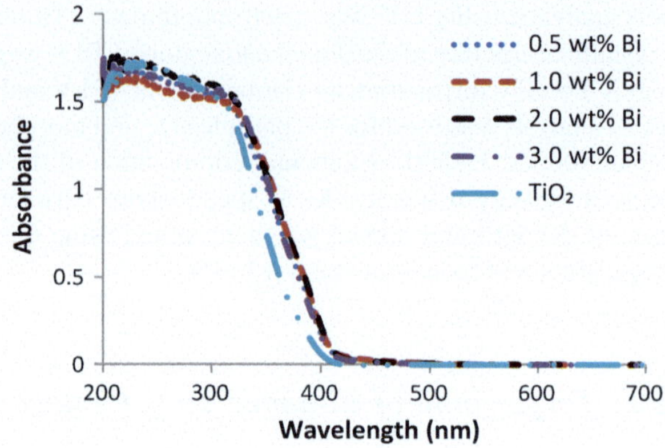

Fig. 1 DRS of undoped TiO_2 and Bi-doped TiO_2 with different dopant concentrations

$$\alpha h\nu = B(h\nu - E_g)^n \tag{1}$$

where $n = 1/2$ for direct allowed transmission; $n = 2$ for indirect allowed transmission; E_g is the band gap in eV; B is the absorption constant; α is the extinction coefficient; h is the Planck's constant (J.s); ν is the light frequency (s^{-1}). The values thus obtained using Eq. 1 are given in Table 1. The band gap energy of P-25 TiO_2 is observed to be 3.0 eV. However, the band gap energies of all the dopants are lower than that of the undoped TiO_2. The lowest band gap energy of 2.88 eV is obtained for the optimum dopant concentration of 1 wt% Bi-doped TiO_2. Further increase in dopant concentration beyond 1 wt% does not make any positive impact towards the absorption spectra, which indicates that no more vacancies are left onto the surface of TiO_2 for the incorporation of Bi, which leads to agglomeration of Bi particles (Kaur et al. 2016). Further, the reduction in band gap from 3.0 to 2.88 eV, band gap narrowing, might be due to the varying lattice parameters and presence of trap states within the conduction and valence bands from electronic Perturbation (Zhang 2011; Kaur et al. 2016).

Table 1 Crystallite size and band gap for different dopant concentrations

Dopant concentration (wt%)	Band gap (eV)	Crystallite size (nm)
0 (TiO_2)	3.00	32.71
0.5	2.92	26.19
1	2.88	23.65
2	2.90	26.23
3	2.94	31.48

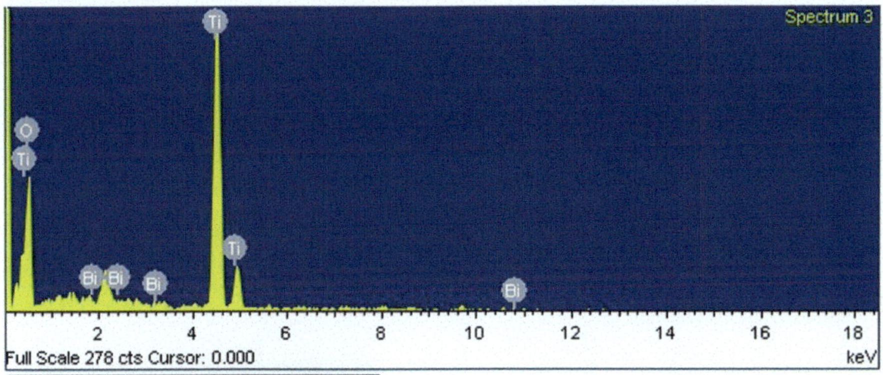

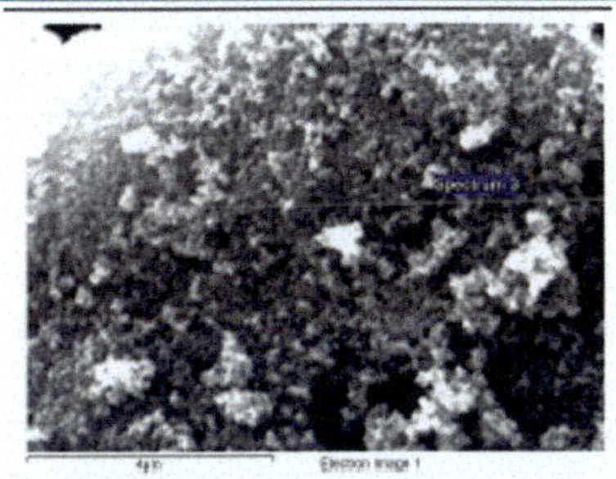

Element	Weight%	Atomic%
O K	47.24	73.11
Ti K	51.81	26.78
Bi M	0.95	0.11

Fig. 2 SEM-EDS of 1 wt% Bi-doped TiO_2

3.2 SEM-EDS Analysis

To see the morphology and composition of the dopant, SEM-EDS analysis was performed. SEM-EDS analysis of the as-prepared 1 wt% Bi-doped TiO_2 (optimum dopant concentration) is given in Fig. 2, which indicates the presence of strongly bonded Bi particles on to the surface of TiO_2. Figure 2 represents the elemental weight and atomic percentages in the prepared dopant and also shows the representative peaks of Bi, Ti and O atoms. The results indicate that the metal ions are well dispersed onto the surface of TiO_2 and no other impurity is present in the prepared dopant. The observed results of elemental composition are also in accordance with the prepared dopant concentration (1 wt%). Further, the morphology of Bi-doped TiO_2 nanoparticles can be observed from SEM image, which shows the uniform distribution of the particles.

3.3 BET Surface Area Analysis

This is a well-known fact that an increase in specific surface area enhances the photocatalytic activity (Lorret et al. 2009), and is considered as one of the important properties of photocatalyst. Therefore, the specific surface area of optimum dopant

(1 wt% Bi-doped TiO_2) and an undoped TiO_2 was determined using Quantachrome Novawin BET surface area analyzer. The result reveals that the specific surface area of optimum-doped TiO_2 is 20.48% higher than undoped TiO_2. The observed values of specific surface area are 41.175 m^2/g and 34.175 m^2/g, respectively, for 1 wt% Bi-doped TiO_2 and undoped TiO_2. The decrease in crystallite size as observed from Table 1 could be the possible reason for enhanced specific surface area.

3.4 XRD Analysis

To determine the crystallite size and to make certain about any changes in the phase composition of TiO_2 after doping, XRD studies were performed. The XRD patterns of doped and undoped TiO_2 are given in Fig. 3a. No significant change in the XRD spectra was observed, which shows that Bi-doping was successfully introduced into the surface of TiO_2 without altering its basic properties, i.e. the anatase form of TiO_2 does not shift to the rutile form. Further, the crystallite size of the prepared dopant may closely relate with the specific surface area, which could affect the photocatalytic activity (Kaur et al. 2016). The crystallite size was determined using Debye–Scherrer equation (Huang et al. 2007). The observed values of crystallite size for different dopant concentrations varying from 0.5 to 3 wt% along with undoped TiO_2 are given in Table 1. The crystallite size of all the dopants is observed to be lower than that of the undoped TiO_2, which could be because of the broadening of diffraction peaks with increase in dopant concentration. The minimum crystallite size of 23.65 nm is obtained for the optimum dopant concentration of 1 wt%, which is in close agreement with the average particle size as obtained from TEM image shown in Fig. 3b. Beyond the optimum dopant concentration of 1 wt%, the crystallite size increases with increases in dopant concentration up to 3 wt%, which could be because of the fact that doping concentration up to a certain value effectively inhibits the TiO_2 grain growth probably by staying at grain boundaries (Ganesh et al. 2012) but increasing the dopant concentration beyond the optimum level causes the agglomeration of the crystallites (Kaur et al. 2016).

3.5 Photolysis, Adsorption and Photocatalytic Studies Under Different Light Conditions

Initially, the degradation studies of phenol were performed under different light conditions such as UV, sunlight and dark. Experiments were performed with and without TiO_2 catalyst at a solution pH of 6. Figure 4 represents the comparison of the results of adsorption (with catalyst in dark), photolysis and photocatalytic studies. It is observed from Fig. 4 that only 5.7% degradation of phenol takes place in about 180 min when the reaction was carried out using TiO_2 in the absence of

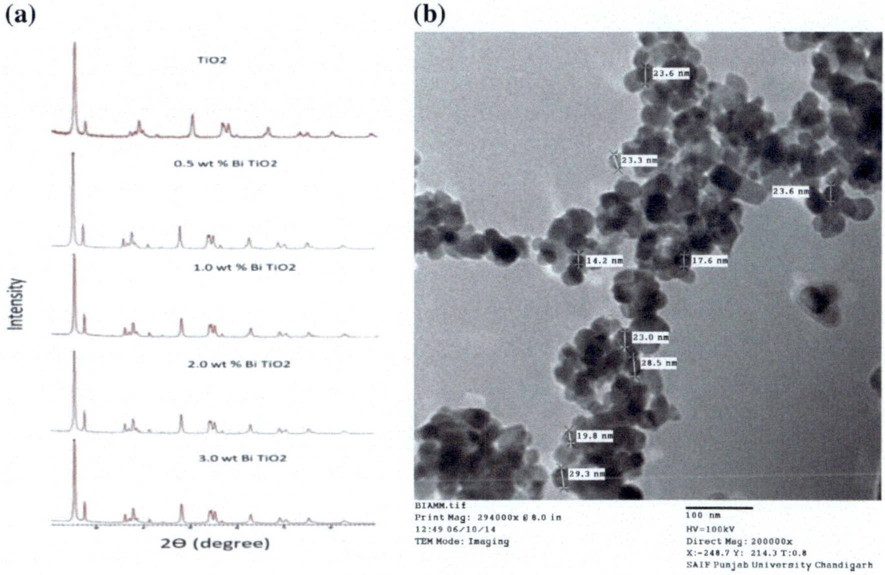

Fig. 3 **a** XRD spectra of Bi-doped TiO_2 and undoped TiO_2 **b** TEM image for 1 wt% Bi-doped TiO_2

light. This small degradation could be because of the adsorption of phenol over the surface of TiO_2 (Sraw et al. 2014). Further, a negligible decrease (2% approx.) in the concentration of phenol was observed when the reactions were carried out in the absence of TiO_2 catalyst under UV and sunlight conditions. Experiments were also conducted using TiO_2 catalyst under UV and sunlight conditions. About 83.51 and 67.79% degradation of phenol was observed when the reactions were carried out using TiO_2 under UV and sunlight conditions, respectively, at 180 min of irradiation time. This is because of the fact that upon illuminating TiO_2, electron-hole pairs were generated because of the shifting of electrons from the valence band to the conduction band (Daneshvar et al. 2003). The above experiments depict that both photocatalyst and UV light are needed for the effective degradation of phenol.

3.6 *Effect of Catalyst (TiO_2) Loading*

Experiments were conducted to determine the effect of catalyst loading (TiO_2) on the photocatalytic degradation of phenol. Figure 5a represents the photocatalytic degradation efficiency of TiO_2 nanoparticles under UV light for different catalyst loading varying from 0.05 to 2 g/L. It is observed from Fig. 5a that the phenol degradation increases with increase in catalyst loading up to 0.5 g/L. However, a decline in the degradation efficiency of TiO_2 nanoparticles was observed beyond the cata-

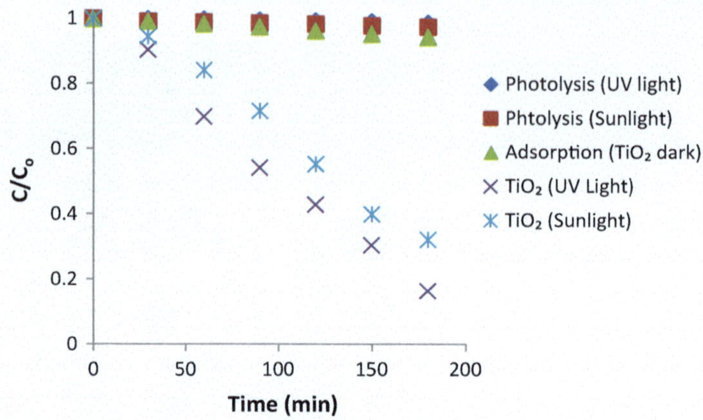

Fig. 4 Photolysis, adsorption and photocatalytic studies under different light conditions

lyst loading of 0.5 g/L. The enhancement in photocatalytic activity at lower level of catalyst concentration could be attributed to the fact that more reactive sites will result with increase in catalyst loading. However, at higher level of catalyst concentration, agglomeration of the catalyst particles may result with increase in catalyst loading because of which part of the catalyst surface becomes unavailable for the absorption of photon. Further, an increased opacity of the suspension at the higher concentration levels may result in reduction of light penetration and the availability of active sites and hence reduces the degradation efficiency of photocatalyst (Huang et al. 2008). The reduction in photocatalytic activity beyond an optimum concentration may also be attributed to the scattering of light due to increased opacity of suspension (Kaur et al. 2016). The maximum degradation of phenol resulted at an optimum catalyst loading of 0.5 g/L.

3.7 Effect of Dopant Concentration

Experiments were also conducted to determine the photocatalytic efficiency of the prepared dopants (Bi-doped TiO_2) on the degradation of phenol under sunlight conditions. Figure 5b represents the degradation of phenol as a function of irradiation time for different dopant concentrations varying from 0.5 to 3 wt%. It is observed from Fig. 5b that as the dopant concentration increases from 0.5 to 1 wt%, the degradation of phenol increases from 67.79 to 85.9% under sunlight conditions. The enhanced photocatalytic efficiency of the Bi-doped TiO_2 could be because of the high surface area of Bi-doped TiO_2, which may provide more number of active sites resulting in the absorption of more reactive species (Toor et al. 2013). However, the % degradation of phenol decreases from 85.9 to 64.37% with further increase in dopant concentration beyond 1 wt%. The decrease in photocatalytic activity of Bi-doped

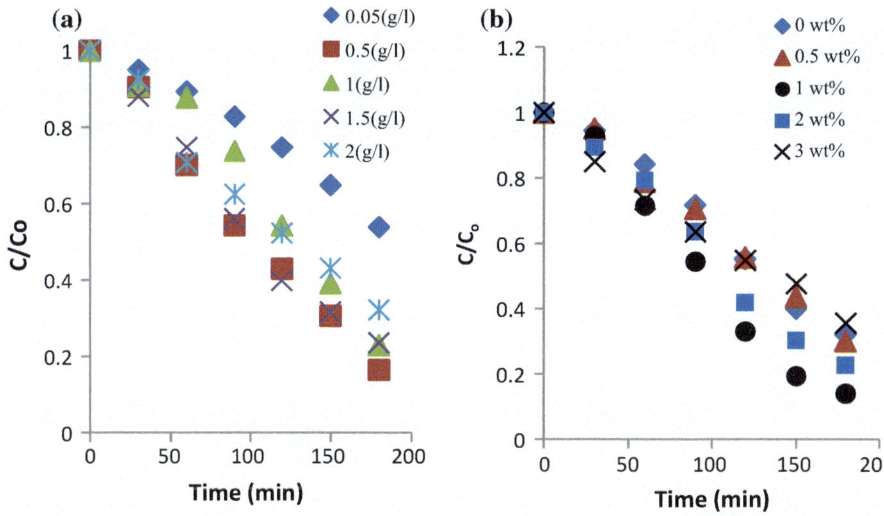

Fig. 5 **a** Effect of catalyst (TiO_2) loading on degradation of phenol under UV light **b** effect of dopant concentration on phenol degradation under sunlight conditions

TiO_2 beyond an optimum dopant concentration of 1 wt% could be attributed to the fact that with increase in dopant concentration, the recombination rate of the charge carriers increases exponentially (Toor et al. 2013). Therefore, the maximum degradation of phenol was obtained using 1 wt% Bi-doped TiO_2.

3.8 Effect of Calcination Temperature

In order to study the influence of the calcination temperature on the photocatalytic activity of Bi-doped TiO_2, the optimum dopant (1 wt% Bi-doped TiO_2) was calcined at different temperatures ranging between 300 and 600 °C for 2 h. Figure 6 represents the degradation of phenol using optimum dopant calcined at various temperatures as a function of irradiation time. Figure 6 depicts that the photocatalytic activity of Bi-doped TiO_2 (1 wt%) increases with increase in calcination temperature from 300 to 500 °C which could be attributed to the large surface area, improved particle size and crystallinity (Tan et al. 2011). However, any further increase in calcination temperature decreases the photocatalytic activity of the optimum dopant. The decrease in photocatalytic activity may be attributed to the fact that at temperatures above 500 °C, the anatase phase of TiO_2 photocatalyst shifts to rutile form having less photocatalytic activity compared to the anatase phase (Huang et al. 2007). Therefore, an optimum calcination temperature of 500 °C, at which maximum photocatalytic activity (85.9%) of 1 wt% Bi-doped TiO_2 towards degradation of phenol has been obtained, is used throughout this study.

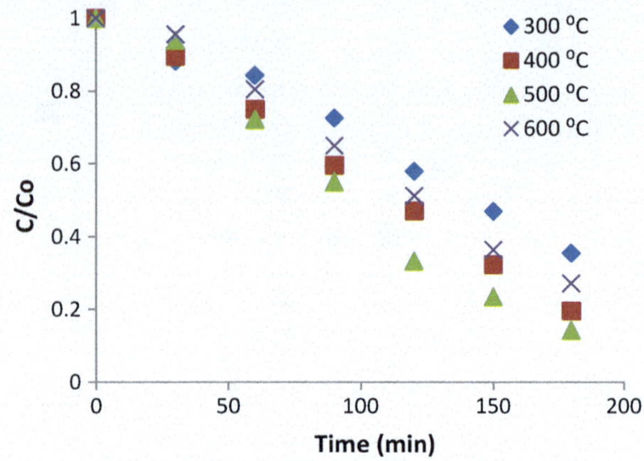

Fig. 6 Effect of calcination temperature on phenol degradation

4 Conclusions

Bismuth-doped TiO_2 nanoparticles were successfully prepared using surface impregnation technique. XRD results showed that the doping does not alter the basic characteristic of TiO_2. Further, the crystallite size as obtained from XRD data for 1 wt% Bi-doped TiO_2 is 23.65 nm compared to that of 32.71 nm for undoped TiO_2 which is also in accordance with the average particle size (22.54 nm) obtained from TEM image. EDS analysis reveals that the Bismuth metal ions are successfully incorporated on to the surface of TiO_2 and the elemental composition as observed from the EDS is in accordance with the prepared optimum dopant concentration. The prepared dopant, as well as undoped TiO_2 was successfully employed for the degradation studies of phenol under UV and sunlight conditions. The results reveal that the absorption edge of Bi-doped TiO_2 has been shifted to the lower energy levels making the process more economical compared to undoped TiO_2. The maximum degradation of 85.9% was obtained using optimum dopant concentration of 1 wt% under sunlight conditions compared to 67.79% using undoped TiO_2 under similar conditions. The maximum degradation of phenol was obtained at an optimum catalyst loading of 0.5 g/L, dopant concentration of 1 wt% and calcination temperature of 500 °C.

Acknowledgements The financial support received from DST PURSE grant-II, TEQIP-II and TEQIP-III of Dr. SSB University Institute of Chemical Engineering and Technology, Panjab University, Chandigarh is gratefully acknowledged.

References

Ahmed S, Rasul MG, Martens WN et al (2010) Heterogeneous photocatalytic degradation of phenols in wastewater: a review on current status and developments. Desalination 261:3–18. https://doi.org/10.1016/j.desal.2010.04.062

Bagwasi S, Tian B, Zhang J et al (2013) Synthesis, characterization and application of bismuth and boron co-doped TiO_2: a visible light active photocatalyst. Chem Eng J217:108–118. https://doi.org/10.1016/j.cej.2012.11.080

Chandraboss VL, Natanapatham L, Karthikeyan B et al (2013) Effect of bismuth doping on the ZnO nanocomposite material and study of its photocatalytic activity under UV-light. Mater Res Bull 48(10):3707–3712. https://doi.org/10.1016/j.materresbull.2013.05.121

Daneshvar N, Salari D, Khataee AR (2003) Photocatalytic degradation of azo dye acid red 14 in water: investigation of the effect of operational parameters. J Photochem Photobiol A Chem 157:111–116. https://doi.org/10.1016/S1010-6030(03)00015-7

Devi LG, Kavitha R (2013) A review on non metal ion doped titania for the photocatalytic degradation of organic pollutants under UV/solar light: Role of photogenerated charge carrier dynamics in enhancing the activity. Appl Catal B Environ 140–141:559–587. https://doi.org/10.1016/j.apcatb.2013.04.035

Ganesh I, Gupta AK, Kumar PP, Sekhar PSC et al (2012) Preparation and characterization of Ni-doped materials for photocurrent and photocatalytic applications. Sci World J 2012:1–16. https://doi.org/10.1100/2012/127326

Huang YU, Xuxu Z, Zhongyi YIN et al (2007) Preparation of nitrogen-doped TiO_2 nanoparticle catalyst and its catalytic activity under visible light. Chin J Chem Eng 15:802–807. https://doi.org/10.1016/S1004-9541(08)60006-3

Huang M, Xu C, Wu Z et al (2008) Photocatalytic discolorization of methyl orange solution by Pt modified TiO_2 loaded on natural zeolite. Dyes Pigments 77(2):327–334. https://doi.org/10.1016/j.dyepig.2007.01.026

Kaur T, Sraw A, Wanchoo RK et al (2016) Visible-light induced photocatalytic degradation of fungicide with Fe and Si doped TiO_2 nanoparticles. Mater Today Proc 3(2):354–361. https://doi.org/10.1016/j.matpr.2016.01.020

López R, Gómez R (2012) Band-gap energy estimation from diffuse reflectance measurements on sol-gel and commercial TiO_2: a comparative study. J Sol-Gel Sci Technol 61:1–7. https://doi.org/10.1007/s10971-011-2582-9

Lorret O, Francová D, Waldner G et al (2009) W-doped titania nanoparticles for UV and visible-light photocatalytic reactions. Appl Catal B Environ 91:39–46. https://doi.org/10.1016/j.apcatb.2009.05.005

Sraw A, Wanchoo RK, Toor AP (2014) Optimization and kinetic studies for degradation of insecticide monocrotophos using LR grade and P25 TiO_2 under UV/sunlight conditions. Environ Prog Sustain Energy 33:1201–1208. https://doi.org/10.1002/ep.11909

Tan YN, Wong CL, Mohamed AR (2011) An overview on the photocatalytic activity of nano-doped-TiO_2 in the degradation of organic pollutants. ISRN Mater Sci 2011:1–18. https://doi.org/10.5402/2011/261219

Teh CM, Mohamed AR (2011) Roles of titanium dioxide and ion-doped titanium dioxide on photocatalytic degradation of organic pollutants (phenolic compounds and dyes) in aqueous solutions: a review. J Alloys Compd 509:1648–1660. https://doi.org/10.1016/j.jallcom.2010.10.181

Toor AP, Yadav N, Wanchoo RK (2013) Enhancement in photocatalytic activity of nano-TiO_2 photocatalyst by carbon doping. Mater Sci Forum 757:271–284. https://doi.org/10.4028/www.scientific.net/MSF.757.271

Turki A, Guillard C, Dappozze F et al (2015) Phenol photocatalytic degradation over anisotropic TiO_2 nanomaterials: kinetic study, adsorption isotherms and formal mechanisms. Appl Catal B Environ 163:404–414. https://doi.org/10.1016/j.apcatb.2014.08.010

Wang J, Jing L, Xue L et al (2008) Enhanced activity of bismuth-compounded TiO_2 nanoparticles for photocatalytically degrading rhodamine B solution. J Hazard Mater 160:208–212. https://doi.org/10.1016/j.jhazmat.2008.02.103

Zhang D (2011) Chemical synthesis of Ni/TiO_2 nanophotocatalyst for UV/visible light assisted degradation of organic dye in aqueous solution. J Sol-Gel Sci Technol 58:312–318. https://doi.org/10.1007/s10971-010-2393-4

Zhou M, Yu J, Cheng B (2006) Effects of Fe-doping on the photocatalytic activity of mesoporous TiO_2 powders prepared by an ultrasonic method. J Hazard Mater 137:1838–1847. https://doi.org/10.1016/j.jhazmat.2006.05.028

Degradation of 4-Chlorophenol Using Homogeneous Fenton's Oxidation Process: Kinetic Study

Sudha Minz, Renu Gupta and Sangeeta Garg

Abstract Degradation of 4-chlorophenol (4-CP) has been studied experimentally using Fenton's oxidation process. To catalyze hydrogen peroxide (H_2O_2), ferrous sulphate heptahydrate ($FeSO_4 \cdot 7H_2O$) was chosen as the source of ferrous ions. Batch tests for degradation of 4-CP were conducted at room temperature and the effect of various parameters such as initial pH, reaction time, concentration of H_2O_2 and ferrous ions on the removal efficiency of 4-CP has been studied. The degradation of 4-CP has been reported in terms of 4-CP removal and reduction in total organic carbon (TOC) and chemical oxygen demand (COD). The study revealed that with increase in pH, Fe^{2+} ions and H_2O_2 dosage, the degradation, TOC removal and COD removal first increased and thereafter decreased. The optimal conditions were observed at pH 5 with 6 and 0.3 mM of H_2O_2 and Fe^{2+} concentrations, respectively. At optimum conditions, degradation of 4-CP was 90%, TOC and COD removal were 54% and 45%, respectively, within 30 min. Kinetic study for the degradation reaction of 4-CP was also done to analyse the order of reaction. The kinetic studies result revealed second-order reaction mechanism for the degradation of 4-CP using Fenton's oxidation process.

Keywords 4-chlorophenol · Fenton's reagent · Hydroxyl radicals · Degradation · Reaction kinetics

1 Introduction

Chlorophenols (CPs) have been widely used as disinfectants and preservative agents in wood, paint, vegetable fibre and leather industries (Zhang et al. 2014). Chlorophenols are mostly utilized as synthesis intermediates in the manufacturing of herbicides,

S. Minz · R. Gupta (✉) · S. Garg
Department of Chemical Engineering, Dr B R Ambedkar National Institute of Technology, Jalandhar, India
e-mail: bansalr@nitj.ac.in

S. Minz
e-mail: minz.sudha@gmail.com

S. Garg
e-mail: gargs@nitj.ac.in

A. K. Agnihotri et al. (eds.), *Sustainable Engineering*, Lecture Notes in Civil Engineering 30, https://doi.org/10.1007/978-981-13-6717-5_21

fungicides, pesticides, insecticides, pharmaceuticals and dyes (Zhou et al. 2008). At the same time, these compounds are very toxic at concentrations higher than 2 mg/L and also have high chemical oxygen demand, which considerably decreases the oxygen concentration and create problems for aquatic life (De Luis et al. 2009; Duan et al. 2014). CPs constitutes a particular group of priority toxic pollutants listed by the US EPA in the Clean Water Act (Titus et al. 2004). Therefore, large amounts of its discharge to the environment are of great concern due to toxicity and suspected carcinogenicity. The limiting permissible concentration of CPs in drinking water should not exceed 10 $\mu g\ L^{-1}$ (Zhou et al. 2014). So, the developments of efficient methods for CPs removal are becoming an important field of study.

Several technologies are available to remove industrial organic wastes, such as biological, thermal and chemical treatments. Conventional biological processes represent an environmentally friendly way of treatment with reasonable costs; however, they are not adequate to treat non-biodegradable wastewater. Thermal treatments present many drawbacks, such as emission of other hazardous compounds. Chemical treatments require a post-treatment, which increases the cost (Liotta et al. 2009). Advanced oxidation technologies (AOPs) are one of the promising technologies among all the available technologies.

Advanced oxidation processes (AOPs) such as Fenton's reagent, ozone, UV, UV/H_2O_2, UV/Fenton and ultrasound have been applied by various researchers for wastewater treatment (Song and Xiao 2005; Kang et al. 2002; Hong and Zeng 2002; Benitez et al. 2000; Brillas et al. 1998; Sauleda and Brillas 2001; Ho and Bolton 1998; Gondrexon et al. 1999). In AOP, the hydroxyl radicals ($^{\bullet}OH$) are generated in solution and these are responsible for the oxidation and mineralization of the organic pollutants to water and carbon dioxide (Ghaly et al. 2001). Among various AOPs, Fenton's oxidation has been used extensively as Fenton's reagent components are easy to handle, environmental benign, require low capital cost, easy operation, and accompanied with non-toxic by-products (Duan et al. 2014). During Fenton's reaction, hydroxyl radicals produced act as a strong oxidizing agent and is able to oxidize organic compounds under ambient conditions. The hydroxyl radical produced by Fenton's reaction can rapidly react with the organic pollutant. The general mechanism using Fenton's reagents, via which the hydroxyl radicals are produced, is a number of cyclic reactions, which utilize the Fe^{2+} or Fe^{3+} ions as a catalyst to decompose the H_2O_2. These ions are regenerated in their original state at the end of the cyclic reactions according to the following scheme of reactions (Kitis et al. 1999; Medien and Khalil 2010):

$$Fe^{2+} + H_2O_2 \xrightarrow{k_1} Fe^{3+} + OH^- + \cdot OH \tag{1}$$

$$Fe^{2+} + \cdot OH \xrightarrow{k_2} Fe^{3+} + OH^- \tag{2}$$

$$Fe^{3+} + H_2O_2 \xrightarrow{k_3} Fe^{2+} + OH^+ + \cdot OOH \tag{3}$$

$$Fe^{2+} + \cdot OOH \xrightarrow{k_4} Fe^{3+} + OOH^- \tag{4}$$

$$Fe^{3+} + \cdot OOH \xrightarrow{k_5} Fe^{2+} + H^+ + O_2 \quad (5)$$

$$H_2O_2 + \cdot OH \xrightarrow{k_6} \cdot OOH + H_2O \quad (6)$$

$$\cdot OH + \cdot OOH \xrightarrow{k_7} H_2O + O_2 \quad (7)$$

Inspite of many advantages, the Fenton's process produces a huge amount of ferric hydroxide $Fe(OH)_3$ sludge which requires additional separation and disposal (Anotai et al. 2012). This disadvantage can be solved by reusing recovered iron sludge in the municipal wastewater treatment for binding of sulphides with the aim to minimize smell and corrosion and to remove phosphates by adsorption (Sarfert et al. 1994). Fenton's oxidation is an important AOP technology that has attracted the interest of researchers due to efficient and fast degradation of organic pollutants. Fenton's oxidation process has been used for the degradation of Direct Blue 71 azo dye, phenol and monochlorophenols (o-CP and m-CP) and other organic pollutants such as di-isopropanolamine, cresols, etc. (Ertugay and Acar 2017; Kang et al. 2002; Khamaruddin et al. 2011; Zazo et al. 2009)

In the present study, the degradation of 4-CP has been studied using homogeneous Fenton's oxidation process. Different parameters that affect the oxidation of 4-CP such as initial pH, reaction time, concentration of hydrogen peroxide (H_2O_2) and ferrous ions (Fe^{2+}) have been optimized. The degradation of 4-CP has been studied in terms of degradation efficiency, COD and TOC removal (%). The kinetic study for the degradation of 4-CP has also been made by fitting the experimental data to first and second-order kinetic model.

2 Materials and Method

2.1 Reagents

All the chemicals used in this study are of AR grade. The organic pollutant 4-chlorophenol (4-CP), potassium ferricyanide, 4-aminoantipyrine were supplied by Loba Chemicals Limited, Mumbai, India. The Fe^{2+} ion solutions were prepared using sulphate heptahydrate ($FeSO_4 \cdot 7H_2O$). The hydrogen peroxide (H_2O_2) (30% w/w) and $FeSO_4 \cdot 7H_2O$ used in the study were purchased from HiMedia Laboratories Pvt. Ltd. Mumbai, India and RFCL Limited Mumbai, India. All the solutions were prepared using double-distilled water.

2.2 Preparation of 4-CP Solution

The 4-CP solution was prepared by dissolving it in double-distilled water to produce a stock solution of 1000 mg/L. This stock solution was diluted in accurate proportions to produce solutions of different initial concentrations.

2.3 Degradation Studies

Batch experiments of 4-CP (100 mg/L) were performed by adding ferrous sulphate solution to the 4-CP solution. The initial pH of 4-CP solution is 5.0. After that, the pH was adjusted to the desired value by the addition of few drops of 0.1 N H_2SO_4 and 0.1 N NaOH. Following the pH adjustment, H_2O_2 was added quickly to the 4-CP solution with continuous stirring. The samples were taken out from the conical flasks periodically, the reaction was stopped by adding 1 M NaOH immediately, and after that, the samples were analysed within 15 min after the reaction. The concentration of the 4-CP in the reaction mixture at different parameters was studied and found in terms of degradation efficiency, COD and TOC removal (%). The efficiency of the Fenton's oxidation process was evaluated by monitoring 4-CP degradation using 4-aminoantipyrine spectrophotometric method at the λ_{max} = 500 nm using UV/Vis spectrophotometer (LaMotte L0504060). Chemical oxygen demand (COD) measurements were made as per APHA method using a MAC COD digester (Model: COD-439) (Clesceri et al. 1999). TOC measurements was done using the total organic carbon analyzer (model: ASI-V and TOC-V CPN, Shimadzu, Japan). The experiments were repeated to study the effect of various parameters such as concentration of hydrogen peroxide, ferrous ions and reaction time.

The degradation efficiency (%), COD removal (%) and TOC removal (%) of 4-CP were calculated as follows:

$$\text{Degradation efficiency (\%)} = \left(1 - \left(\frac{C_t}{C_o}\right)\right) \times 100 \tag{8}$$

where C_0 is the initial concentration of 4-CP and C_t is the concentration at any time t (min)

$$\text{COD Removal (\%)} = \left(1 - \frac{COD_t}{COD_0}\right) \times 100 \tag{9}$$

where COD_t concentration at any time t, and COD_0 is initial concentration at time zero

$$\text{TOC Removal (\%)} = \left(1 - \frac{TOC_t}{TOC_0}\right) \times 100 \tag{10}$$

where TOC_t concentration at any time t and TOC_0 is initial concentration at time zero.

2.4 *Kinetic Study*

Kinetic study was done as kinetics provides useful information about the mechanism and rate of chemical reaction, which helps to run a chemical reaction successfully by the way of selecting optimum conditions. The kinetic study also helps to study the factors, which influence the rate of reaction like reactant concentration, oxidant concentration and composition of the reaction mixture. First- and second-order reaction kinetics were used to investigate the oxidation of 4-CP (Sun et al. 2007; Ertugay and Acar 2017).

First-order kinetics

$$\frac{dC}{dt} = -k_1C \tag{11}$$

$$C_t = C_o + e^{-k_1t} \tag{12}$$

Second-order kinetics

$$\frac{dC}{dt} = -k_2C^2 \tag{13}$$

$$\frac{1}{C_t} = \frac{1}{C_o} + k_2t \tag{14}$$

where C is the concentration of 4-CP, k_1 and k_2 represent the apparent kinetic rate constant of first- and second-order reaction, respectively, C_0 is the initial concentration of 4-CP and C_t is the concentration of 4-CP at any time t.

3 Results and Discussion

3.1 *Effect of PH*

The pH value of the solution has a great impact on the degradation of 4-CP. The pH value influenced the generation of hydroxyl radicals and the oxidation efficiency. The effect of pH variation on the degradation (%) of 4-CP by Fenton's process is shown in Fig. 1. It was observed that the degradation of 4-CP increased from pH 1–3 and after that, it decreased. Similarly, the COD and TOC removal (%) of 4-CP has increased with increase in pH from 1 to 3 and it decreased on further increase in pH from 4 to 7. The maximum

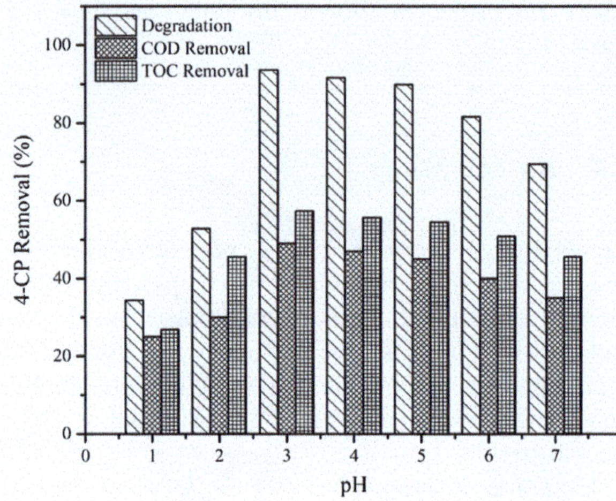

Fig. 1 Effect of pH on the degradation of 4-CP, COD and TOC removal by Fenton's oxidation process. Conditions: [4-CP] = 100 mg/L, [H_2O_2] = 6 mM, [Fe^{2+}] = 0.3 mM, time 40 min

degradation of 93.6% was obtained for 4-CP at pH = 3. The COD removal (%) and TOC removal (%) at pH 3 was observed to be 49% and 57.3%, respectively.

At pH below 3, the degradation of 4-CP was observed low because of low decomposition of hydrogen peroxide by Fe^{2+} due to the stable oxygen concentration. Thus, the hydrogen peroxide stays stable probably because it solvated a proton to form oxonium ion (H_3O^{2+}) and an oxonium ion made hydrogen peroxide electrophilic to enhance its stability and reduced its reactivity with ferrous ions (Kwon et al. 1999). When pH was increased above 3 or 5, ferrous and ferric oxy-hydroxides were formed which inhibited the reaction between Fe^{2+} and H_2O_2 and only a small amount of the $^{\bullet}OH$ was generated thus reducing the degradation of 4-CP (Ghaly et al. 2001). When the pH of the solution was increased from 1 to 3, there is a sharp increase in degradation of 4-CP from 36 to 90%. A sharp increase in COD removal (from 20 to 50%) and TOC removal (from 30 to 60%) was also observed. With further increase in pH the degradation as well as COD and TOC removal decreased. But the decrease in 4-CP removal (%) was low as compared to initial increase in degradation COD and TOC removal rate (Fig. 1). The degradation of 4-CP, COD removal and TOC removal at pH 5 was 90, 45 and 54.6% and this pH was considered as the optimum pH for further studies.

3.2 *Effect of Reaction Time*

In order to study the effect of reaction time on the degradation of 4-CP, a series of experiments were conducted at different reaction time separately and the results are shown in Fig. 2. It was observed that the degradation of 4-CP increased with increase in reaction time from 4 to 30 min. After 30 min of reaction time, the increase in the % degradation of 4-CP was not seen much. The degradation of 4-CP was also increased

Fig. 2 Effect of reaction time on the degradation of 4-CP. Conditions: [4-CP] = 100 mg/L, $[H_2O_2] = 6$ mM, $[Fe^{2+}] = 0.3$ mM, pH = 5

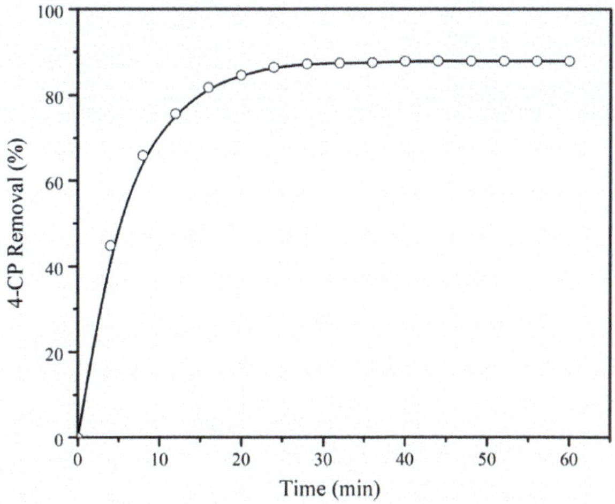

up to 90% with increase in time from 4 to 30 min. After that, the change in degradation was almost negligible. Thus, the optimum degradation of 4-CP could be achieved within 30 min of reaction time.

3.3 *Effect of Hydrogen Peroxide Concentration*

The concentration of H_2O_2 was varied from 3 to 12 mM to study the effect of H_2O_2 dose on the degradation of 4-CP. Due to the generation of $^{\bullet}OH$ radicals H_2O_2 plays a crucial role in the degradation of organic compounds. The effect of hydrogen peroxide concentration on the degradation of 4-CP is shown in Fig. 3. It was observed that the degradation of 4-CP increased by increasing the concentration of H_2O_2 from 3 mM to 6 mM. Further increase in concentration of H_2O_2 above 6 mM the degradation efficiency decreased. The maximum degradation of 4-CP (90%), COD removal (45%) and TOC removal (54.4%) were observed at H_2O_2 concentration of 6 mM. The increase in degradation of 4-CP up to a certain limit by increasing concentration of H_2O_2 can be explained by the effect of additionally produced $^{\bullet}OH$ radicals. The addition of H_2O_2, exceeding 6 mM of aqueous solution decreased the degradation of 4-CP. This may be due to the scavenging of generated hydroxyl radicals, which occurred at large quantities of H_2O_2 and also the unused portion of H_2O_2 during the Fenton's process contributed to COD and hence excess amount was not recommended (Babuponnusami and Muthukumar 2014; Lin and Lo 1997). Therefore, an optimum dosage of H_2O_2 is necessary for the degradation of 4-CP in Fenton's oxidation process as excess of H_2O_2 is not encouraged.

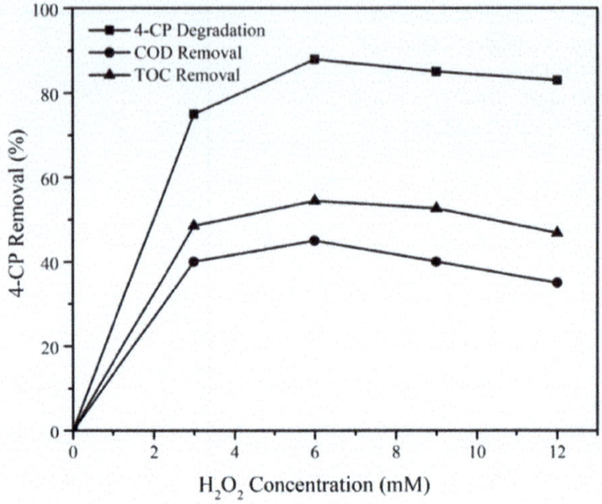

Fig. 3 Effect of hydrogen peroxide concentration on the degradation of 4-CP. Conditions: [4-CP] = 100 mg/L, $[Fe^{2+}] = 0.3$ mM, pH = 5 and time 30 min

3.4 *Effect of Ferrous Ions Concentration*

Ferrous ions catalyze H_2O_2 to generate $^{\bullet}OH$ radicals. Larger the generation of $^{\bullet}OH$ radicals, greater is the degradation rate. The degradation of 4-CP as a function of ferrous ions added was investigated and the results are shown in Fig. 4. In the degradation study of 4-CP, the concentration of Fe^{2+} ions were varied from 0.1 to 1 mM keeping other parameters constant. It was found that the degradation of 4-CP increased with increase in the ferrous ions concentration up to 0.3 mM of $FeSO_4$aqueous solution and after that the removal (%) decreased. The degradation of 4-CP was observed to be maximum (90%) at 0.3 mM concentration of Fe^{2+} ions. The COD removal (%)

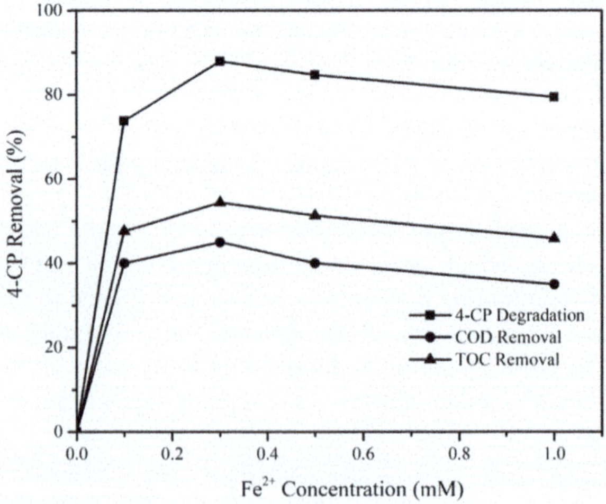

Fig. 4 Effect of ferrous ions concentration on the degradation of 4-CP. Conditions: [4-CP] = 100 mg/L, $[H_2O_2] = 6$ mM, pH = 5 and time 30 min

and TOC removal (%) was also maximum at (i.e. 45 and 54.4%) at 0.3 mM Fe^{2+} ions concentration. The optimum concentration of Fe^{2+} ions is necessary because at lower Fe^{2+} ions concentrations, the degradation of 4-CP was slower due to insufficient amount of Fe^{2+} ions as a catalyst to catalyze H_2O_2 for $^{\bullet}OH$ radicals generation. Addition of higher concentration of Fe^{2+} ions may lead to scavenging of $^{\bullet}OH$ radicals due to reaction of $^{\bullet}OH$ radicals with H_2O_2 or Fe^{2+} ions present in the solution. (Meriç et al. 2004). Also increase in concentration of Fe^{2+} ions, may lead to unutilized quatity of iron salt in the solution which will result in increase in solid contents and requires further removal (Zhou et al. 2012). Thus, it was required to establish the optimum loading of Fe^{2+} ions to mineralize 4-CP.

3.5 *Reaction Kinetics Study*

In the present study, first-order and second-order kinetics were used to study the reaction kinetics of 4-CP by the Fenton's oxidation process. The kinetic study for the oxidation of 4-CP was done by varying time and keeping other parameters constant (initial 4-CP concentration 100 mg/l, $[H_2O_2] = 6$ mM, $[Fe^{2+}] = 0.3$ mM and pH $= 5$). The regression analysis based on the first- and second-order reaction kinetics for the degradation of 4-CP was carried on and the results are shown in Fig. 5. The results indicated that the correlation coefficient R^2 value is 0.9669 and k_1 (min^{-1}) is 0.0884 for first-order reaction kinetics and R^2 is 0.9895 and k_2 ($l\ mg^{-1}\ min^{-1}$) is 0.0026 for second-order kinetics (Fig. 5a, b). The results observed in the present study are in accordance with the study of Ghosh et al. 2012. Thus, the kinetic study indicated that the degradation of 4-CP followed second-order reaction kinetics as it fitted the experimental data more closely than first order.

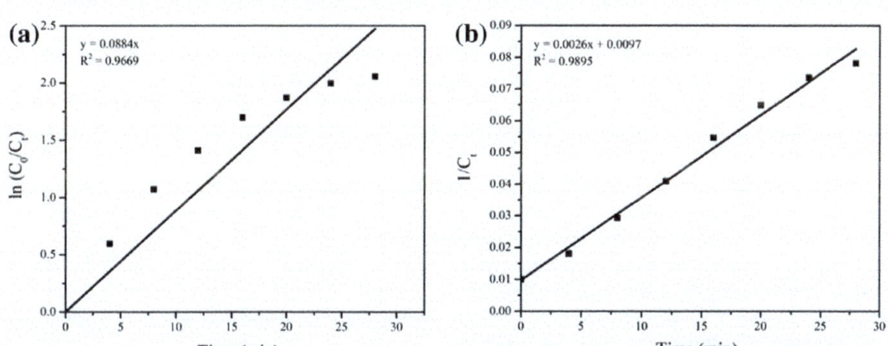

Fig. 5 First and second-order reaction kinetics for the removal of 4-CP by Fenton's oxidation process **a**, **b** at $[4\text{-CP}] = 100$ mg/L, $[H_2O_2] = 6$ mM, $[Fe^{2+}] = 0.3$ mM, pH $= 5$ and time 30 min

4 Conclusions

The degradation of 4-CP by Fenton's oxidation process was investigated. The investigation showed that Fenton's oxidation process is very effective for the removal of 4-CP from aqueous solution. At optimized conditions of $H_2O_2 = 6$ mM, $Fe^{2+} = 0.3$ mM and pH = 5, 90% of 4-CP was degraded in 30 min. The COD removal and TOC removal at optimized conditions were 45% and 54.4%, respectively. The present study indicates that the degradation of 4-CP followed second order reaction kinetics. Thus, Fenton's oxidation process has proved to be an efficient method due to high degradation in short reaction time.

References

Anotai J, Chen CM, Bellotindos LM, Lu MC (2012) Treatment of TFT-LCD wastewater containing ethanolamine by fluidized bed Fenton technology. Bioresour Technol 113:272–275

Babuponnusami A, Muthukumar K (2014) A review on Fenton and improvements to the Fenton process for wastewater treatment. J Environ Chem Eng 2(1):557–572

Benitez JF, Heredia JB, Acero JL, Rubio JF (2000) Rate constants for the reactions of ozone with chlorophenols in aqueous solutions. J Hazard Mater 79(3):271–285

Brillas E, Sauleda R, Casado J (1998) Degradation of 4-chlorophenol by anodic oxidation, electro-Fenton, photoelectro-Fenton and peroxi-coagulation processes. J Electrochem Soc 145(3):759–765

Clesceri LS, Greenberg AE, Eaton AD (1999) Standard methods for the examination of water and wastewater, 20th edn. American Public Health Association (APHA), Washington

De Luis A, Lombrana JI, Varona F, Menendez A (2009) Kinetic study and hydrogen peroxide consumption of phenolic compounds oxidation by Fenton's reagent. Korean J Chem Eng 26(1):48–56

Duan F, Yang Y, Li Y, Cao H, Wang Y, Zhang Y (2014) Heterogeneous Fenton-like degradation of 4-chlorophenol using iron/ordered mesoporous carbon catalyst. J Environ Sci 26(5):1171–1179

Ertugay N, Acar FN (2017) Removal of COD and color from Direct Blue 71 azo dye wastewater by Fenton's oxidation: Kinetic study. Arab J Chem 10:S1158–S1163

Ghaly MY, Hartel G, Mayer R, Hasender R (2001) Photochemical oxidation of p-chlorophenol by UV/H_2O_2 and photo-Fenton process: a comparative study. Waste Manag 21(1):41–47

Ghosh P, Kelapure P, Samanta AN, Ray S (2012) Determination of reaction rate constant for p-chlorophenol and nitrobenzene reacting with OH• during oxidation by Fe(II)/H_2O_2 system. Int J Chem Technol Res 4:116–123

Gondrexon N, Renaudin V, Petrier C, Boldo P, Bernis A, Gonthier Y (1999) Degradation of pentachlorophenol aqueous solutions using a continuous flow ultrasonic reactor: experimental performance and modelling. Ultrason Sonochemistry 5(4):125–131

Ho TFL, Bolton JR (1998) Toxicity changes during the UV treatment of pentachlorophenol in dilute aqueous solution. Water Res 32(2):489–497

Hong APK, Zeng Y (2002) Degradation of pentachlorophenol by ozonation and biodegradability of intermediates. Water Res 36(17):4243–4254

Kang N, Lee DS, Yoon J (2002) Kinetic modeling of Fenton oxidation of phenol and monochlorophenols. Chemosphere 47(9):915–924

Khamaruddin PF, Bustam MA Omar AA (2011) Using Fenton's reagents for the degradation of diisopropanolamine: effect of temperature and pH. In: International conference on environment and industrial innovation, vol 12, pp 12–17

Kitis M, Adams CD, Daigger GT (1999) The effects of Fenton's reagent pretreatment on the biodegradability of nonionic surfactants. Water Res 33(11):2561–2568

Kwon BG, Lee DS, Kang N, Yoon J (1999) Characteristics of p-chlorophenol oxidation by Fenton's reagent. Water Res 33(9):2110–2118

Lin SH, Lo CC (1997) Fenton process for treatment of desizing wastewater. Water Res 31(8):2050–2056

Liotta LF, Gruttadauria M, Carlo GD, Perrini G, Librando V (2009) Heterogeneous catalytic degradation of phenolic substrates: catalysts activity. J Hazard Mater 162(2–3):588–606

Medien HA, Khalil SM (2010) Kinetics of the oxidative decolorization of some organic dyes utilizing Fenton-like reaction in water. J King Saud Univ Sci 22(3):147–153

Meriç S, Kaptan D, Ölmez T (2004) Color and COD removal from wastewater containing reactive black 5 using Fenton's oxidation process. Chemosphere 54(3):435–441

Sarfert F, Jekel M, Wichmann K (1994) Treatment and utilization of water work sludge's in Germany. Water Supply 12(1):6–8

Sauleda R, Brillas E (2001) Mineralization of aniline and 4-chlorophenol in acid solution by ozonation catalyzed with Fe^{2+} and UVA light. Appl Catal B 29(2):135–145

Song Y, Xiao L (2005) Comparison treatment of various chlorophenols by electro-Fenton method: relationship between chlorine content and degradation. J Hazard Mater 118(1–3):85–92

Sun JH, Sun SP, Fan MH, Guo HQ, Qiao LP, Sun RX (2007) A kinetic study on the degradation of p-nitroaniline by Fenton oxidation process. J Hazard Mater 148(1–2):172–177

Titus MP, Molina VG, Banos MA, Gimenez J, Esplugas S (2004) Degradation of chlorophenols by means of advanced oxidation processes: a general review. Appl Catal B 47(4):219–256

Zazo JA, Casas JA, Mohedano AF, Rodriguez JJ (2009) Semicontinuous fenton oxidation of phenol in aqueous solution. A kinetic study. Water Res 43(16):4063–4069

Zhang L, Zeng H, Zeng Y, Zhang Z, Zhao X (2014) Heterogeneous Fenton like degradation of 4-chlorophenol using a novel Fe^{III} containing polyoxometalate as the catalyst. J Mol Catal A Chem 392:202–207

Zhou L, Hu J, Zhong H, Li X (2012) Study of phenol using fluidized-bed Fenton process. Chem Eng Res Des 90(3):377–382

Zhou T, Li Y, Ji J, Wong FS, Lu X (2008) Oxidation of 4-chlorophenol in a heterogeneous zero valent iron/H_2O_2 Fenton-like system: kinetic, pathway and effect factors. Sep Purif Technol 62(3):551–558

Zhou S, Zhang C, Hu X, Wang Y, Xu R, Xia C, Zhang H, Song Z (2014) Catalytic wet peroxide oxidation of 4-chlorophenol over Al–Fe, Al–Cu, and Al–Fe–Cu-pillared clays: sensitivity, kinetics and mechanism. Appl Clay Sci 95:275–283

Effectiveness of Remote Sensing and GIS-Based Landslide Susceptibility Zonation Mapping Using Information Value Method

Virender Kumar and Kanwarpreet Singh

Abstract Himalayas, having complex geography and geotechnical conditions, is highly prone to landslide instability. The present study aims at the slope instability prediction along the National Highway-21, from Rohtang to Bara-lacha in Lahaul–Spiti District, Himachal Pradesh, India by employing ArcGIS and remote sensing database. Several factors triggering the landslides were generated on the map by utilizing the available multi-temporal remote sensing data at macroscale of 1:50,000. To find out the degree of susceptibility of the landslides, information value model was utilized. Information values were attributed on the basis of landslides pixel count in each class of a causative factor triggering landslides. The success rate curve was developed by employing an area under curve (AUC) method to check the accuracy of the applied model. The landslide susceptibility map was classified on the basis of their probability values into "low, moderate, high, and very high" susceptibility zones with the help of natural break spatial analyst tool in ArcGIS. The landslide susceptibility zonation map delineated in this paper can be used by various stakeholders like state governments, researchers, and the citizens to develop, and manage the developmental activities in the study area in particular, and in the other similar areas in general.

Keywords Landslides · Himalayas · GIS · Hazard · Himachal Pradesh

1 Introduction

Himalayan Mountains are tectonically active and associated with landslide activities (Anbalagan et al. 2008).

The slope failures are the most common natural hazards in the Lahaul–Spiti district of Himachal Pradesh, India and are responsible for large-scale transformations and

---

V. Kumar (✉) · K. Singh
NIT Hamirpur, Hamirpur, Himachal Pradesh, India
e-mail: sardavk@yahoo.com

K. Singh
e-mail: singhkanwarpreetnit@gmail.com

A. K. Agnihotri et al. (eds.), *Sustainable Engineering*, Lecture Notes in Civil Engineering 30, https://doi.org/10.1007/978-981-13-6717-5_22

massive destruction of life and property in the area. Lahaul is ice-covered land mass and some of its most dramatic glaciers are Chota Shigri, Bara Shigri, Sonapani, and Samundari glaciers (Chandel et al. 2011). Large-scale constructional works like hydroelectricity power projects, roads, buildings, tunnels, etc., cause terrain alteration with environmental destruction of the Himalayas and resulting in landslides in many places. In order to decrease the threat of landslides, it becomes essential to develop a landslide susceptibility map. LSI is an important tool in distinguishing an area into a number of derived classes depending upon their susceptibility to landslide potential activities. Some major factors like "slope, aspect, soil, geological character, drainage, snowmelt, and land use/land cover" play a major role to identify potentially unstable slopes. Ample of literature is available on qualitative and quantitative approaches, which have been proposed for landslide susceptibility zonation (LSZ) (Yin and Yan 1988; Choubey and Litoria 1990; Anbalagan and Singh 1996; Soeters and Van Westen 1996; Aleotti and Chowdhury 1999; Van Westen 2000; Dai and Lee 2002; Lin and Tung 2003; Mathew et al. 2007; Sharma and Kumar 2008; Pourghasemi et al. 2012; Singh and Kumar 2017).

The present paper aims to generate information about the causative factors affecting landslides using remote sensing and other ancillary data and to delineate LSZ map using the information value method. For this purpose, the research area has been chosen in Lahaul–Spiti district of Himachal Pradesh, India. It lies along NH-21 from Rohtang to Bara-lacha la in the central zone of the Himalayas as mentioned in Fig. 1a.

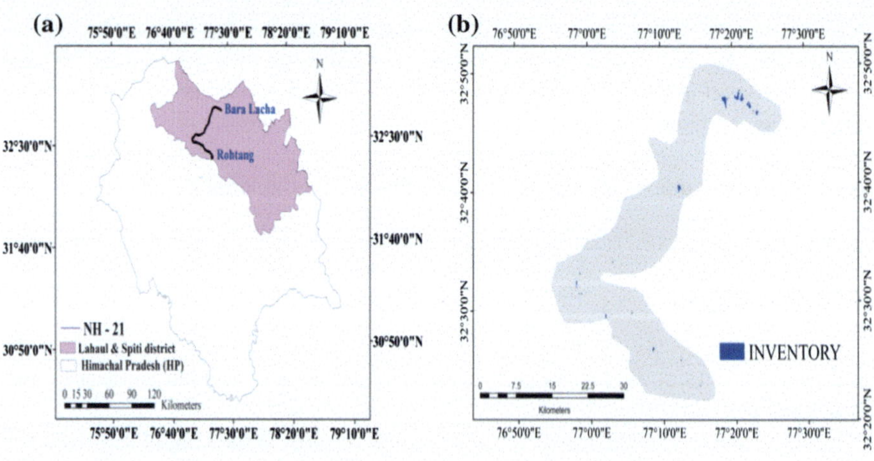

Fig. 1 **a** Study area Rohtang to Bara-lacha La **b** Landslide training dataset

2 Methodology

Two steps are involved in the current investigation (i) generation of landslide inventory map, (ii) application of a bivariate statistical method for developing the LSZ map of the study area.

2.1 *Landslide Inventory Map*

The information related to the location, date of event, type of slope failure, and landslide properties represents the landslide inventory map (Guzettiet al. 2012). The landslide training dataset used in the current modeling is shown in Fig. 1b.

2.2 *Data Source and Data Preparation*

The database used for the delineation of the LSZ map of the research area is mentioned in Table 1. The thematic layers have been generated using ArcGIS (ESRI) environment at scale of 1:50,000.

Various thematic maps like "slope, aspect, drainage, geological structure, and land use/land cover" are developed as shown in Fig. 2 on the basis of the abovementioned database for information value modeling.

2.3 *Bivariate Statistical Analysis*

An assumption was followed for the current assessment of susceptibility that the factors causing slope instability in the past are those, which will cause the same

Table 1 Data type/source for LSZ

Sr. No.	Data type	Data description	Use/purpose
1	Toposheets by the survey of India	1:50,000	Land use land cover
2	DEM	1 m resolution	Slope, aspect, drainage network, and density
3	IRS-P6 LISS III	Spatial resolution 23.5 m	Land use/land cover
4	Geological map	Scale 1:50,000	Geology, Lithology
5	Soil map	Scale 1:50,000	Soil
6	Field data	GPS locations	Landslide locations

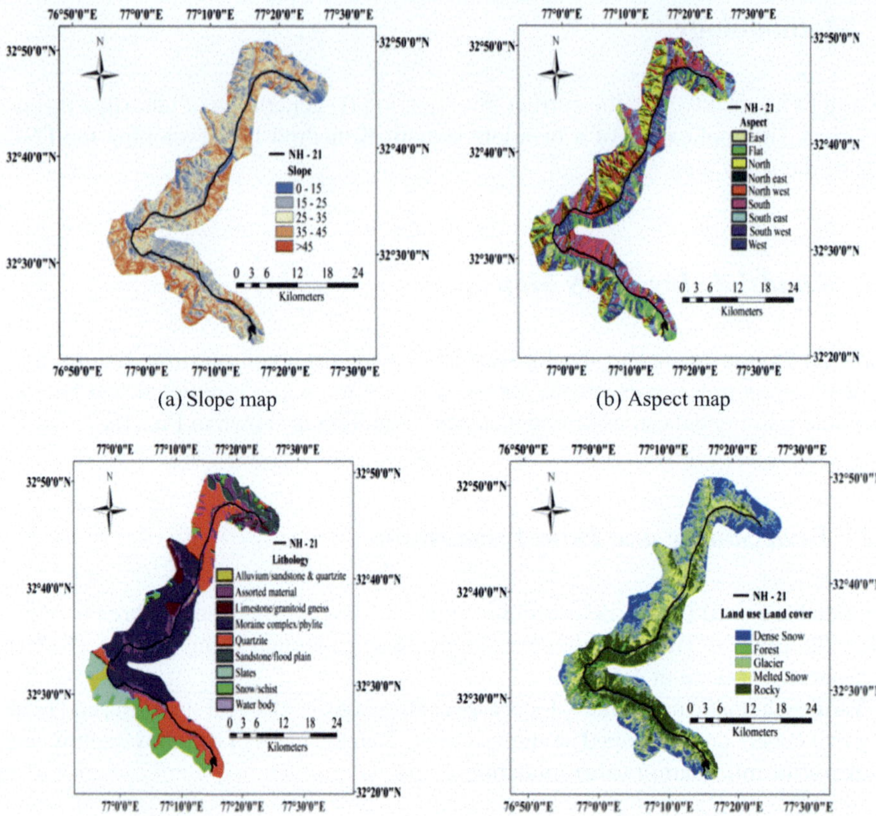

(a) Slope map

(b) Aspect map

(c) Lithology map

(d) Land use/Land cover map

Fig. 2 Maps of various causative factors

phenomena in the upcoming time period at the same region to run information value model (IVM) to compare the landslide pixel count in the training data set with each thematic layer of the landslide triggering agents. The IVM, which is a statistical analysis method proposed by some of the researchers (Yin and Yan 1988), and later on it was modified by one researcher (Van Westen 1993). It was practiced by various researchers for spatial prediction of geological hazards and disaster risk assessment. In this method, the weighted class value is determined on the basis of density of landslides in each causative parameter (Lin and Tung 2003).

However, in actual practice, the information value of each factor can be calculated as

$$\mathrm{I} = \ln \frac{Npix(Si)/Npix(Ni)}{\sum Npi(Si)/\sum Npix(Ni)} \tag{1}$$

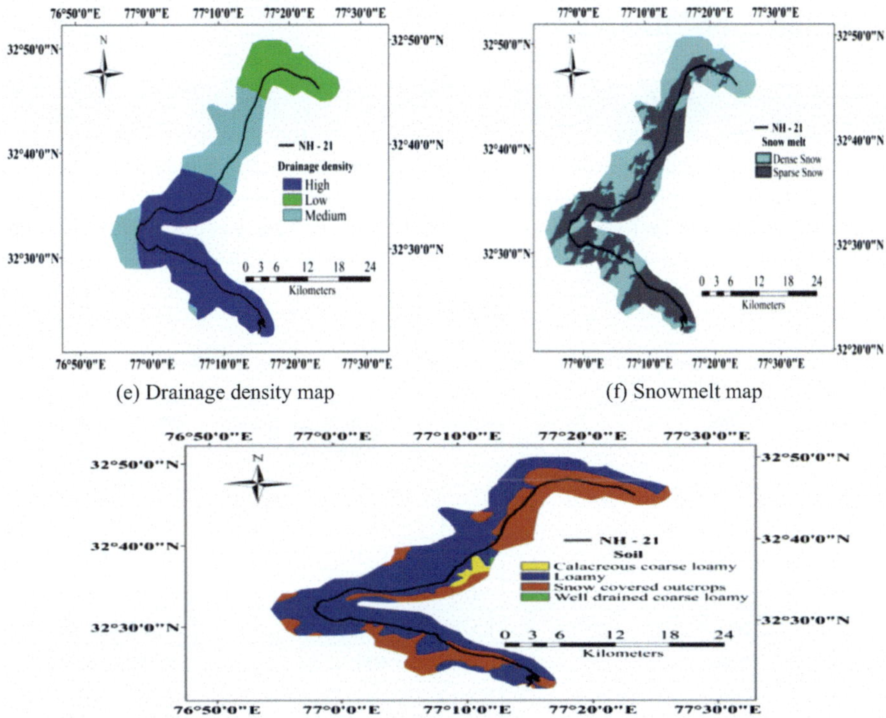

(e) Drainage density map

(f) Snowmelt map

(g) Soil map

Fig. 2 (continued)

where the symbols of Eq. 1 can be denoted as: "I (Information value of a factor class), Npix(Si) (Numbers of pixels of a landslide within class i), Npix(Ni) (Numbers of pixels within class i), $\sum Npix(Si)$ (Number of pixels of a landslide within the entire study area), and $\sum Npix(Ni)$ (Number of pixels within the entire study area)".

The landslide susceptibility index (LSI) for each pixel was calculated by summing up the information values of each factor class as given below:

$$\text{LSI} = \sum\nolimits_{i=1}^{n} I = \sum\nolimits_{i=1}^{n} ln \frac{Npix(Si)/Npix(Ni)}{\sum Npi(Si)/\sum Npix(Ni)} \quad (2)$$

The chance of a landslide occurrence is lesser than average when the information value is less than negative. When the information value is greater than zero, the landslide occurrence likelihood will be equal to average. This indicates that if the information value is greater, then the possibility of the landslide occurrence will be higher.

3 Results and Discussion

Table 2 depicts the information about the landslides relationship with each class of landslide causative factor. The weights of each factor class put their effect on this relationship. The negative information value means that factor class has no contribution for causing slope instability. On the other hand, the positive information value reveals that the factor class may trigger a landslip. Moreover, zero value of the information weight shows that neither landslide is present nor absent in that particular factor class.

Table 2 Pixel count-based information value table

Causative factors	Class	Count	Total no of pixels in landslide	Information value
Slope (Degree)	0–15	115914	591	0.081800877
	15–25	218617	1275	0.140175543
	25–35	264640	1336	0.077500259
	35–45	201424	394	−0.334265706
	>45	83787	139	−0.405812552
Aspect	Flat	96050	400	−0.006088089
	East	89917	106	−0.554186655
	North	104469	98	−0.653412073
	Northeast	107320	114	−0.599426524
	Northwest	101751	494	0.060537551
	South	114724	1035	0.329635358
	Southeast	98390	502	0.082102045
	Southwest	83831	213	−0.220675755
	West	87930	773	0.318391709
LULC	Dense snow	258756	540	−0.306147381
	Glacier	98521	151	−0.440202575
	Melted snow	248205	1496	0.154470356
	Forest	21563	27	−0.52799613
	Rocky	257337	1521	0.145976269
Lithology	Sandstone/Flood plain	64032	149	−0.25886151
	Assorted material	79000	1881	0.751110993
	Snow/Schist	90502	3	−2.105187633
	Quartzite	259782	1033	−0.026159446

(continued)

Table 2 (continued)

Causative factors	Class	Count	Total no of pixels in landslide	Information value
	Alluvium/Sandstone and Quartzite	13885	362	0.790511975
	Moraine complex/Phylite	291991	221	−0.746627903
	Water body	14503	70	0.057989482
	Limestone/Granitoid gneiss	26432	0	0
	Slates	44255	16	−1.067493074
Soil	Snow-covered outcrops	291721	3279	0.425123012
	Calcareous coarse loamy	18650	0	0
	Well-drained coarse loamy	17370	0	0
	Loamy	556641	456	−0.71226106
Drainage density	Low	197700	2770	0.520822388
	Medium	305780	543	−0.376259958
	High	380902	422	−0.581151513
Snowmelt	Sparse snow	534215	942	−0.379315887
	Dense snow	350167	2793	0.276145009

3.1 Landslide Susceptibility Map

Finally, all the thematic layers were converted into 30 m × 30 m grid cells, and then the relation between the causing factors and existing landslide were analyzed. Subsequently, the contribution of each class in the factors for landslide occurrence was calculated. For the delineation of the LSZ map, the information values of each factor class were attributed as mentioned in Table 2 in ArcGIS. The final LSZ map was generated using raster calculator tool from map algebra within ArcGIS. The index values were classified into four susceptibility classes, i.e., “very high, high, moderate and low”. The final LSZ map prepared is shown in Fig. 3.

It was observed that out of the total area of 780.66 km^2 under study, about 19.37% area falls under low susceptibility, 35.36% under moderate, 28.30% under high, and 16.94% falls under very high susceptibility zone.

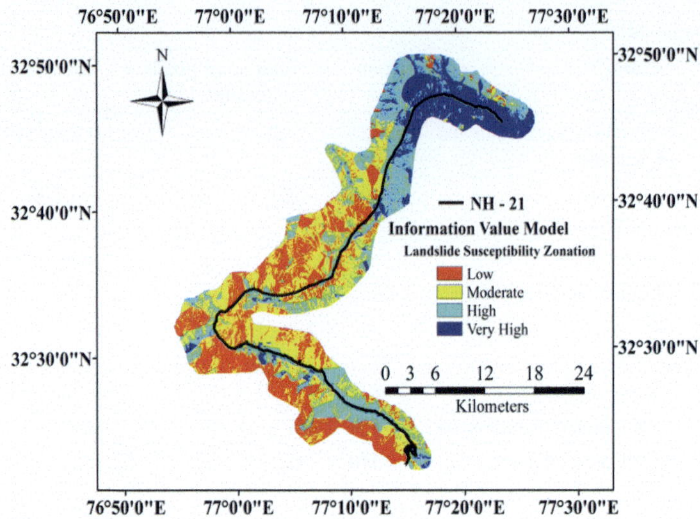

Fig. 3 Landslide susceptibility zonation (LSZ) map

3.2 Validation of Landslide Susceptibility Map

It is an essential task to determine the accuracy of the derived LSZ map of the study area (Chung and Fabbri 2003). In this method, area under curve (AUC) technique was employed to determine the success rate of the LSZ map by tabulating landslide training dataset with LSZ map within ArcGIS. The ability of the probabilistic model can be determined in terms of the accuracy in a precise manner using AUC approach (Feizizadeh et al. 2013). An AUC value within the range of 0.5−1 depicts a random fitness of the model. From Fig. 4 the area under the curve (AUC) comes out to be 0.7271. This overall success rate (72.71%) can be considered as excellent (Hosmer and Lameshow 2000). Hence, the generated map exhibits promising results.

4 Conclusions

The present research is giving an attention on the quantitative analysis of landslides along NH-21, from Rohtang to Bara-lacha la in Lahaul–Spiti district of Himachal Pradesh in India. An ArcGIS (ESRI) and the information value model (IVM) have been used for the generation of the LSZ map. Total seven factor maps, i.e., slope, aspect, geology, soil, land use/cover, drainage density, and snowmelt were generated using GIS software. Standardization of the causative factors has been done in detail by using the information value method in which weights were calculated for each class. The weights of each factor class triggering slope instability influence the results of landslide susceptibility assessment. The weighted rating approach was used in the

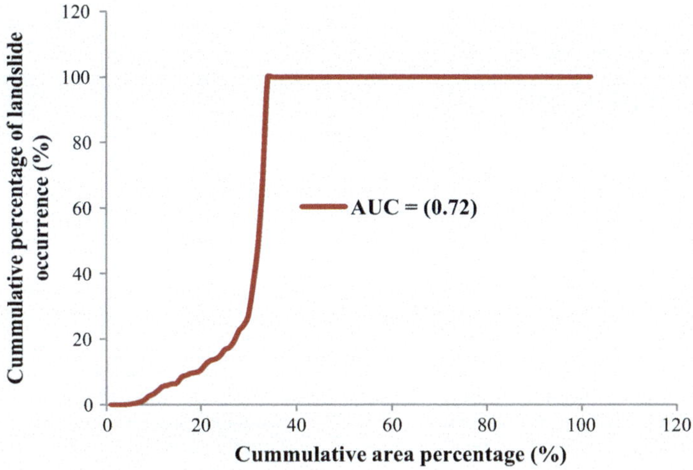

Fig. 4 The success rate curve for validating LSZ map

arithmetic overlay analysis in ArcGIS. After the lookup process in ArcGIS, all of the thematic layers were integrated in raster calculator for the generation of LSZ map. Out of the total study area, about 16.94% fall under the very high susceptible zone, 28.30% under the high susceptible zone, 35.36% under the moderate, and only 19.37% under the low susceptible zone as shown in Fig. 3. The susceptibility map results have been validated with AUC curve, which shows 72.71% success rate indicating that the produced LSZ map has exhibited promising results.

Acknowledgements This paper is a part of the Department of Science and Technology, Ministry of Science and Technology, Government of India sponsored project "GIS-Based Snow Melt Water/Rainfall-Induced Debris Flow Landslide Susceptibility Assessment between Rohtang and Baralacha - La in Himachal Pradesh, India". The authors duly acknowledge the support of DST for carrying out this project at National Institute of Technology, Hamirpur (HP). The authors are thankful to Dr. Bhoop Singh and Dr. A. K. Singh of Department of Science and Technology, Ministry of Science and Technology, for their suggestions from time to time.

References

Aleotti P, Chowdhury R (1999) Landslide Hazard Assessment: Summary Review and New Perspectives. Bulletins of Engineering Geology and the Environment 58(1):21–44

Anbalagan R, Singh B (1996) Landslide hazard and risk assessment mapping of mountainous terrains- a case study from Kumaun Himalaya, India. Eng Geol 43:237–246

Anbalagan R, Chakraborty D, Kohli A (2008) Landslide hazard zonation (LHZ) mapping on meso-scale for systematic planning in mountainous terrain. J Sci Ind Res 67:486–497

Chandel VBS, Brar KK, Chauhan Y (2011) RS & GIS based landslide hazard zonation of mountainous terrains a study from middle Himalayan Kullu district, Himachal Pradesh, India. Int J Geomat Geosci 2:121–132

Choubey VD, Litoria PK (1990) Terrain classification and land hazard mapping in KaLHI-Chakrata area (Garhwal Himalaya), India. ITC J 1:65–68
Chung CJF, Fabbri AG (2003) Validation of spatial prediction models for landslide hazard mapping. Nat Hazards 30(3):451–472
Dai FC, Lee CF (2002) Landslide characteristics and slope instability modelling using GIS, Lantau Island, Hong Kong. Geomorphology 42:213–238
Feizizadeh B, Blaschke T, Nazmfar H et al (2013) Landslide susceptibility mapping for the Urmia Lake basin, Iran: a multi-criteria evaluation approach using GIS. Int J Environ Res 7(2):319–3336
Guzetti F, Mondini AC, Cardinali M et al (2012) Landslide inventory maps: new tools for an old problem. Earth-Sci Rev 112:42–66
Hosmer DW, Lameshow S (2000) Applied logistic regression, 2nd edn. Wiley, London
Lin ML, Tung CC (2003) A GIS-based potential analysis of the landslides induced by the Chi-Chi earthquake. Eng Geol 71:63–77
Mathew J, Jha VK, Rawat GS (2007) Weights of evidence modelling for landslide Hazard zonation mapping in part of Bhagirathi valley, Uttarakhand. Curr Sci 92(5):628–638
Pourghasemi HR, Pardhan B, Gokceoglu C et al (2012) Landslide susceptibility mapping using a spatial multi-criteria evaluation model at Haraz watershed, Iran. Springer, Berlin, pp 23–49
Sharma M, Kumar R (2008) GIS-based landslide hazard zonation: a case study from the Parwanoo Area, lesser and outer Himalaya, H.P., India. Bull Eng Geol Environ 67:129–137
Singh K, Kumar V (2017) Landslide hazard assessment along national highway-154A in Himachal Pradesh, India using information value and frequency ratio. Arab J Geosci 10:539. https://doi.org/10.1007/s12517-017-3315-3
Soeters R, Van Westen CJ (1996) Slope instability recognition, analysis and zonation. In: Turner KT, Schuster RL (eds) Landslides: investigation and mitigation. Special Report No. 247. Transportation Research Board National Research Council, Washington, DC, pp 129–177
Van Westen CJ (1993) Application of geographic information systems to landslide hazard zonation. ITC Publication, vol. 15. International Institute for Aerospace and Earth Resources Survey, Enschede, p 245
Van Westen CJ (2000) The modelling of landslide hazards using GIS. Geophysics 21:241–255
Yin KL, Yan TZ (1988) Statistical prediction model for slope instability of metamorphosed rocks. In: Bonnard C (ed) Proceedings of fifth international symposium in landslides, Lausanne, vol 2. A.A. Balkema, Rotterdam, pp 1269–1272

Modelling and Analysis of Sustainable Green Building Design Using Autodesk Ecotect Analysis

Shristi Khosla and Umesh Sharma

Abstract Buildings have major environmental impacts during their entire life cycle. Their construction and maintenance cause damage to the environment as they consume up to 40% of the total energy resources. It is the need of the hour to promote sustainable development that necessitates the requirements of the present without compromising the ability of future generations. Sustainability in the construction industry is achieved through Green Building approach. Green building concept makes it possible to preserve natural resources by reducing consumption of non-renewable sources of energy and increasing self-recovery of the ecosystem. Keeping in mind, the importance of the energy savings for our future generation and to minimize the impact of buildings on environment degradation, the paper presents a study based on the identification of parameters of sustainable green building design and conceptualizing a model building using Autodesk Ecotect Analysis for the Chandigarh region.

Keywords Sustainability · Green building · Autodesk Ecotect Analysis

1 Introduction

The continuous use of natural resources in the past has placed a tremendous pressure on the environment, thereby depleting the natural resources, polluting the environment, warming the earth, raising sea levels and endangering our biodiversity. Due to an escalating population growth, there is rising concern over having enough resources for development in the present and that for future generations.

The construction and operation of buildings contribute to many environmental impacts, starting from the generation of solid waste during construction to the production of carbon dioxide from occupant energy use. The construction sector in India emits about 22% of the total annual emission of CO_2 resulting from the Indian economy (Buildings and climate change 2007). They also account for 40% CO_2

S. Khosla (✉) · U. Sharma
Civil Engineering Department, Punjab Engineering College, Chandigarh, India
e-mail: shristi24@gmail.com

A. K. Agnihotri et al. (eds.), *Sustainable Engineering*, Lecture Notes in Civil Engineering 30, https://doi.org/10.1007/978-981-13-6717-5_23

emissions, 30% solid wastes and 20% water pollution in the world (Sridhar 2010). One-sixth of all water pumped out of natural flows is consumed in buildings (Reffat 2004). Combined, buildings form one of the major catalysts for environmental degradation on the planet. Thus, green development becomes the world's new agenda to ensure that the human standard of living can be maintained.

Green Building practices can substantially reduce or eliminate negative environmental impacts and improve the existing unsustainable design, construction and operational practices (Mohanty 2012). A green building is an outcome of a design philosophy which focuses on increasing the efficiency of resource use; energy, water, and materials while reducing the impact on human health and the environment during the building's lifecycle, through better design, construction, operation, maintenance and removal. They intend to minimize the waste in construction and demolition processes by recovering materials and reusing or recycling those.

The five major elements of green building design are:

- Sustainable Site Design/Site Efficiency.
- Water Conservation and Quality.
- Energy and Environment.
- Indoor Environmental Quality.
- Conservation of Materials and Resources.

Environment-friendly innovative technologies like energy efficient materials, intelligent gadgets, energy efficient doors and windows, solar water heating and generating power, rain water harvesting are used by many developers in different parts all over the country. Green building is rated on the basis of points system that is given by LEED INDIA and GRIHA. LEED—INDIA is the Indian counterpart of United States Green Building Council's Leadership in Energy and Environmental Design (LEED). It is led by the Indian Green Building Council (IGBC) in India. The LEED India Green Building Rating System is a voluntary, consensus-based, market-driven building rating system based on the existing proven technology. It evaluates environmental performance from a whole building perspective over a building's life cycle. Green rating for integrated habitat assessment (GRIHA) Council is mandated to promote the development of buildings and habitats in India through GRIHA. It is an independent platform for the interaction on scientific and administrative issues related to sustainable habitats in the Indian subcontinent. It was founded by TERI (The Energy and Resources Institute, New Delhi) with support from MNRE (Ministry of New and Renewable Energy, Government of India). GRIHA is a rating tool that helps people assesses the performance of their building against certain nationally acceptable benchmarks.

In the present study, an attempt is being made to identify the green parameters in the region of Chandigarh. The parameters thus selected are integrated in a model building and is analysed using Autodesk Ecotect Analysis software.

2 Methodology

The building survey was done in the vicinity of Chandigarh. The study area has a humid subtropical climate characterized by a seasonal rhythm: very hot summers, mild winters, unreliable rainfall and great variation in temperature. The prototype was thereby designed taking into consideration the climatic conditions.

Paryavaran Bhavan, Chandigarh was selected for the detailed study. The building is an attempt by the Chandigarh Administration to construct an energy efficient building in the Chandigarh. The building is targeted to achieve Leadership in energy and environment design (LEED) India Platinum Rating and Green rating for integrated habitat assessment (GRIHA) five-star rating.

Courtyard—The building's form was basically derived as a concrete block as shown in Fig. 1 that has self cast shadow to protect itself from solar heat and the entire form of the building has been developed keeping in mind the solar activity in the vicinity by increasing the set back as we go to lower floors. The building is oriented along east–west direction to have maximum exposure to the North light.

Earth air tunnel (EAT) to provide space conditioning—Earth air tunnel system comprises of buried pipes at a depth of 4 m below the ground level. At a depth of 4 m, soil temperature remains constant throughout the year. Air is sucked from one end of the tunnel at ambient temperature and is allowed to pass through the tunnel. The system is 69% efficient for Chandigarh region but the efficiency drops during the humid summer period. As for energy-saving potential is concerned, EAT can easily provide savings up to 30% as compared to the conventional system.

Sustainable Building Materials—Walls with fly ash bricks and lightweight and heat-insulating AAC blocks is provided along with entire concreting, mortars and plasters with PPC (fly ash cement). Locally available stone floorings, claddings, terrazzo flooring and flooring with broken marble and Kota stone is used. High

Fig. 1 Central courtyard

Fig. 2 Solar fixtures

reflectance terrace tiles and PUF insulations are provided for low heat ingress. Low VOC paints are used for better indoor air quality.

Renewable energy fixtures—The external lightning is solar powered and entire hot water generation is done using solar panels. Renewable energy systems are installed on site with solar photovoltaic cells to meet total energy demand. The solar fixtures used are shown in Fig. 2.

3 Modelling and Analysis

The important parameters of green building design are as follows:

- Building orientation
- Passive design analysis
- Solar energy analysis

A model is then conceptualized incorporating these parameters for energy efficiency using Autodesk Ecotect analysis software. Building orientation refers to the way a building is situated on a site and the positioning of windows, rooflines and other features. Ecotect software analyses the model for best orientation using the sun path, which helps in deciding the facade for providing openings in a building. The specifications of model building are as follows:

One rectangular office space—7 * 12 m
Window size = 2400 * 1500 mm
Door size = 1500 * 2300 mm
Windows VLT = 60%

Figure 3 depicts the sun path and shadow analysis for the model building in the region of Chandigarh.

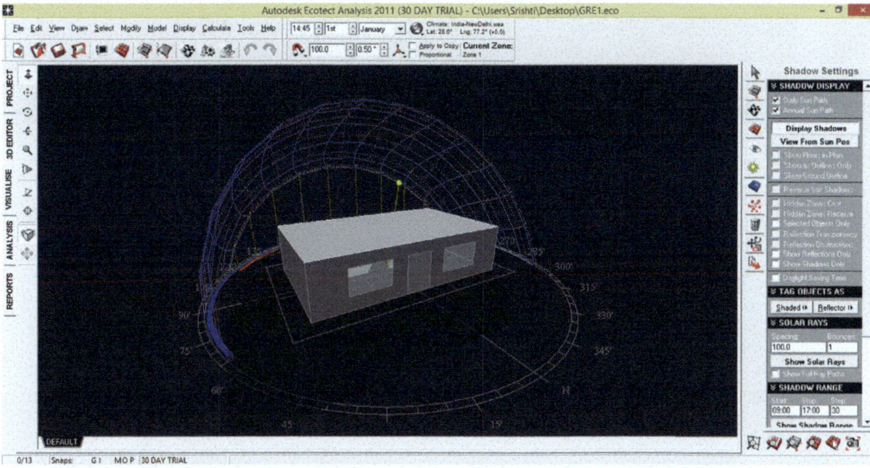

Fig. 3 Shading analysis

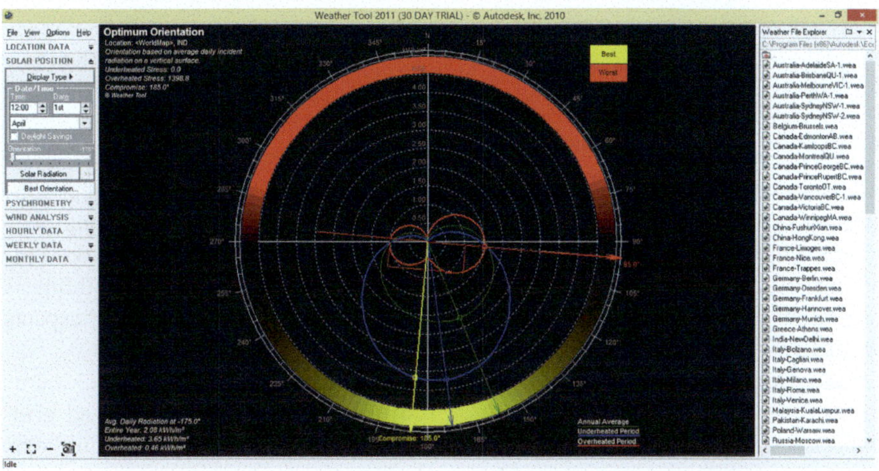

Fig. 4 Optimum orientation

On the basis of sun path, the optimum orientation for the building to be constructed in Chandigarh is determined. Figure 4 depicts the optimum orientation for the building. It is observed that longer facade should be oriented at 185° from north along the NE–SW axis for maximum gain through sunlight for lighting in the rooms.

Once the orientation for the building is fixed, passive design analysis is performed to investigate the enhanced comfort by using some indirect methods of providing high level of comfort and the comparison is done with the base case say conventional case without the use of passive design strategies.

The selected passive design techniques are:

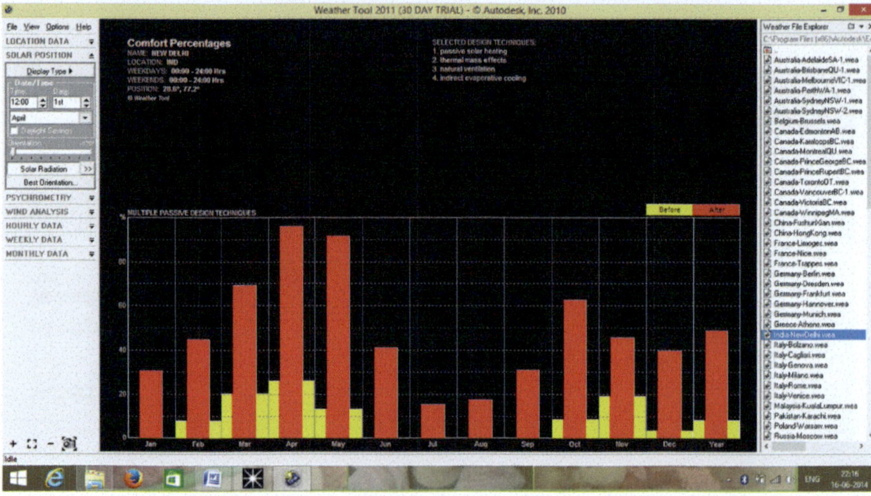

Fig. 5 Passive design analysis

i. Passive solar heating
ii. Natural ventilation
iii. Indirect evaporative cooling

Figure 5 depicts the results obtained for passive design analysis in Ecotect software. It is observed that enhanced comfort is of the order of 40–50% during peak months when using passive design techniques along with proper orientation. Orientation itself improves natural ventilation and indirect cooling as the building is modelled by the direction of sun path. Thus, orientation is one of the major aspects for achieving energy savings without any incremental cost.

Solar Energy Analysis—The residential and commercial sectors cover 73.9 km^2 areas in the city out of the total area of 114 km^2. Rooftop solar PV-based grid-connected system may be well quite feasible in the city. The commercial buildings, government buildings, markets etc., have very large roof areas which are not being used. The grid-connected solar PV systems of 100–500 kW capacities are technically feasible in commercial buildings while 25–50 kW capacity systems might be feasible in the residential sector.

The average rooftop area of residential buildings in Chandigarh is about 320 m^2 (State of Environment Report, Chandigarh). Solar PV panels can be easily installed for this area.

Figure 6 depicts the annual average solar irradiation in the region of Chandigarh. It has been calculated using solar electricity handbook. This handbook shows the monthly solar irradiation for a selected town or city. For the region of Chandigarh, it can be seen that average solar irradiation for the year is 5.22 $kWh/m^2/day$. Annually, this value turns up to $5.22 * 365 = 1905$ kWh/m^2

Fig. 6 Value of solar irradiation

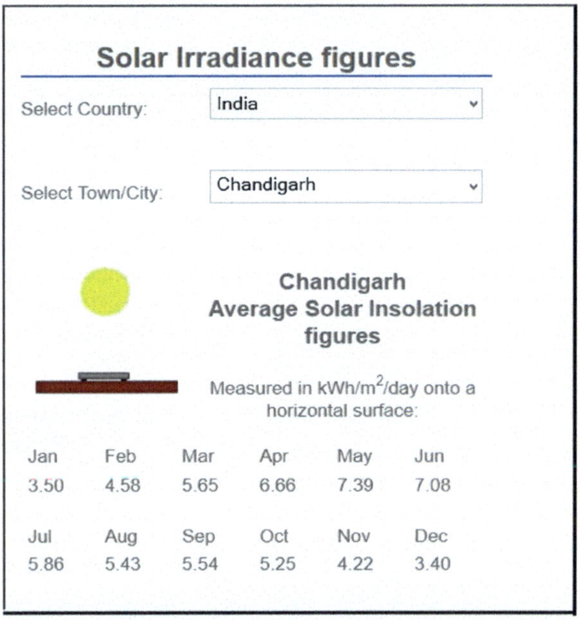

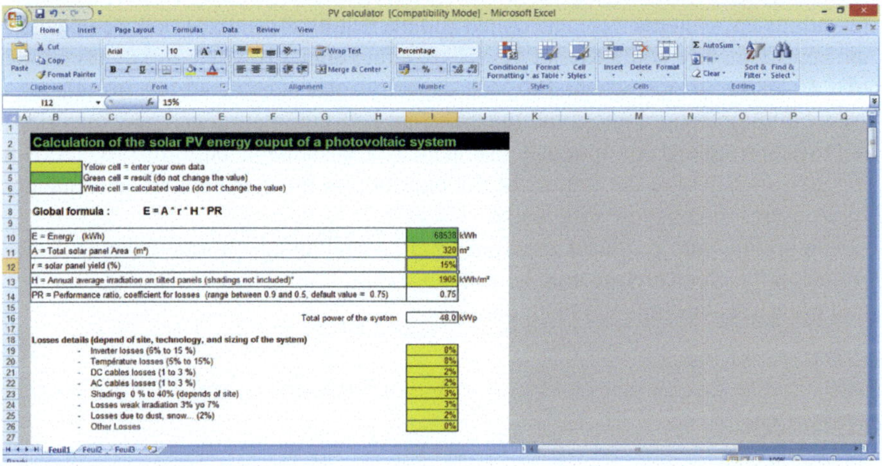

Fig. 7 Calculation of solar PV energy output

On the basis of simulation program using RETScreen software, the performance of various PV systems is calculated. The values for the Chandigarh profile is entered into the simulation software and other required values such as roof and building type are entered for simulation. It can be observed from Fig. 7 that the total output power of the PV system if installed is 48 kWp based on the values of solar irradiation.

It is clear from the analysis that Chandigarh has high potential of solar energy, as the annual solar irradiation can generate up to 76.58 MU of electricity annually.

4 Conclusions

Sustainable development involves efforts in effective utilization of available resources to ensure that we sustain our human and natural environments now and for future generations. The study was carried out to identify the green parameters through different case studies. The green building parameters were identified and following deductions are made:

- Paryavaran Bhavan, Chandigarh—The profile of the building is designed with inward stepping floors from the top floor till the first floor so as to cast self-shadows on the walls to achieve thermal comfort during extreme summer and in turn, reduce the load on the cooling system of the building.
- For a rectangular office space of 7 * 12 m, with longer facade towards north-east–south-west direction, the passive design analysis results show enhanced comfort of the order of 40–50% during the peak months in the Chandigarh region.
- On the basis of simulation program using RETScreen software, the performances of various PV systems were calculated and the results show that 48kWp can generate approximately 76.58 MU of electrical output annually. Chandigarh has a total of 1.10 lakh rooftops, where 400 MW capacity solar plants could be installed.
- New building technologies, and in particular, ICT automation and new materials, are to be introduced continuously to enhance the sustainable building process with the goal of reducing the impact of the building on the surrounding environment by using resources more efficiently.
- Green architecture produces environmental, social and economic benefits. Environmentally, green architecture helps reduce pollution, conserve natural resources and prevent environmental degradation.

References

Huovila P, Ala-Juusela M, Melchert L, Pouffary S (2007) Buildings and climate change. UNEP

Sridhar KS (2010) India infrastructure report 2010

Mohanty B (2012) Low carbon green growth roadmap for Asia and the Pacific. UNESCAP Publication

Reffat R (2004) Sustainable development of buildings and environment. In: Proceedings of second international conference on development and environment. Assiut University, Egypt

Aeration Efficiency Evaluation of Modified Small Parshall Flume Using M5P and Adaptive Neuro-Fuzzy Inference System

Sangeeta, Subodh Ranjan and N. K. Tiwari

Abstract This paper studies the potential of modeling techniques for predicting the aeration efficiency of Modified Small Parshall flume. The experiments were performed in the hydraulic laboratory of Civil Engineering Department, NIT, Kurukshetra. Parshall flume is used to measure the flow rate in irrigation and sewer system. The aeration is nothing but transfer of oxygen from atmosphere to water. Where 100 experimental observations were collected. Out of 100 observations, arbitrarily, 67 were taken as training data and the remaining 33 as testing data. The input data table consists of discharge (m^3/s), temperature (°C), depth (m), and the ratio of width (m) and depth (m) of the modified small Parshall flume, whereas aeration efficiency was taken as the output. Three quantitative standard performance evaluation measures, the coefficient of correlation (CC), Nash Sutcliff (NS), and root mean square error (RMSE) have been performed to compare the results of modeling technique. The comparison of results using M5P and Adaptive Neuro-Fuzzy Inference System (ANFIS) was made out and M5P was found to be the best.

Keywords Aeration efficiency · Modified small Parshall flume · M5P · ANFIS · CC · RMSE · NS

1 Introduction

Many pollution changes of water can adversely affect the aquatic life but none is generally accepted to be more important and common than the reduction of dissolved oxygen. Some extents in all water bodies are receiving industrial and domestic wastes that contain decomposable organic matter. Major sources of pollution in water are

Sangeeta (✉) · S. Ranjan · N. K. Tiwari
Department of Civil Engineering, NIT Kurukshetra, Kurukshetra, Haryana, India
e-mail: Sangeeta_31602506@nitkkr.ac.in

S. Ranjan
e-mail: sranjan_nitk@yahoo.com

N. K. Tiwari
e-mail: nand_nitk@rediffmail.com

A. K. Agnihotri et al. (eds.), *Sustainable Engineering*, Lecture Notes in Civil Engineering 30, https://doi.org/10.1007/978-981-13-6717-5_24

sediment dispersion, biocides, detergent fertilizers, pesticides, mining products, mineral oils, organic wastes, radioactive materials, and industrial wastes, etc. Therefore, water pollution control affords have been directed largely toward the maintenance of concentration of dissolved oxygen above the minimum that is required by aquatic life. Dissolved oxygen concentration is one of the best indicators of overall water quality. In most of the natural water systems, it requires 5–6 mg/l of dissolved oxygen concentration to support a diverse aquatic population. Oxygen solubility in water is directly proportional to the partial pressure in the gas phase. Its solubility decreases as temperature goes on increasing. Dissolved oxygen enters the water by absorption directly from the air or by photosynthesis of aquatic fauna. To deal with water pollution-related problem, wastewater has to be treated before release into any water system. Aeration is one of the major important steps for treatment of polluted water. Aeration induced dissolved oxygen into the water system, which is important for the survival of flora and fauna in the water bodies. If sewage is discharged into rivers without aeration, aquatic life cannot survive. A minimum of 4 ppm of oxygen concentration is necessary for the sustainable life of aquatic bodies.

Hydraulic structures such as stepped spillway, nozzle orifice, or free overflow construction are the better answer for improving water quality. In the river system, aeration can be done by stepped spillway or free overfall spillway to increase the amount of dissolved oxygen. In the straight flow canals, these types of alternative would not be preferred, so drop structures such as baffled block, chutes, weirs, and cascades are favored. Small Parshall flume provides a good solution for straight prismatic channels with a low and negligible slope. Parshall flume is composed of a converging section, ending with a diverging outlet and a flat floor, named as throat with upward slope section and downward slope floor. Parshall flume is used in several projects such as sewage treatment plants, watershed monitoring, dam seepage, stream gauging, industrial discharge monitoring, and irrigation canals. Parshall (1928) developed and studied the improved venturi flume, which operates successfully with relatively small loss of head. For free flow, head loss in standard weir is approximately four times than that of the flume. Cox et al. (2013) developed a new rating equation more concerned with the supercritical flow in large Parshall flumes. Singh et al. (2014) studied different throat width Parshall flumes and their test results of the experiments gave relationship of throat width with discharge. Dursun (2016) studied the aeration efficiency of a small 3 in. Parshall flume and Modified Venturi flumes and concluded that Parshall and Venturi flumes are effective for oxygen transfer for channels having highly low slopes. Dursun et al. (2016) conducted experiments on oxygen transfer efficiency at baffled chutes and concluded that baffle blocks had significantly higher aeration efficiency values than those without chute channels. Baylar et al. (2009) performed experiments on water aeration processes in Hydraulic structures and their results concluded that stepped cascades were very efficient means of aeration because of strong turbulent mixing, the large residence time and the substantial air bubble entrainment. Baylar et al. (2009) performed modeling on air entrainment rate and aeration efficiency of weirs using ANN approach from which they concluded that multi nonlinear regression model was found to be better than the simple linear regression model in the computation of aeration efficiency.

Recently, software techniques are applied to construct models. Here, M5P and ANFIS technique is modeled for predicting oxygen transfer efficiency. Both these techniques have an advantage of developing input and output relationship.

2 Oxygen Transfer Efficiency (E)

Oxygen is an extremely volatile matter with an air–liquid transfer rate that is influenced entirely by the liquid phase. Lewis and Whitman (1924) advised that two smooth and regular layer lies on each side of the air–liquid interface. When the oxygen concentration over time in a package of water as the package passes through a hydraulic structure can be shown as

$$\frac{d(m)}{dt} = V\frac{d(c)}{dt} = K_L\frac{A}{V}(C_s - C), \tag{1}$$

where

$\frac{d(m)}{dt}$	Interface oxygen transfer rate,
$\frac{d(c)}{dt}$	Change of concentration rate,
K_L	Coefficient of liquid bulk film,
C_S	Constant oxygen saturation concentration at prevailing ambient conditions, and
$\frac{A}{V}$	Specific surface area/volume.

From the charts and equations, saturation concentration of deionized water can be calculated which is a resemblance for natural water. It is often different from that of deionized water due to the impact of water impurities. C is the factual concentration of oxygen of the water at instantaneous time (t). Mostly in hydraulic structures, water surface has free turbulence and the greater quantity of air is transferred during the flow. So, in the above equation, source and sink of oxygen is not assigned to the water channel.

The aeration transfer efficiency (E) given by Gulliver et al. (1990)

$$\mathrm{E} = \frac{Cd - c}{Cd - cu} = 1 - e^{-\int_u^d K_L}\frac{A}{V}dt = 1 - \frac{1}{r}, \tag{2}$$

where

U and D	Subscripts showing u/s and d/s, respectively.
R	Oxygen deficient ratio.

Conditions:

E > 1	Downstream water has become supersaturated (i.e., Cd > Cu)
E = 1	No oxygen transfer through interface.

Aeration efficiency is very responsive to temperature changes and investigators have commonly employed a correction factor for temperature. For hydraulic structures, Adelman and Gulliver (1990), formulated a relationship for temperature correction,

$$(1 - E_{20})^{F} = 1 - E \tag{3}$$

where

E Aeration transfer efficiency at the measured water temperature,
E_{20} Aeration transfer efficiency at 20 °C and
F Exponent for aeration efficiency equation is calculated by Eq. 2,

$$F = 1 + 2.103 * 10^{-2} * (T - 20) + 8.261 * 10^{-2} * (T - 20)^{2} \tag{4}$$

where

T Water temperature in °C.

3 Factors Affecting Aeration Efficiency Rate

The oxygen transfer rate is directly proportional to an interfacial surface contact area (A) and oxygen deficient (r); and adversely proportional to a liquid film thickness (Y_t). Those factors which affect these parameters will also affect aeration efficiency. Some of these factors are described below briefly. The oxygen transfer rate in Parshall flume is affected by water temperature, water quality and discharge (m^3/s), etc.

(1) Temperature of water oxygen transfer efficiency is heavily responsible for water temperature variation, and investigators have formulated various temperature correction factors. Mostly used temperature correction factor was given by Gameson et al. (1958). Arrhenius type of water temperature corrections is used by Holler (1970) in their theory. Gulliver et al. (1990) used the study of Levich (1962), Hinte (1955), and Azbel (1981) for air transfer simulation and gave the relationship of aeration efficiency with temperature correction as shown in Eqs. 1 and 2.
(2) Quality of Experimental water.
(3) Due to the presence of an organic substance, dissolved solvents and suspended solids, quality of water are adversely affected. They have been reasonable for impact on aeration processes. These solvents affect hydrodynamic characteristics, inhibiting film at the air–liquid interface and reduce surface tension. The solvent content and salt concentration are maintained in this experiment. So that, the aeration processes are not affected by the presence of any type of chemical and pollutants.

(4) Parshall Flume discharge Parshall flume consists of u/s converging section, d/s diverging section, and middle throat. In this, two different type of flow occurs on the bases of water head; namely free flow and submerged flow. For different flume, different flow equations and methods of flow calculation are used. In Parshall Flume under free flow conditions depend on point of measurement at U/S, discharge is calculated by the following formula:

$$Q = CH^{n} \tag{5}$$

where

Q Discharge through the flume (m^3/s)
C Coefficient of discharge
H U/S head measured at 2/3 converging section of the throat and
n exponent of the head, H.

4 Adaptive Neuro-Fuzzy Interference System (ANFIS)

Multi-modeling numerical computing surface is provided by MATLAB (matrix laboratory). Although MATLAB is planned mainly for numerical computing but additional work are also done as multi-domain simulation. The computing system based on computer software is ANFIS, which have different training and testing data as input to give a meaningful result as output. ANFIS works as an efficient hybrid learning method that consists of gradient descent and the least square method. The least square method is really the important driving force that gives fast training, whereas the gradient descent leads to slowly change the underlying function of membership, for the least squares approach.

As the outcome, ANFIS will mostly give a satisfying outcome right after the starting epoch of training that is after the initial employment of the least square method. For output, a connection of all input is done in the layer of interface. Several approaches for supplier classification have developed in the literature, including system depend on fuzzy logic. The important point for a fuzzy logic approach is the requirement to manage unclear in the situation. Scholar tries to construct effective models for analysis of quantitative aspects as to convert human opinion about meaningful category into significant results. Baylar et al. (2008) performed ANFIS software modeling to compute air entrainment rate of weir by taking three inputs to two different ANFIS model.

Different membership functions (MF) includes many shape and size as trapezoidal, triangular, Gaussian, generalized bell-shaped, and many others. In this study, approaches were done on triangular, generalized bell-shaped model, and Gaussian function. The output of these functions is compared. All membership functions are briefly described below.

Triangular:

$$\mu \mathrm{X(a)} = \frac{\{a - x\}}{\{b - y\}}, (x \leq a \geq y) \tag{6}$$

$$= \frac{\{z - a\}}{\{z - y\}}, (z \leq a \geq y)$$

$$= 0 \tag{7}$$

Gaussian:

$$\mu_{\mathrm{x}}(\mathrm{a}) = \mathrm{G}\frac{1}{1 + (\frac{(a-zj)}{x})^2} \tag{8}$$

Bell shaped:

$$\mu_{\mathrm{xi}}(\mathrm{a}) = \mathrm{G}\frac{1}{1 + \{\frac{(a-zi)}{xi}\}^{2xi}} \tag{9}$$

$$\mu_{\mathrm{yi}}(\mathrm{b}) = \mathrm{G}\frac{1}{1 + \{\frac{(a-zij)}{xj}\}^{2yi}} \tag{10}$$

where i and j = 1, 2, 3, 4…

In the present study, ANFIS was applied to training and testing data set. Modeling was done using the fuzzy logic toolbox of MATLAB 2013. ANFIS computing technique does not have any stick rules. The Trimf membership function gave better results for every input (Table 1).

M5P in 1992, Quinlam (1992) represented his idea of "MODEL TREE" that he called it as M5, as an innovative method for dealing with quantitative problems. Model trees are constructed as a taking all decision as the tree and linear regression function at their leaves. For better performance of M5P on a dataset and reduce tree size, Wang and Witten (1997) had done some changes which developed a new version of M5P. M5P is electing attributes that minimize intra-subset differences in the clause

Table 1 Specifications used in the development of ANFIS model

Specification	TRIMF	GAUSSMF	GBELLMF
Number of nodes	193	193	193
Linear parameters	81	81	81
Nonlinear parameters	48	24	36
Total number of parameter	129	105	117
Number of training data pairs	67	67	67
Number of fuzzy rules	81	81	81

Table 2 Optimum parameters used for user-defined parameters of M5P model

Batch size	Minimum number of instances	Unpruned
100	4.0	TRUE
100	4.0	FALSE

values of cases that move down each branch. The differences are determined by the standard deviation of the data that deliver to the node from substructure through their branches by computing the expected decrease in error as an outcome of testing each value at a node. The values which gave more reduction in expected error are chosen. The division stops if the attributions of all that reach a node differ shortly or only less quantity of instant remained.

The calculation of standard deviation reduction (SDR) is done as

$$\text{SDR} = \text{Sd(T)} - \sum_{i}^{j} \frac{[[T_i]]}{[[T]]} * \text{sd}[[T_i]] \tag{11}$$

where

T Set of values that deliver to node
T_i ith outcome of the potential set
sd Standard deviation.

The division in the M5 model tree has line regression (LR) function at the leaves, which forecast continuous numerical values. Advantage of model tree above regression is that tree formed is much smaller and does not involve many variables (Ajmera and Goyal (2012). Another benefit of tree model is its calculation efficiency (Table 2).

5 Methodology

For computing aeration efficiency, experiments were done at NIT, Kurukshetra hydraulic laboratory. The test was conducted in a steel rectangular channel with 0.25 m wide, 0.30 m deep and 3 m long. Different structures of Parshall flume models was fixed to the channel. Flow in the channel was regulated by value where 2HP centrifugal pump which could supply a maximum discharge of 6 l/s.

Dataset consist of 100 observations, from which 67 observations were taken randomly as training data and the remaining 33 were used as testing data. Input consisted of discharge (m^3/s), temperature (°C), ratio of width to depth and throat width (m) while aeration efficiency (E_{20}) was taken as output.

6 Results and Discussion

To compare the results of ANFIS and M5P, statistical performance parameters CC, NS, and RMSE values were determined as shown in Table 4. Table 3 shows the performance of modelling technique as discussed. A substantial increase in predictive accuracy of M5P approach shows that it can be performed to predict the aeration efficiency. Figures 1 and 2 shows, actual and predicted dissolved oxygen transfer efficiency for training and testing dataset using ANFIS and M5P. Figure 3 indicates the relative performance of the above-applied methods on test data.

Table 3 Details of performance evaluation of training and test data set

Variables	Minimum	Maximum	Mean	Standard deviation
Training data				
Dicharge (m^3/s)	0.001	0.005	0.003	0.001
Depth (m)	0.034	0.217	0.109	0.046
Oxygen deficit ratio (r)	1.04	4.669	1.465	0.522
Throat width/length	0.25	0.333	0.298	0.039
E_{20}	0.042	0.73	0.47	0.126
Test data				
Dicharge (m^3/s)	0.001	0.005	0.003	0.001
Depth (m)	0.034	0.212	0.109	0.046
Oxygen deficit ratio (r)	1.038	2.79	1.439	0.383
Throat width/length	0.25	0.333	0.296	0.04
E_{20}	0.042	0.577	0.244	0.122

Table 4 The detail of performance evaluation using M5P and ANFIS

	Training data			Test data		
	CC	N.S	RMSE	CC	N.S	RMSE
UNPRUNED M5P	0.984	0.969	0.022	0.991	0.969	0.020
PRUNED M5P	0.999	0.999	0.002	0.999	0.970	0.021
TRIMF	0.992	0.970	0.002	0.993	0.975	0.019
GAUSSMF	0.999	0.999	0.001	0.852	0.725	0.096
GBELLMF	0.999	0.999	0.001	0.952	0.907	0.166

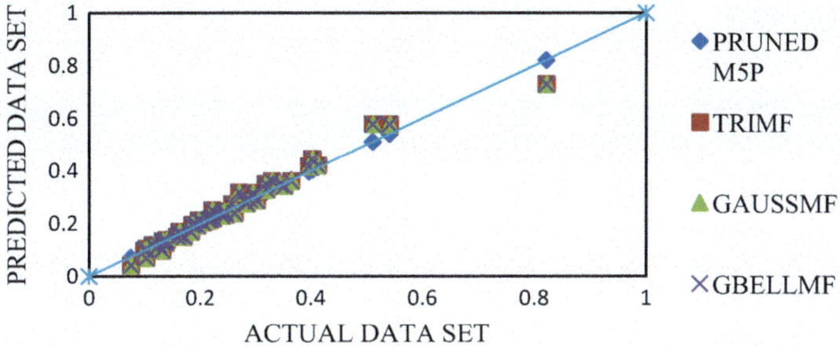

Fig. 1 Actual and predicted aeration efficiency (E_{20}) of training data set using ANFIS and M5P

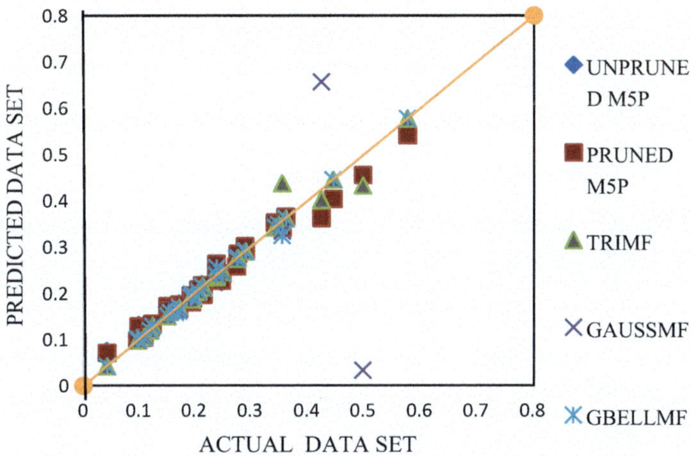

Fig. 2 Actual and predicted aeration efficiency (E_{20}) of test data set using ANIFIS and M5P

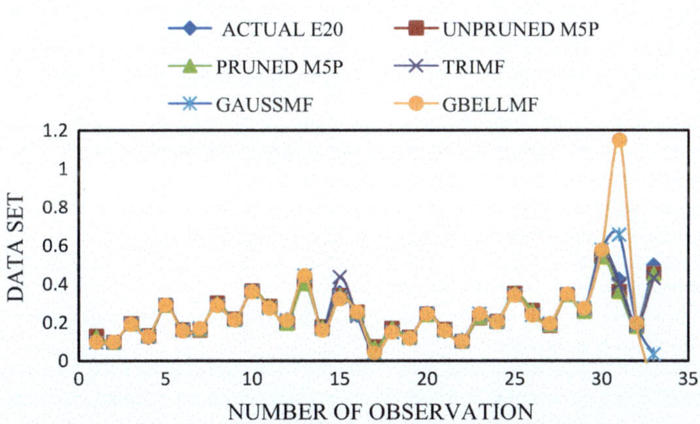

Fig. 3 Number of observation versus software modeling technique (ANFIS and M5P)

7 Conclusions

In this paper, the potential of M5P (PRUNED AND UNPURNED) and ANFIS (TRIMF, GAUSSMF, and GBELLMF) were compared for predicting oxygen transfer efficiency. From the comparison of different performance evaluation parameters, it has been found that M5P (PRUNED) approach work well than the ANFIS. M5P can be efficiently worked in the estimation of oxygen transfer efficiency of small modified Parshall flume.

References

Adelman SJ, Gulliver AF (1990) An elemental abundance analysis of the superficially normal a star vega. Astrophys J 348:712–717

Ajmera TK, Goyal MK (2012) Development of stage–discharge rating curve using model tree and neural networks: an application to Peachtree Creek in Atlanta. Expert Syst Appl 39(5):5702–5710

Azbel MY (1981) Quantum delta-dimensional Landauer formula. J Phys C Solid State Phys 14(9):L225

Baylar A, Hanbay D, Ozpolat E (2008) An expert system for predicting aeration performance of weirs by using ANFIS. Expert Syst Appl 35(3):1214–1222

Baylar A, Hanbay D, Batan M (2009) Application of least square support vecto machines in the prediction of aeration performance of plunging overfall jets from weirs. Expert Syst Appl 36(4):8368–8374

Cox AL, Thornton CI, Abt SR (2013) Supercritical flow measurement using a large Parshall flume. J Irrig Drain Eng 139(8):655–662

Dursun OF (2016) An experimental investigation of the aeration performance of Parshall flume and venturi flumes. KSCE J Civ Eng 20(2):943–950

Dursun OF, Talu MF, Kaya N, Alcin OF (2016) Length prediction of non-aerated region flow at baffled chutes using intelligent nonlinear regression methods. Environ Earth Sci 75(8):680

Fromm H (1955) Untersuchungen zum Marienleben des Priesters

Gameson ALH, Vandyke KG, Ogden CG (1958) The effect of temperature on aeration at weirs. Water Water Eng (British) 753(62):489

Gulliver JS, Thene JR, Rindels AJ (1990) Indexing gas transfer in self-aerated flows. J Environ Eng 116(3):503–523

Hanbay D, Baylar A, Batan M (2009) Prediction of aeration efficiency on stepped cascades by using least square support vector machines. Expert Syst Appl 36(3):4248–4252

Levich VG (1962) Physicochemical hydrodynamics. Prentice Hall

Lewis WK, Whitman WG (1924) Principles of gas absorption. Ind Eng Chem 16(12):1215–1220

Parshall RL (1928) The improved Venturi flume. CER; 47/52–54

Quinlan JR (1992) November. Learning with continuous classes. In: 5th Australian joint conference on artificial intelligence, vol 92, pp 343–348)

Singh J, Mittal SK, Tiwari HL (2014) Discharge relation for small Parshall flume in free flow condition. Int J Res Eng Technol 3(4):317–321

Wang Y, Witten IH (1997) Inducing model trees for continuous classes. In: Proceedings of the ninth European conference on machine learning. Wernher, vol 52. Turun yliopiston kustantamo

Typical Sites in the Proposed Smart City Ludhiana

Manish Bhutani and Sanjeev Naval

Abstract Ground response analysis for a specific site plays an important role in designing earthquake-resistant structures as several conditions at the surface may occur due to the amplification of rock/hard layers at the bed level. In this study, 1D EQL ground response investigation has been conducted for sites of Ludhiana city using DEEPSOIL software. The analysis has been carried out by considering Sikkim 2011 and Uttarkashi 1999 earthquake. Geotechnical data have been collected from various organizations and Shear wave velocity estimated using already available correlation between SPT-N value and Vs. The sites were characterized on the basis of average SPT 'N' values as per the recommendations of National Earthquake Hazards Reduction Program (NEHRP). The peak ground acceleration for different sites has been calculated and found to vary between 0.173 and 0.254 g. It has been observed that sites of Ludhiana can amplify due to which site-specific seismic ground motion analysis should be adopted for RCC structures having high important value.

Keywords Equivalent linear ground response analysis · Input motion · SPT-N value · Soil characterization · Shear wave velocity · Peak ground acceleration

1 Introduction

Earthquake is a natural event which may cause tremendous loss of life and property. Earthquake can be explained as it is the process of rupture at the source, which causes movement of seismic waves through underlying rock. Seismic waves mostly travel deeply in rock through several kilometers surrounding as compared to rock individual having thrust up to shallow depth of few meters in soil layers but still it plays a significant role in identifying the earthquake motion characteristics. Because of the moments of earthquake motion, the equivalent motion is considerably improved at

M. Bhutani (✉) · S. Naval
CE, DAVIET, Jalandhar, Punjab, India
e-mail: manish_0220@yahoo.com

S. Naval
e-mail: sanjeevnaval@gmail.com

A. K. Agnihotri et al. (eds.), *Sustainable Engineering*, Lecture Notes in Civil Engineering 30, https://doi.org/10.1007/978-981-13-6717-5_25

the ground surface. This happens due to the presence of local soil layers above the bedrock beneath the site of interest.

Ground response analysis is considered as one of the most important and commonly encountered problems in Civil Engineering. Ground response analysis is determined to study the effects of the soils on the rock motion available at bed surface. It is used to predict natural periods, evaluate ground motion amplification, provides, evaluate liquefaction potential, and to access the forces induced due to an earthquake which causes instability of structures supported on earth. The determination of the soil amplifications in ground response due to the local region is very complex problem to the structural designers. It is so important that all structure such as superstructure and sub structure should be designed for earthquake including rupture mechanics including nearest fault to the site of interest.

In northern India, Naval and Chandan (2017) and Siddhartha et al. (2017) carried out a lot of research, which has been undertaken for estimating surface PGA values. Deterministic Seismic Hazard Analysis of Jalandhar revealed that the maximum value of Peak Ground Acceleration is 0.454 g and it is much higher value as compared to the values published in IS 1893 (Part 1): 2001 by Bureau of Indian Standards. According to the National Centre for Seismology (NCS), Ludhiana featured among other 29 cities highly vulnerable to earthquakes. After carrying out Deterministic Seismic Hazard Analysis of Ludhiana city by dividing the whole city into $0.025° \times 0.025°$ grids and using ground motion equation developed by the National Disaster Management Authority (NDMA), the Peak Ground Acceleration obtained ranged from 0.10 to 0.392 g. So adequate measures must be adopted for the safety of structures falling in the high seismic hazard zone.

In this study, 1-D earthquake ground response investigation has been conducted on selected 05 sites of Ludhiana City using DEEPSOIL software using Sikkim 2011 and Uttarkashi 1999 earthquake. Geotechnical data have been collected from various organizations and Shear wave velocity was calculated using already available correlation between SPT-N value and Vs. The sites were characterized on the basis of average SPT 'N' values as per the recommendations of National Earthquake Hazards Reduction Program (NEHRP). The peak ground acceleration for different sites of Ludhiana city has been calculated and found to vary between 0.173 and 0.254 g. Desai and Choudhury (2015) and Phanikanth et al. (2011) performed the site specific ground response analysis for the sites of Mumbai, whereas Gupta et al. (2017) also conducted the equivalent linear ground response analysis for the sites of Haryana region.

2 Study Region

Ludhiana being an industrial center of northern India, is the largest city in the state, with an area of 310 km^2 and an estimated population of about 3 million (Census 2011). It lies under seismic zone IV as per the seismic zoning map of India. The earthquake database in India is yet to be completed and zoning offers a preliminary guide to earthquake hazard for a particular region. Ludhiana district being central core of Punjab, is confined between latitude 30°33′N and 31°01′N and longitude 75°25′E and 76°27′E. The boundary of the Ludhiana district has been formed by Satluj in the North with Jalandhar and Hoshiarpur districts. In eastern and southern region, Ropar and Fatehgarhsahib districts marks limits of the Ludhiana district. The western border is adjoining Moga and Ferozpur districts. Ludhiana region is occupied with Indo-Gangetic alluvium with key drains are Satluj and its tributaries and Budha nala.

The dissimilarities in the characteristics of soil profile are much more pronounced due to the regional climatic differences. The soil of this sector has established under semi-arid situation. The soil is sandy loam to clayey with normal reaction.

3 Site Characterization

In this study, 1D EQL ground response analysis of five sites of Ludhiana city is carried out. The field borehole data of the sites were collected. The sites selected in Ludhiana region for GRA has been shown in Fig. 1. SPT-N values are found at 1.5 m intervals. The geologic strata contain shallow silt and clay layer (up to 3 m) at top with layer of sand at the bottom up to 20 m depth.

The details of SPT test of the five sites of Ludhiana region have been presented in Table 1a–e. The laboratory tests results on disturbed and undisturbed samples provide the information regarding thickness of subsoil strata, standard penetration test values, and index properties of the soil deposits. The SPT site effects that represent seismic ground response characteristics are usually incorporated as amplification factors in seismic codal provisions (e.g. NEHRP 2001, UBC 97, IBC 2000 and EC8

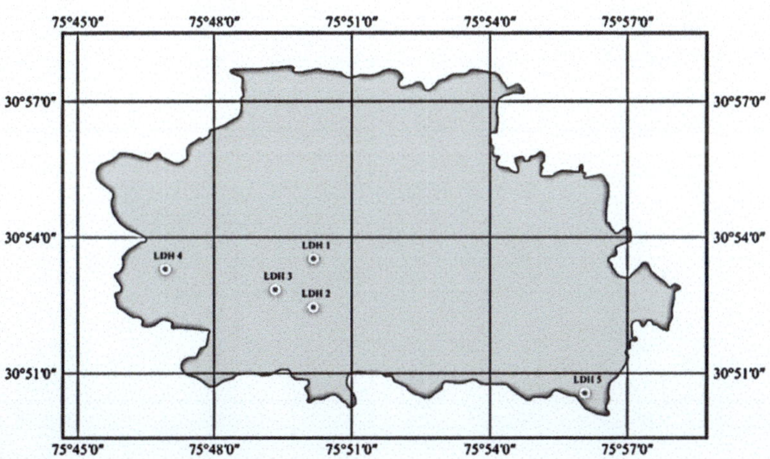

Fig. 1 Location of bore hole in Ludhiana district

Table 1 Soil profile with shear wave velocity at the investigated sites (LDH-01 to LDH-05)

Layer no.	Depth (m)	Soil type	Thickness (m)	SPT-N value	Vs (m/s)
(a) South model gram (LDH-01)					
1	1.50	SP	1.50	6	178.48
2	3.00		1.50	9	212.47
3	4.50		1.50	12	240.45
4	6.00		1.50	16	272.11
5	7.50		1.50	19	292.98
6	9.00		1.50	22	312.05
7	10.50		1.50	27	340.77
8	12.00		1.50	30	356.57
9	13.50		1.50	38	394.72
10	15.00		1.50	39	399.15
11	16.50		1.50	42	412.08
12	18.00		1.50	47	432.50
13	20.00		2.00	51	447.96
(b) Urban phase I (LDH-02)					
1	1.50	SP	1.50	5	165.02
2	3.00		1.50	9	212.47
3	4.50		1.50	12	240.45
4	6.00		1.50	15	264.67
5	7.50		1.50	19	292.98
6	9.00		1.50	23	318.05
7	10.50		1.50	27	340.77
8	12.00		1.50	31	361.63
9	13.50		1.50	35	381.00
10	15.00		1.50	39	399.15
11	16.50		1.50	43	416.27
12	18.00		1.50	45	424.48
13	20.00		2.00	48	436.43
(c) South model gram (LDH-03)					
1	1.50	SP	1.50	10	222.32
2	3.00		1.50	9	212.47
3	4.50		1.50	14	256.93
4	6.00		1.50	18	286.25
5	7.50		1.50	21	305.87
6	9.00		1.50	25	329.68

(continued)

Table 1 (continued)

Layer no.	Depth (m)	Soil type	Thickness (m)	SPT-N value	Vs (m/s)
7	10.50		1.50	29	351.41
8	12.00		1.50	33	371.48
9	13.50		1.50	39	399.15
10	15.00		1.50	41	407.83
11	16.50		1.50	43	416.27
12	18.00		1.50	45	424.48
13	20.00		2.00	48	436.43
(d) Punjab Mata Nagar (LDH-04)					
1	1.50	SP	1.50	7	190.71
2	3.00		1.50	13	248.87
3	4.50		1.50	17	279.30
4	6.00		1.50	21	305.87
5	7.50		1.50	24	323.94
6	9.00		1.50	27	340.77
7	10.50		1.50	30	356.57
8	12.00		1.50	33	371.48
9	13.50		1.50	37	390.22
10	15.00		1.50	40	403.52
11	16.50		1.50	43	416.27
12	18.00		1.50	47	432.50
13	20.00		2.00	49	440.32
(e) Kangawal (LDH-05)					
1	1.50	SP	1.50	7	190.71
2	3.00		1.50	13	248.87
3	4.50		1.50	17	279.30
4	6.00		1.50	21	305.87
5	7.50		1.50	24	323.94
6	9.00		1.50	27	340.77
7	10.50		1.50	30	356.57
8	12.00		1.50	33	371.48
9	13.50		1.50	37	390.22
10	15.00		1.50	40	403.52
11	16.50		1.50	43	416.27
12	18.00		1.50	47	432.50
13	20.00		2.00	49	440.32

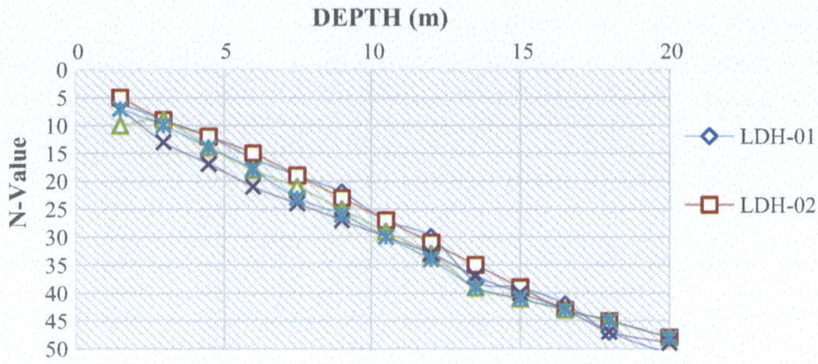

Fig. 2 Variation of N-value with depth for Ludhiana sites (LDH-01 to LDH-05)

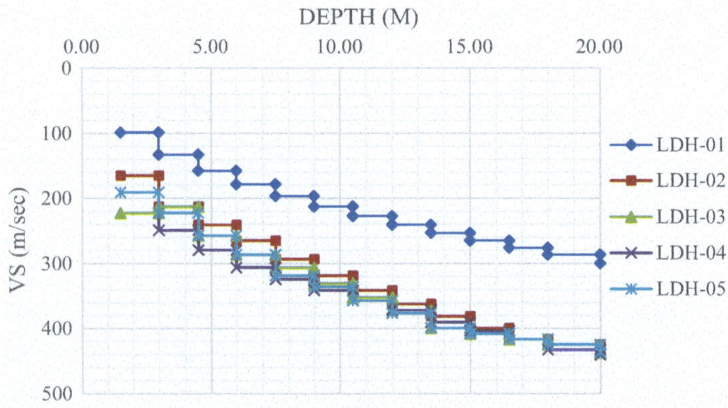

Fig. 3 Variation of Vs with depth for Ludhiana sites (LDH-01 to LDH-05)

2003). So, that site effects can be accounted for while designing. These factors are based on average shear wave velocity of top 30 m of the soil profile (Vs30). It is a general recommendation to use the actual shear wave velocity of the bedrock in site classification. However, due to non-accessibility of Vs profiles, sites have been classified using N30 values as per the recommendations of NEHRP. The present NEHRP provisions categorize soils into the classes A, B, C, D, E and F based on average N-value of the profile (BBSC 2000).

For the calculation of average N-value (N_{30}) for the soils of Ludhiana region, the following equation can be used:

$$N_{30} = \frac{\sum_{i=1}^{n} di}{\sum_{i=1}^{n} \frac{di}{Ni}} \tag{1}$$

where N_{30} = average SPT-N value of the soil for 30 m depth, Ni = SPT-N value of soil layer and di = depth of soil layer. It has been witnessed that all five sites of Ludhiana city falls in the D category with the average N-value for the soil varies from 17.85 to 22.77 with lowest and extreme values observed at LDH-02 and LDH-

04 sites. The thickness of the deposits is so adjusted that the extreme frequency of particular layer can transmit is always beyond 25 Hz. The engineering bedrock is usually expected to be the topmost deposit with a shear wave velocity of soil (V_s) ≥ 760 m/s as per the provisions of NEHRP (Puri and Jain 2016). Figure 2 shows the variation of N-value with depth for the sites of Ludhiana (LDH-01 to LDH-05).

Therefore, for the present study, for N-value greater than 50 has been considered as a refusal for bedrock for 15 cm penetration or N-value greater 100 for 30 cm penetration. Hence, for this analysis, the boreholes bored up to refusal have been considered. Solid rock has been modeled as an elastic half-space while considering the density of soil 2.5 g/cc with damping of 2% and shear wave velocity (Vs) of 760 m/s.

4 Dynamic Soil Properties

Shear wave velocity (Vs) is being the most important input parameters to represent the stiffness of the soil layers. The Earthquake motion factors at the top of ground is achieved using one-dimensional ground response investigation allowing only the upward movement, the stiffness of various layers of soil is represented by shear wave velocity (dynamic properties). As shear wave velocity profile is not available for the study region, it has been calculated using SPT-N values through available correlations. In the present study, the relation, $\mathbf{V_s = 82.6N^{0.43}}$, suitable for all types of soil is used. This equation was developed by considering the data of more than 200 borelogs and shear wave velocity profiles at more than 80 boreholes available at several locations for Delhi (Hanumantharao and Ramana 2008). For the Ludhiana sites (LDH-01 to LDH-05), shear wave velocity varies from 165.02 to 222.32 m/s at the top layer of 1.5 m thickness and from 436.43 to 447.96 m/s at 20 m depth. The variation of Vs with depth for the sites of Ludhiana is shown in Fig. 3.

The maximum and minimum value of Shear Wave velocity at the top layer is achieved at LDH-03 and LDH-02 sites, respectively. Sites LDH-02, LDH-03 and LDH-05 show minimum Shear Wave velocity at 20 m depth whereas, LDH-01 site exhibits maximum Shear Wave velocity at 20 m depth.

5 Input Motions

For the present analysis, two earthquake motions Sikkim earthquake motion (2011) and Uttarkashi earthquake motion (1991) having PGA values 0.152 and 0.253 g have been applied at the bedrock level to study the soil effects.

The time history of 1991 Uttarkashi earthquake and for 2011 Sikkim earthquake considered in this study is shown in Fig. 4 and the earthquake characteristics of these motions are given in Table 2.

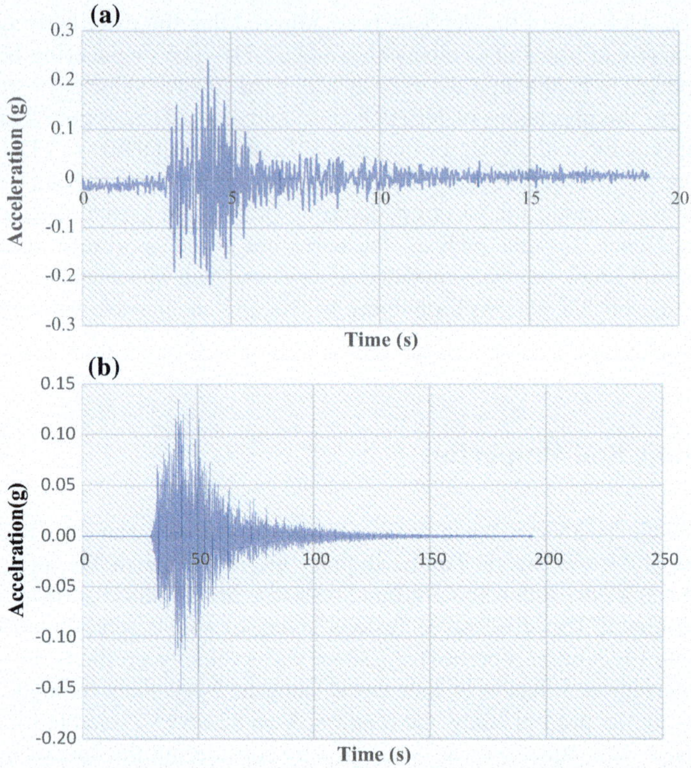

Fig. 4 Acceleration (g) versus time graph of Sikkim (2011) and Uttarkashi (1991) motion. **a** Acceleration versus time graph of Sikkim. **b** Acceleration versus time graph of Uttarkashi earthquake (1991)

Table 2 Strong motion characteristics considered in the present study

S. no	Seismic motion	Date of occurrence	Location of epicenter	Recording station	Magnitude	Distance from source (km)	PGA (g)
1	Sikkim earthquake	18 September 2011	27.723°N 88.064E	Gangotak	6.8	68	0.152
2	Uttarkashi earthquake	19 October 1991	30.837N 78.984E	Uttarkashi	7.0	34	0.253

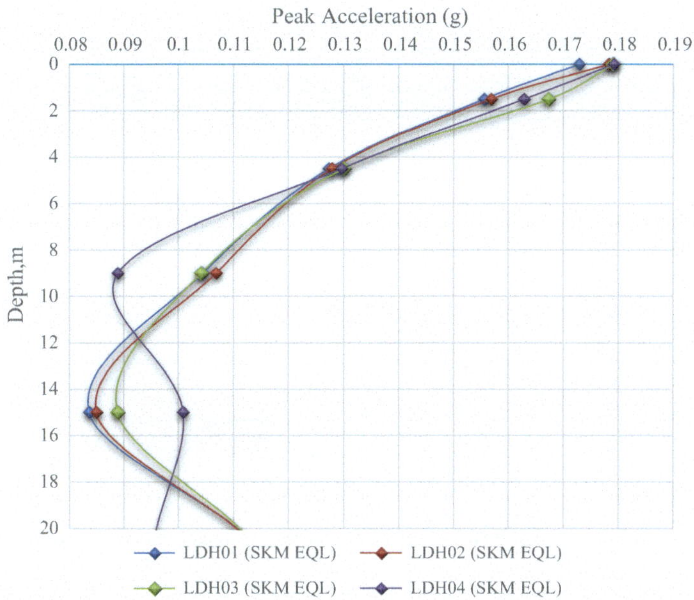

Fig. 5 PGA variations with depth due to 2011 Sikkim earthquake for all sites of Ludhiana city

6 Results and Discussion

EQL ground response investigation has been conducted to study the effect of local ground conditions for the sites of Ludhiana city using DEEPSOIL v6.1 software (Hashash et al. 2016). The ground surface acceleration with time history was computed at all five locations in response to the 1999 Uttarkashi and 2011 Sikkim earthquake applied at the bedrock. The variation of PGA with depth at all the locations is presented in Fig. 5. It has been observed that under the influence of Sikkim earthquake all the sites under consideration show surface PGA value varies from 0.173 to 0.183 g with maximum surface PGA at LDH5 and minimum surface PGA at LDH1 (Fig. 5). Similarly after carrying out euivalent linear GRA for sites LDH1 to LDH5 with strong input motion of Uttarkashi earthquake having PGA 0.24 g, the surface PGA was observed to vary from 0.241 to 0.254 g (Fig. 6).

When using Uttarkashi earthquake motion, LDH2 site was observed to have maximum surface PGA and minimum surface PGA was observed at LDH4 site under consideration. The average surface PGA for various sites under consideration comes out to be 0.179 g and 0.248 g against input motion of Sikkim earthquake having PGA 0.153 g and Uttarkashi Earthquake having PGA 0.24 g, respectively.

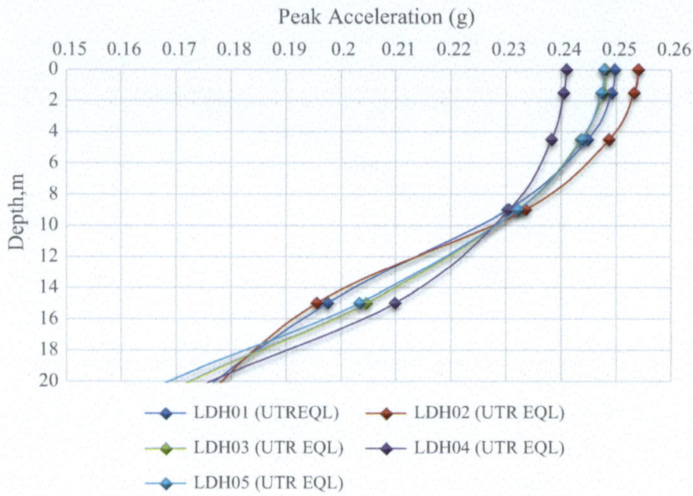

Fig. 6 PGA variation with depth due to 1991 Uttarkashi earthquake for all sites of Ludhiana city

7 Conclusion

EQL ground response investigation has been conducted to study the effect of local ground conditions for the sites of Ludhiana city using DEEPSOIL v6.1 software. The following points were observed:

1. The maximum value of Shear Wave Velocity in top layer and bottommost layer was found to be 222.32 m/s at LDH3 and 440.32 m/s at LDH4, respectively..
2. The minimum value of Shear Wave Velocity in top layer and bottommost layer was found to be 98.33 m/s at LDH1 and 299.52 m/s at LDH1, respectively.
3. Two input bedrock motions Sikkim and Uttarkashi earthquakes having PGA of 0.152 g and 0.253 g, respectively, were used for carrying out Equivalent Linear GRA.
4. The maximum value of PGA on surface was found to be 0.183 g at LDH5 and 0.254 g at LDH2 for input bedrock motions of Sikkim and Uttarkashi earthquakes, respectively.
5. The minimum value of PGA on surface was found to be 0.173 g at LDH1 and 0.241 g at LDH4 for input bedrock motions of Sikkim and Uttarkashi earthquakes, respectively.

References

BBSC (2000) NEHRP recommended provisions for seismic regulations for new buildings and other structures. Building Seismic Safety Council, USA

Desai SS, Choudhury D (2015) Site specific seismic ground response study for nuclear power plant and ports in Mumbai. National Hazards 16:1–13

Gupta R et al (2017) Equivalent linear ground response analysis of typical sites in state of Haryana using spectrum compatible ground motions. In: Sixth Indian young geotechnical engineers conference 6IYGEC2017, 10–11 March 2017. NIT Trichy, India

Hanumantharao C, Ramana GV (2008) Dynamic soil properties for microzonation of Delhi, India. J Earth Syst Sci 117:719–730

Hashash YMA et al (2016) DEEPSOIL—1-D wave propagation analysis program for geotechnical site response analysis of deep soil deposits version 6.1, user manual and tutorial. University of Illinois, UC

Nath RR, Jakka RS (2012) Effect of bedrock depth on site classification. In: 15th world conference on earthquake engineering 15WCEE. Lisbon, Portugal, 24–28 Sept 2012

Naval S, Chandan K (2017) Deterministic seismic hazard analysis for proposed smart city, Ludhiana (India). Electron J Geotech Eng 22:4255–4270

Phanikanth VS et al (2011) Equivalent-linear seismic ground response analysis of some typical sites in Mumbai. J Geotech Geol Eng. https://doi.org/10.1007/s10706-011-9443-8

Puri N, Jain A (2016) Deterministic seismic hazard analysis for the state of Haryana, India. Ind Geotech J 46:164–174

Siddhartha S et al (2017) Deterministic seismic hazard analysis of Chandigarh city. In: Sixth Indian young geotechnical engineers conference 6IYGEC2017, 10–11 March 2017. NIT Trichy, India

Synthesis of rGO/TiO_2 Nanocomposite for the Efficient Photocatalytic Degradation of RhB Dye

Vikash Kumar, Ajay Bansal and Renu Gupta

Abstract In this work, the successful synthesis of reduced graphene oxide–TiO_2 (rGO/T) photocatalysts with various concentrations of GO (5, 10, 15, and 20 wt%) has been reported. The synthesis of rGO/T photocatalysts was done by the photo-assisted reduction of GO to rGO using ultraviolet radiation followed by the hydrothermal deposition of rGO onto the TiO_2. The synthesized photocatalysts were characterized by field-emission scanning electron microscopy (FE-SEM), X-ray diffraction (XRD), and energy dispersive X-ray spectroscopy (EDXs) techniques. The photocatalysts were used for the photodegradation of Rhodamine B (RhB) dye. The photocatalytic degradation of RhB dye of 10 ppm was successfully carried out under UV light, as well as solar light irradiation over rGO/T photocatalysts. Here, the influence of GO over the photoactivity of TiO_2 was explored. It was found that the synthesized rGO/T photocatalysts exhibit enhanced photocatalytic efficiency as compared to unmodified TiO_2. The rate of photodegradation of RhB dye was successfully fitted by Langmuir–Hinshelwood kinetic model.

Keywords TiO_2 · Graphene oxide · rGO/T · Photocatalysis · Rhodamine B

V. Kumar · A. Bansal (✉) · R. Gupta
Department of Chemical Engineering, Dr. B. R. Ambedkar National Institute of Technology, Jalandhar 144011, India
e-mail: bansala@nitj.ac.in

V. Kumar
e-mail: vikash.jaiswal15@gmail.com

R. Gupta
e-mail: bansalr@nitj.ac.in

A. K. Agnihotri et al. (eds.), *Sustainable Engineering*, Lecture Notes in Civil Engineering 30, https://doi.org/10.1007/978-981-13-6717-5_26

1 Introduction

The discharge of a huge quantity of colored dyes from various industries such as textile, plastic, paper, printing, pharmaceutical, and food industry into water bodies are the life-threatening health hazard to aquatic as well as human life. These dyes being toxic in nature are imposing a serious threat to the environment (Chang et al. 2009). Rhodamine B (RhB) is one of the very common water-soluble organic dyes and is extensively used as a colorant in the textile and food industries, as well as a biological stain in biomedical laboratories. RhB has been banished to use in the food industry for many years due to its suspected carcinogenic nature. However, with the development of industry and the illegal discharge, RhB still has the chances to enter the aliment chain to hazard human health. It is therefore essential to remove these dyes from dye effluents prior to their discharge into the receiving water bodies (Rani et al. 2016; Shen and Gondal 2017; Yu et al. 2004; Li et al. 2014a, b; Carneiro et al. 2016). To date, several treatment methods such as chemical oxidation, coagulation, biological treatment, ozonolysis, and adsorption have been extensively explored for the removal of organic pollutants from industrial effluents, but all such methods suffer from a number of drawbacks such as very low removal efficiency, generation of secondary pollutant, high chemical consumption, and high cost of treatment (Carneiro et al. 2016; Oseghe et al. 2014). Photocatalysis using semiconductor nanocomposites is involved in a group of waste treatment methods called advanced oxidation processes (AOPs) (Carneiro et al. 2016). AOPs are recommended when wastewater components have a high chemical stability and/or low biodegradability. A chemical wastewater treatment using AOPs can produce the complete mineralization of pollutants to CO_2, water, and inorganic compounds (Thiruvenkatachari et al. 2008).

Titanium dioxide (TiO_2) has gained intensive attention due to their strong oxidizing power, physicochemical stability, non-toxicity, low-cost production and eco-friendly nature (Friedmann et al. 2010) (Krejčíková et al. 2012; Nakata and Fujishima 2012; Gnanasekaran et al. 2015). However, the photocatalytic activity of TiO_2 is strongly influenced by its structural properties such as morphology, crystalline phase, particle size, surface hydrophilicity, and oxygen vacancy concentration (Shao et al. 2013). The efficiency of the photocatalytic reaction of TiO_2 gets significantly reduced owing to its wide band gap energy of 3.2 eV and high electron-hole recombination rate. Thus, restricting the electron-hole pair recombination and extending the light absorption to the longer wavelength are the key factors to improve the photocatalytic activity of TiO_2 (Chen and Mao 2007; Abdel-Messih et al. 2013; Yadav et al. 2017). Consequently, several attempts have been made to improve the performance of TiO_2 such as modifying it with carbonaceous materials like CNTs, Fullerene, graphene, or graphene oxide. Doping or immobilization of TiO_2 over the surface of carbon extends the absorption range of TiO_2 to the visible region. In addition, the nonpolar, nonreactive, and nontoxic natures of carbon materials play a significant role in the treatment of wastewater containing complex toxic organic compounds (Fang et al. 2017).

Graphene oxide (GO), a two-dimensional sp^2 hybridized carbonaceous material, which has been extensively investigated because of its unique structure, large surface area (2630 m^2/gm), high mechanical strength, and good physicochemical stability. Benefiting from the large surface area, superior electronic, thermal, and optoelectronic properties, GO has gained great attention in various fields such as catalysts, sensors, supercapacitors, and energy devices (Dubey et al. 2014; Lin et al. 2015; Lorestani et al. 2015; Li et al. 2014a, b; Štengl et al. 2013). The synthesis of GO-TiO_2 composite is regarded as a significantly important approach for improving the performance of photocatalysts. The hybridization of GO-TiO_2 can reduce the recombination of photogenerated electron-hole pairs, extend the light absorption to the visible light region, and accelerate the transfer rate of charge carriers (Lv et al. 2012; Kumar et al. 2015; Hamandi et al. 2017).

To explore the influence of GO over photocatalytic activity of TiO_2, the studies have been made on the degradation of RhB dye using GO-supported TiO_2 nanocomposites. The effects of various parameters such as light source, catalyst dosage, and type of catalysts have also been investigated.

2 Methodology

2.1 *Preparation of Graphene Oxide*

GO was prepared by the oxidation of natural graphite powder according to modified Hummer's method. Typically, 2 gm of graphite powder and 1 gm of sodium nitrate were added to 80 ml of concentrated H_2SO_4 in a 2000 mL beaker maintained at 0 °C under continuous agitation. After 10 min, 12 gm of $KMnO_4$ was added slowly to the above reaction mixture by keeping the reaction temperature below 20 °C. Successively, the reaction system was transferred to a 40 ± 3 °C water bath and vigorously stirred for about 2 h. At the end of 2 h, 40 ml of deionized water (dH_2O) was added slowly to the reaction mixture causing violent effervescence and an increase in temperature. The solution is maintained for 15 min at 90 ± 3 °C under agitation. The excess $KMnO_4$ was removed by the addition of 6 ml of 30% H_2O_2. The yellow bright solution was filtered and washed three–four times with 1400 ml of dH_2O followed by 1:10 HCl aqueous solution. The resulting solid was dried in vacuum at 100 °C.

2.2 *Preparation of TiO_2*

In a typical synthesis, 5 ml of titanium (IV) isopropoxide (TTIP) was mixed dropwise in a vigorously stirred 6 ml isopropanol. The resulting mixture was stirred for 15 min and followed by the dropwise addition of 75 ml of dH_2O. The addition of dH_2O to the above solution resulted in a solution containing a white precipitate. Then, the temperature knob was adjusted so that the temperature of the solution becomes 80 °C. After stirring at 80 °C for 1 h 2.5 ml of acetic acid was added and continued to stir for 5 h. Then, 1 ml of nitric acid was added for peptidization and the resulting solution was further stirred for 2 h and cooled naturally to room temperature. The as-formed solution was oven dried at 60 °C. The dried powder was crushed in the mortar to make it fine powder, and then this powder was finally annealed at 500 °C for 3 h.

2.3 Synthesis of rGO-TiO_2 Photocatalyst

In a typical synthesis process, 500 mg of TiO_2 np powder was dispersed in absolute ethanol solution by vigorous stirring for 1 h exposed under UV radiation to store the electrons inside the TiO_2 np. Then, GO with different mass ratios (5, 10, 15, and 20 wt%, respectively) was added to the TiO_2 np solution. The GO-TiO_2 suspension was stirred under UV exposure for another 2 h in order to reduce GO to rGO. The resulting dispersions were subjected to hydrothermal treatment in an autoclave for 2 h. During the hydrothermal process, the partial reduction of GO to reduced graphene oxide (rGO) and the deposition of rGO onto the TiO_2 were simultaneously achieved. Finally, the resulted nanocomposites were recovered by centrifugation, followed by washing and drying in oven 100 °C to get the 5-rGO/T, 10-rGO/T, 15-rGO/T, and 20-rGO/T nanocomposites.

2.4 Photocatalyst Characterization

X-Ray Powder Diffraction analysis was performed at room temperature with X'Pert Pro Diffractometer equipped with X'Celerator solid-state detector using Cu-K α radiation ($\lambda = 1.54060$ Å) to observe the crystalline structure and phase composition of the photocatalysts synthesized. The surface morphology and size of the samples were investigated by Field-Emission Scanning Electron Microscope (Carl Zeiss Ultra Plus) coupled with Energy dispersive X-ray spectroscopy (EDXs).

2.5 Photocatalytic Activity Test

The photocatalytic activities of synthesized photocatalysts were evaluated by photodegradation of RhB. The reaction was carried out in batch mode at the initial RhB concentration of 10 ppm under UV and solar light, irrespective of pH of the dye solution. In all the experiments, the volume of dye solution used was 100 ml. Before the irradiation, the suspensions were magnetically stirred in dark for 30 min to reach the adsorption–desorption equilibrium of RhB dye. The RhB concentration after various time intervals were estimated using the following equation:

$$\%\,\text{RhB Degradation} = \frac{C_o - C_t}{C_o} \times 100\%$$

where C_o is the initial RhB concentration and C_t is the concentration after various time intervals (t).

3 Results and Discussion

3.1 FE-SEM

The field-emission scanning electron micrographs of the samples synthesized are depicted in Fig. 1. FE-SEM was done to inspect the topography of the samples synthesized. It is observed from Fig. 1a that GO has a layered structure with cascades of folds, which is due to the introduction of oxidizing functional groups between interlamination of graphite. The different degrees of wrinkles are also observed over the sheets of graphene oxide due to the absence of functional groups. Figure 1b shows agglomerated particles of TiO_2 having spherical morphology and rough surfaces. Figure 1c–f shows the micrographs of 5-rGO/T, 10-rGO/T, 15-rGO/T, and 20-rGO/T, respectively. In Fig. 1c–f, the images of graphene oxide-modified TiO_2 does not show a clear-cut presence of the rGO flakes, however, it shows the uniform distribution of TiO_2 nanoparticles. The uniform distribution of TiO_2 nanoparticles is due to very small amount of rGO with respect to TiO_2 and intercalation of rGO inside the TiO_2 matrix. The FE-SEM investigation reveals that the crystallites are of nanometer size ranging from 36 to 56 nm.

3.2 EDXs

Figure 2 shows EDXs spectra of the prepared rGO/TiO_2 photocatalysts with varying concentration of rGO and GO sample. From the EDXs data, the existence of main elements such as C, O, and Ti has been observed. There is no impurity peak observed in the EDXs spectra. This confirms that the prepared samples are in pure form. The elemental composition of the GO, TiO_2, 5-rGO/T, 10-rGO/T, 15-rGO/T, and 20-rGO/T are depicted in Table 1.

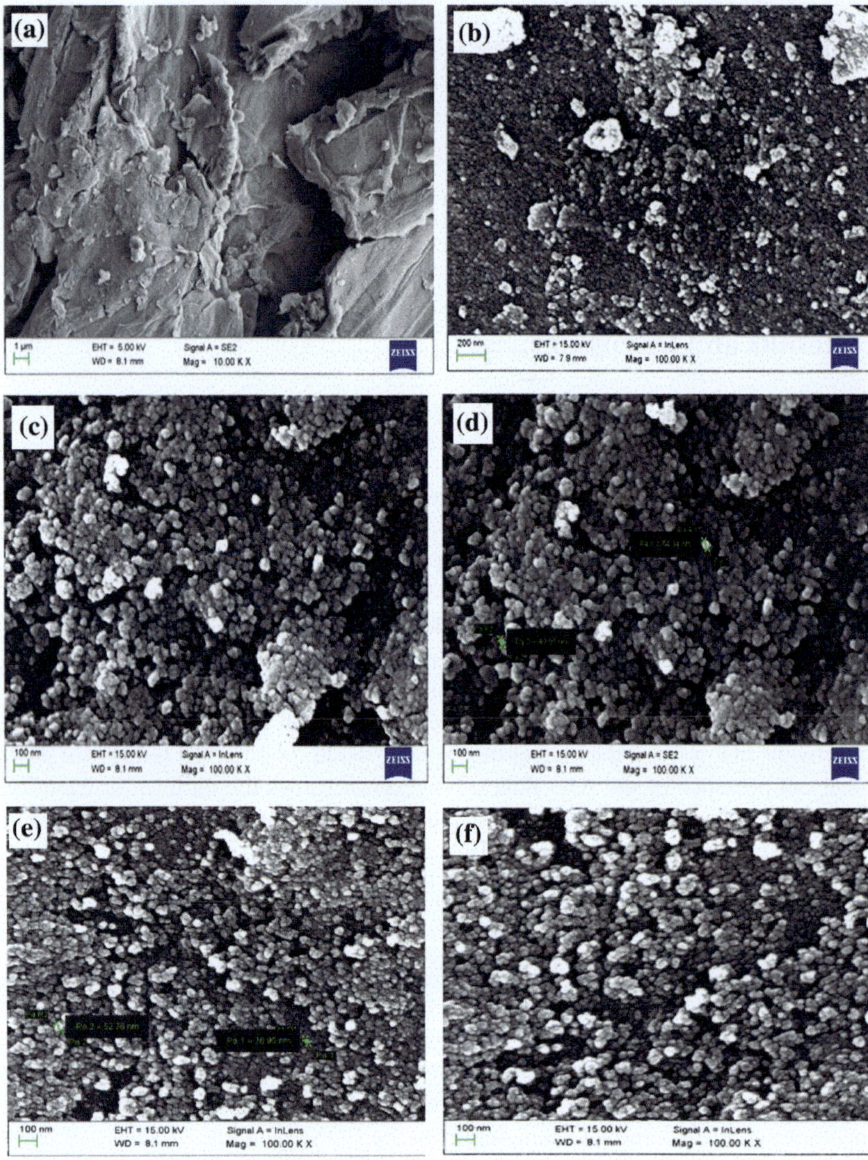

Fig. 1 FE-SEM images of different nanocomposite photocatalysts **a** GO, **b** TiO_2 (synthesized), **c** 5-rGO/T, **d** 10-rGO/T, **e** 15-rGO/T, **f** 20-rGO/T

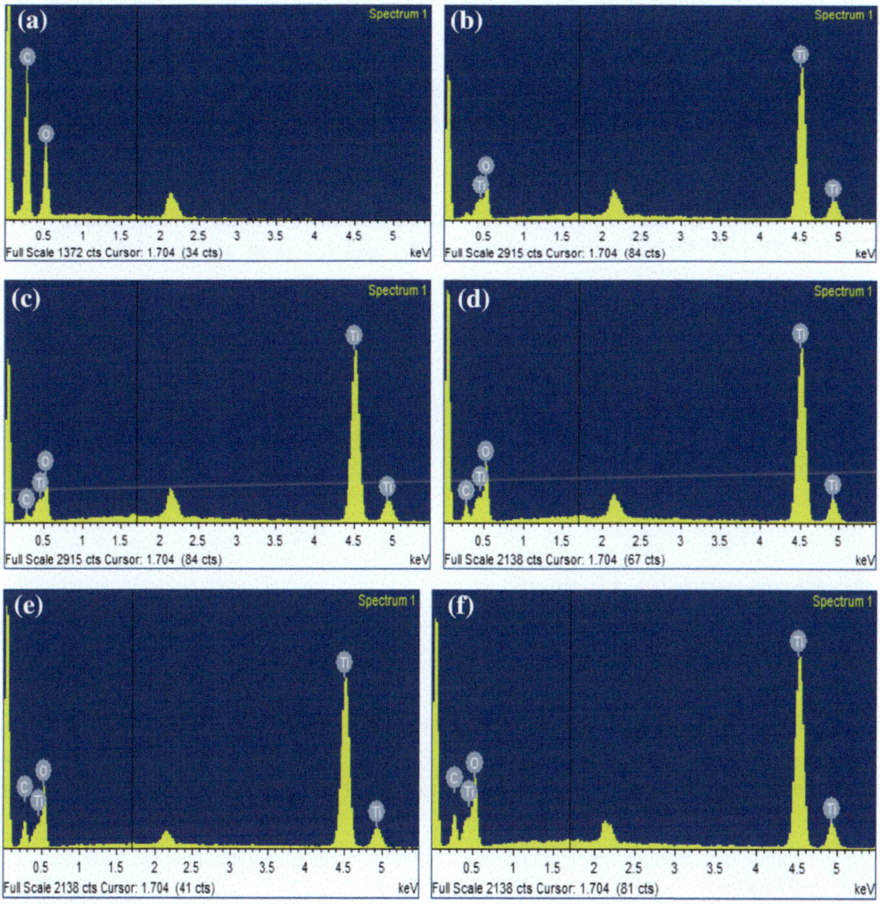

Fig. 2 EDXs elemental microanalysis of **a** GO, **b** TiO_2, **c** 5-rGO/T, **d** 10-rGO/T, **e** 15-rGO/T and **f** 20-rGO/T

Table 1 Surface composition from EDXs elemental microanalysis

Sample		Element wt%			
		C	O	Ti	Total
(a)	GO	60.28	39.72	–	100
(b)	TiO_2	–	34.89	65.11	100
(c)	5-rGO/T	2.33	33.92	63.75	100
(d)	10-rGO/T	5.37	37.17	57.46	100
(e)	15-rGO/T	8.06	38.54	53.40	100
(f)	20-rGO/T	10.07	39.27	50.66	100

3.3 XRD Analysis

The XRD pattern of sample photocatalysts is shown in Fig. 3. The broad peaks indicate either particles of very small crystalline size, or the semi-crystalline nature. A sharp diffraction peak of GO appears at $2\theta = 11.14°$ with an interlayer spacing of 7.93 Å. The diffraction pattern of TiO_2 shows major peaks at $2\theta = 25.58$, 37.20, 48.19, 54.13, 55.36, 62.89, 69.01, 70.70, 75.35 and corresponds to the planes (101), (004), (200), (105), (211), (204), (116), (220) and (215) of anatase TiO_2, respectively (Li et al. 2014a, b). It can be observed that the intensity of the peaks attributed to rutile phase that is very small in TiO_2. The effect of rGO in TiO_2 nanocomposite found negligible and is nearly same diffraction pattern appears in all 5-rGO/T, 10-rGO/T, 15-rGO/T and 20-rGO/T photocatalysts. Since the existence of rGO in TiO_2 nanocomposite did not affect the crystalline phase of TiO_2 as shown in the XRD patterns, such interaction could be responsible for the enhanced photocatalytic performance of rGO/T.

3.4 Photocatalytic Analysis

3.4.1 Effect of Light Source

Experiments were performed under UV light source (15 W UV tube) as well as in solar light without any photocatalyst. Initially, RhB solution (10 ppm, 100 ml) was poured in a Petri dish (diameter 190 mm) and kept under illumination of UV light

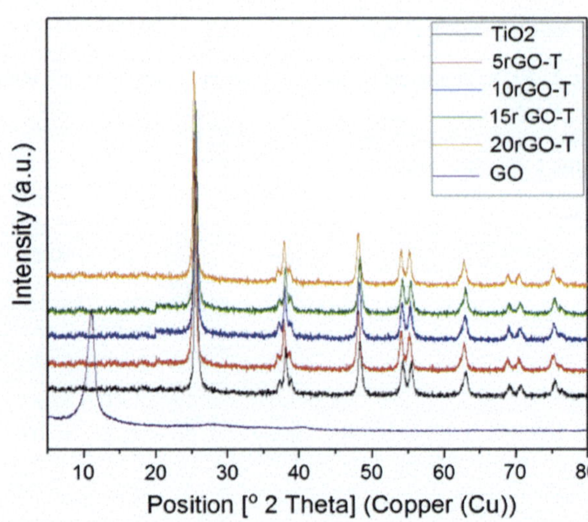

Fig. 3 XRD patterns of GO, TiO_2, 5-rGO/T, 10-rGO/T and 20-rGO/T

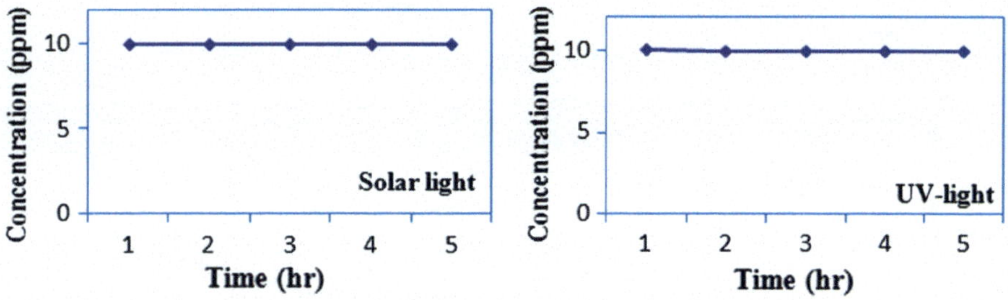

Fig. 4 Effect of light source on dye solution without any catalyst

as well as solar light. It was observed that there was no change in the concentration of RhB dye both under UV light as well as solar light (Fig. 4).

3.4.2 Catalyst Doses

The experiments were conducted to study the effect of catalyst dose for the removal of RhB dye. The best suitable dose of catalyst was determined on the basis of degradation efficiency of the unmodified TiO_2 photocatalyst. In these experiments, three different doses (25, 50, and 100 mg) of TiO_2 were taken for the same volume of dye solution (100 ml) of 10 ppm concentration. The degradation of dye was observed under both UV light as well as solar light. It can be seen from Fig. 5 that percentage degradation is high for the high value of catalyst concentration for the same time interval (3 h). The percentage degradation under UV light was obtained as 84.4%, 92.4%, and 99% for the amount of catalyst 25 mg, 50 mg, and 100 mg, respectively; whereas, the percentage degradation under solar light was 29%, 34%, and 50% for the amount of TiO_2 25 mg, 50 mg, and 100 mg, respectively, which is very less. Among all three doses, 50 mg dose was selected as the best suitable dose as there is no far difference in the degradation efficiency of 50 and 100 mg dose. As we know irradiation time is also an important parameter; by seeing the above result, it can be inferred that prolonged irradiation may lead to 100% degradation of dye.

3.4.3 Effect of Type of Catalyst Under UV Light

The photocatalytic activity of all four types of photocatalysts with variable ratio of rGO to TiO_2 synthesized was tested to determine the optimal catalyst for RhB degradation. The catalysts 5-rGO/T, 10-rGO/T, 15-rGO/T, and 20-rGO/T with rGO content 5, 10, 15, and 20 wt%, respectively, were set under examination keeping other parameters constant. 50 mg of each sample was taken in 100 ml of the dye solution. The results are depicted in Fig. 6. The % degradation in 2.5 h for 5-rGO/T, 10-rGO/T, 15-rGO/T, and 20-rGO/T was found to be 97%, 95%, 97%, and 100%, respectively.

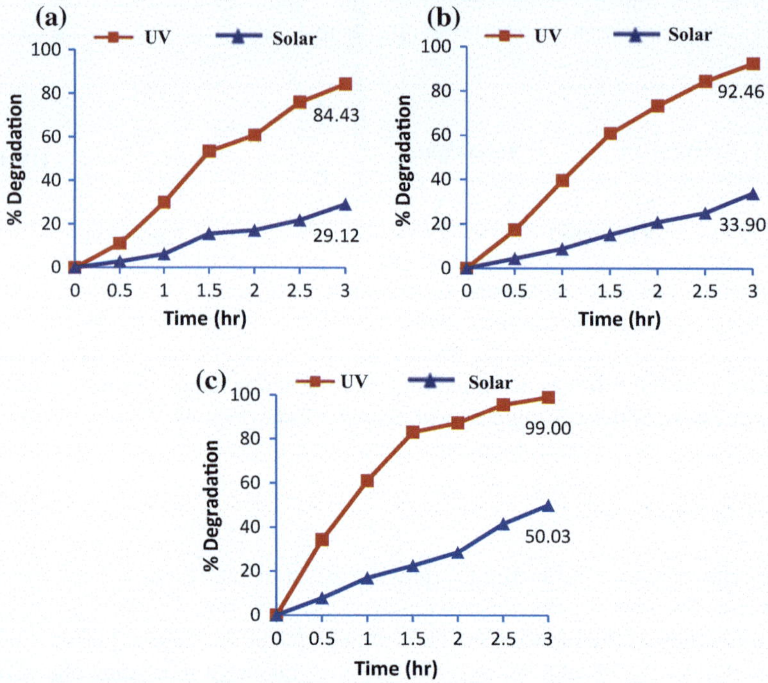

Fig. 5 Effect of catalyst dose on photocatalytic degradation of RhB (TiO_2 dose per 100 ml dye solution: **a** 25 mg, **b** 50 mg, **c** 100 mg)

The activity of TiO_2 has increased by 12–18% after modifying it with GO. The % degradation is nearly the same for 5-rGO/T and 15-rGO/T; however, there is slightly decrease in % degradation for 10-rGO/T may be due to the agglomeration of the nanocomposite. The maximum was observed for 20-rGO/T nanocomposite, which might be due to the adsorption of RhB dye as the catalyst has large surface area support. From the above results, it can be concluded that TiO_2 loaded with the lowest concentration of rGO (i.e., 5-rGO/T) is better among all four photocatalysts because increase in the concentration of rGO is not significantly changing the % degradation.

3.4.4 Effect of Type of Catalyst Under Solar Light

The activities of all four types of photocatalysts synthesized (5-rGO/T, 10-rGO/T, 15-rGO/T, and 20-rGO/T) were investigated under solar light keeping all the parameters the same as in under UV light. The results are shown in Fig. 7. It has been observed that the trend of degradation of RhB dye under solar light is the same as in UV light. The % degradation of RhB dye using 5-rGO/T, 10-rGO/T, 15-rGO/T and 20-rGO/T photocatalysts are 70%, 64%, 77% and 81%, respectively, whereas that of

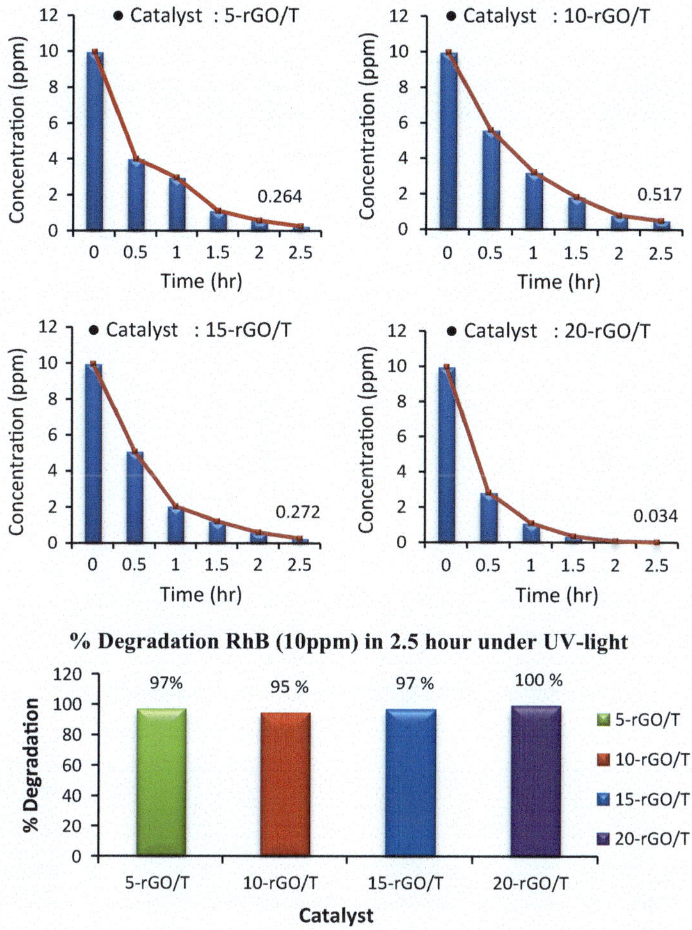

Fig. 6 Effect of type of catalyst on photocatalytic degradation of RhB (catalyst dose 50 mg/100 ml of dye solution) under UV light

bare TiO_2 is only 25%. The tremendous increase in the photocatalytic performance of rGO/TiO_2 nanocomposite can be attributed to increase in light absorption capacity and the reduction of electron-hole pair recombination in TiO_2. The characteristic feature of GO as good electron acceptor provides a flexible platform to electrons in the conduction band of TiO_2, where electrons are transferred over the surface of GO and thereby providing more time for photocatalytic reaction.

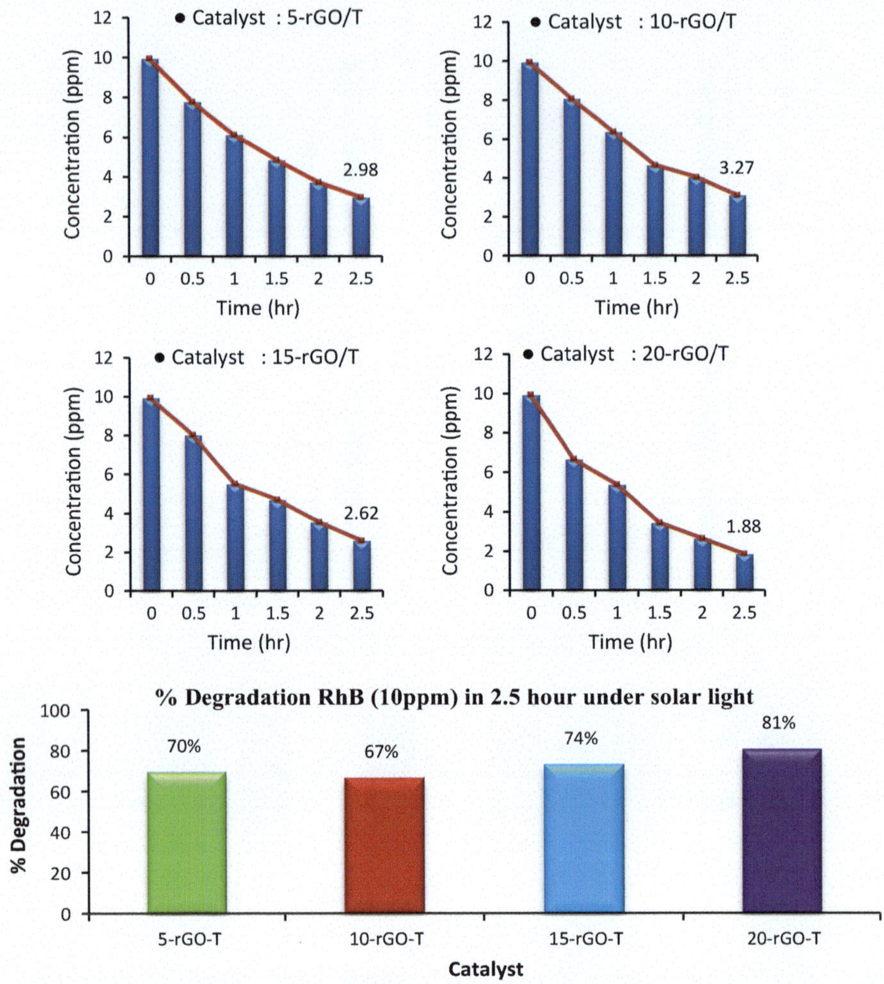

Fig. 7 Effect of type of catalyst on photocatalytic degradation of RhB (catalyst dose 50 mg/100 ml of dye solution) under solar light

3.5 Kinetic Study of Degradation of RhB Dye

The selection of a reaction system that operates in the safest and most efficient manner can be the key to the economic success or failure of a chemical process, so the kinetics study for any reaction becomes more important and necessary factor to study. The rate of reaction depends upon the number of parameters; the most important is the nature of the species involved in the reaction, concentrations of reactants and weight

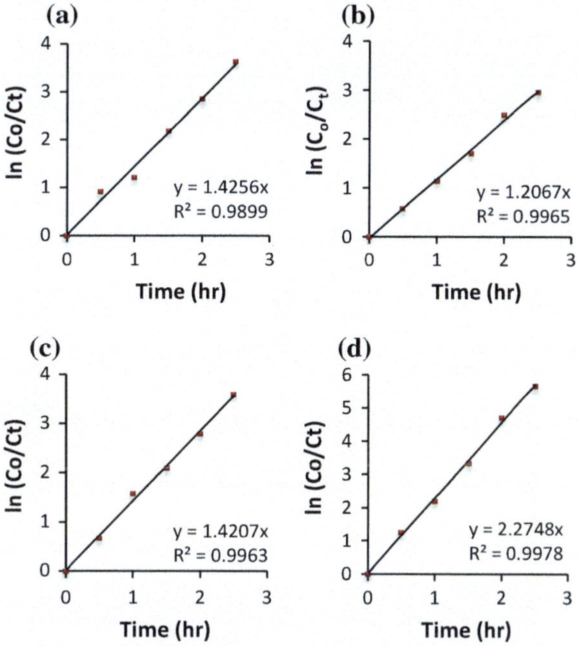

Fig. 8 Kinetic degradation profile for **a** 5-rGO/T, **b** 10-rGO/T, **c** 15-rGO/T and **d** 20-rGO/T photocatalyst under UV light

of catalyst loading in the reacting system. The plot of ln (C/C_o) versus time data for all the four types of photocatalysts is plotted and shown in Figs. 8a–d and 9a–d for UV light and solar light irradiation, respectively. The straight line with $R^2 = 0.99$ shows the best fit and indicates that the photocatalytic degradation followed the pseudo-first-order kinetics according to L-H (Langmuir-Hinshelwood) law (Li et al. 2014a, b).

Table 2 shows the pseudo-first-order reaction rate constant for all the samples in UV light and solar light. It also shows the half-life period for the degradation of RhB for all the photocatalysts and their correlation coefficient. The unmodified TiO_2 has the maximum half-life time ($t_{1/2}$) 0.923 h and 5.638 h under UV light and solar light respectively, which is almost 2–4 times higher than that of 5-rGO/T. After analyzing the above results, it has been found that 5-rGO/T photocatalyst showed the best fit for the pseudo-first-order reaction kinetics.

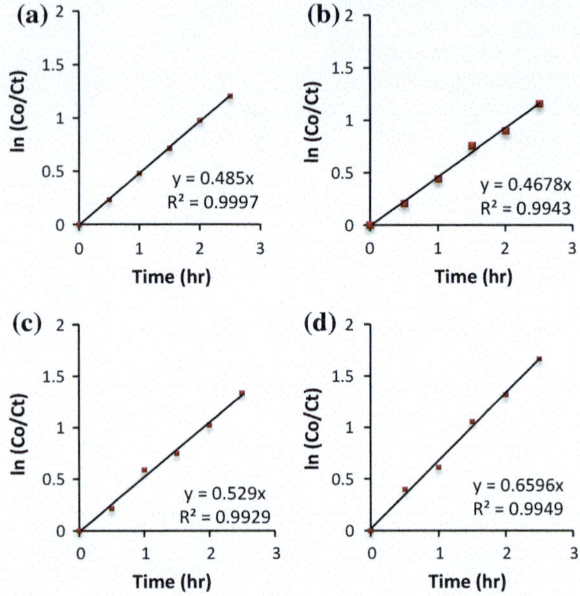

Fig. 9 Kinetic degradation profile for **a** 5-rGO/T, **b** 10-rGO/T, **c** 15-rGO/T and **d** 20-rGO/T photocatalyst under solar light

Table 2 Kinetic degradation coefficients for different photocatalysts

Type of sample	Under UV light			Under solar light		
	Rate const. K (h^{-1})	Half-life $t_{1/2}$ (h)	Correlation coefficient (R^2)	Rate const. K (h^{-1})	Half-life $t_{1/2}$ (h)	Correlation coefficient (R^2)
Pure TiO_2	0.751	0.923	0.988	0.123	5.638	0.9682
5-rGO/T	1.426	0.486	0.989	0.485	1.429	0.999
10-rGO/T	1.206	0.574	0.996	0.467	1.484	0.994
15-rGO/T	1.420	0.488	0.996	0.529	1.310	0.992
20-rGO/T	2.274	0.304	0.998	0.659	1.051	0.994

4 Conclusions

In this work, TiO_2 is selected as a base material for the degradation of RhB dye and was modified to enhance its photocatalytic activity. TiO_2 np was synthesized by sol–gel method. GO was prepared by the oxidation of natural graphite powder according to the modified Hummers method. The synthesis of rGO/T photocatalyst was successfully done by the photo-assisted reduction of GO using ultraviolet radiation followed by hydrothermal deposition of rGO onto the TiO_2. The synthesized photocatalysts were characterized by different physical techniques, such as X-ray diffraction, FE-SEM, and EDXs. FE-SEM analysis reflected the morphology of GO as layered structure and that of TiO_2 and rGO/T as spherical structure. The XRD spectra confirmed crystalline nature of TiO_2 and rGO/T, with anatase being the main crystalline phase. The introduction of GO over the TiO_2 np reflected no change in the crystalline phase of GO modified TiO_2 nanocomposites. The results of photocatalytic degradation of RhB dye showed an improvement in the degradation efficiency of rGO/T photocatalysts by almost two–three times that of pure TiO_2. From the experimental data, it has been concluded that 5-rGO/T having 5% GO content is best among all the photocatalysts synthesized. The rate of photodegradation of RhB was successfully fitted by Langmuir–Hinshelwood kinetic model. RhB dye followed the pseudo-first-order reaction kinetics for photodegradation. This study explored the feasibility of using sunlight as the main light source for photocatalytic degradation of organic dye.

References

Abdel-Messih M, Ahmed M, El-Sayed A (2013) Photocatalytic decolorization of Rhodamine B dye using novel mesoporous SnO_2–TiO_2 nano mixed oxides prepared by sol–gel method. J Photochem Photobiol A 260:1–8. https://doi.org/10.1016/j.jphotochem.2013.03.011

Carneiro J, Samantilleke A, Parpot P et al (2016) Visible light induced enhanced photocatalytic degradation of industrial effluents (Rhodamine B) in aqueous media using TiO_2 nanoparticles. J Nanomater 2016:1–13. https://doi.org/10.1155/2016/4396175

Chang W, Shen Y, Xie A, Tong W (2009) Preparation of Al_2O_3-supported nano-Cu_2O catalysts for the oxidative treatment of industrial wastewater. Russ J Phys Chem A 83:2308–2312. https://doi.org/10.1134/s0036024409130214

Chen X, Mao S (2007) Titanium dioxide nanomaterials: synthesis, properties, modifications, and applications. ChemInform. https://doi.org/10.1002/chin.200741216

Dubey P, Tripathi P, Tiwari R et al (2014) Synthesis of reduced graphene oxide–TiO_2 nanoparticle composite systems and its application in hydrogen production. Int J Hydrogen Energy 39:16282–16292. https://doi.org/10.1016/j.ijhydene.2014.03.104

Fang W, Xing M, Zhang J (2017) Modifications on reduced titanium dioxide photocatalysts: a review. J Photochem Photobiol C 32:21–39. https://doi.org/10.1016/j.jphotochemrev.2017.05.003

Friedmann D, Mendive C, Bahnemann D (2010) TiO2 for water treatment: parameters affecting the kinetics and mechanisms of photocatalysis. Appl Catal B 99:398–406. https://doi.org/10.1016/j.apcatb.2010.05.014

Gnanasekaran L, Hemamalini R, Ravichandran K (2015) Synthesis and characterization of TiO_2 quantum dots for photocatalytic application. J Saudi Chem Soc 19:589–594. https://doi.org/10.1016/j.jscs.2015.05.002

Hamandi M, Berhault G, Guillard C, Kochkar H (2017) Reduced graphene oxide/TiO_2 nanotube composites for formic acid photodegradation. Appl Catal B 209:203–213. https://doi.org/10.1016/j.apcatb.2017.02.062

Krejčíková S, Matějová L, Kočí K et al (2012) Preparation and characterization of Ag-doped crystalline titania for photocatalysis applications. Appl Catal B 111–112:119–125. https://doi.org/10.1016/j.apcatb.2011.09.024

Kumar K, Kumar G, Reddy K (2015) Rapid microwave synthesis of reduced graphene oxide-supported TiO_2 nanostructures as high performance photocatalyst. Mater Today Proc 2:3736–3742. https://doi.org/10.1016/j.matpr.2015.07.204

Li D, Zheng H, Wang Q et al (2014a) A novel double-cylindrical-shell photoreactor immobilized with monolayer TiO_2-coated silica gel beads for photocatalytic degradation of Rhodamine B and

Methyl Orange in aqueous solution. Sep Purif Technol 123:130–138. https://doi.org/10.1016/j.seppur.2013.12.029

Li Y, Yan J, Su Q et al (2014b) Preparation of Graphene–TiO_2 nanotubes/nanofibers composites as an enhanced visible light photocatalyst using a hybrid synthetic strategy. Mater Sci Semicond Process 27:695–701. https://doi.org/10.1016/j.mssp.2014.08.001

Lin W, Liao C, Chang T et al (2015) Humidity sensing properties of novel graphene/TiO_2 composites by sol–gel process. Sens Actuators B Chem 209:555–561. https://doi.org/10.1016/j.snb.2014.12.013

Lorestani F, Shahnavaz Z, Mn P et al (2015) One-step hydrothermal green synthesis of silver nanoparticle-carbon nanotube reduced-graphene oxide composite and its application as hydrogen peroxide sensor. Sens Actuators B Chem 208:389–398. https://doi.org/10.1016/j.snb.2014.11.074

Lv X, Zhang G, Fu W (2012) Highly efficient hydrogen evolution using TiO_2/graphene composite photocatalysts. Procedia Eng 27:570–576. https://doi.org/10.1016/j.proeng.2011.12.489

Nakata K, Fujishima A (2012) TiO_2 photocatalysis: design and applications. J Photochem Photobiol C 13:169–189. https://doi.org/10.1016/j.jphotochemrev.2012.06.001

Oseghe E, Ndungu P, Jonnalagadda S (2014) Synthesis of mesoporous Mn/TiO_2 nanocomposites and investigating the photocatalytic properties in aqueous systems. Environ Sci Pollut Res 22:211–222. https://doi.org/10.1007/s11356-014-3356-z

Rani S, Aggarwal M, Kumar M et al (2016) Removal of methylene blue and rhodamine B from water by zirconium oxide/graphene. Water Sci 30:51–60. https://doi.org/10.1016/j.wsj.2016.04.001

Shao X, Lu W, Zhang R, Pan F (2013) Enhanced photocatalytic activity of TiO_2-C hybrid aerogels for methylene blue degradation. Scientific Reports. https://doi.org/10.1038/srep03018

Shen K, Gondal M (2017) Removal of hazardous Rhodamine dye from water by adsorption onto exhausted coffee ground. J Saudi Chem Soc 21:S120–S127. https://doi.org/10.1016/j.jscs.2013.11.005

Štengl V, Bakardjieva S, Grygar T et al (2013) TiO_2-graphene oxide nanocomposite as advanced photocatalytic materials. Chem Cent J 7:41. https://doi.org/10.1186/1752-153x-7-41

Thiruvenkatachari R, Vigneswaran S, Moon I (2008) A review on UV/TiO_2 photocatalytic oxidation process (Journal Review). Korean J Chem Eng 25:64–72. https://doi.org/10.1007/s11814-008-0011-8

Yadav M, Yadav A, Fernandes R et al (2017) Tungsten-doped TiO_2/reduced Graphene Oxide nanocomposite photocatalyst for degradation of phenol: a system to reduce surface and bulk electron-hole recombination. J Environ Manag 203:364–374. https://doi.org/10.1016/j.jenvman.2017.08.010

Yu D, Cai R, Liu Z (2004) Studies on the photodegradation of Rhodamine dyes on nanometer-sized zinc oxide. Spectrochim Acta Part A Mol Biomol Spectrosc 60:1617–1624. https://doi.org/10.1016/j.saa.2003.09.00

Use of Geosynthetics for Sustainable Pavements

G. R. Harish, K. Madesh, N. Swathi, G. R. Vijayabhaskara Reddy and H. Deepak

Abstract Nowadays, the pavements are subjected to premature failure due to rapid growth in vehicular traffic. Among various modes of failure, rutting and fatigue are the predominant ones. In addition to this, there is scarcity for good quality materials. To address these problems, the engineers are forced to design new/alternate construction materials. Geosynthetic is one among those new materials. In this study, model pavement sections were built in the field under unreinforced and geosynthetic reinforced conditions. For reinforcement, uniaxial geogrid, geotextile, biaxial geogrid and geocell were considered. The reinforcement was provided at the interface of subgrade and granular sub-base course layers. The loading was done by passing a two wheeler over the pavement at an average speed of 20 kmph. Each model pavement section was subjected to 100 vehicular passes. During testing, the rut depth was measured at various locations along the length and width of the pavement at the end of 10 vehicular passes. The provision of geosynthetics was found to be effective in reducing rutting in pavements. Among various forms of geosynthetics, biaxial geogrids and geocell were found to be suitable for pavement application. With the provision of biaxial geogrid and geocell, the rut depth was found to be reduced by 28% and 56%, respectively. In addition, the provision of geosynthetics reduced the surface heaving with geocell being very effective and reduced surface heave results in improved riding quality of the pavements leading to sustainable pavements.

Keywords Pavement · Rutting · Fatigue · Geosynthetics

1 Introduction

India has the second largest road network in the world. Among various categories of road, low volume roads constitute about 80% and most of these roads are unpaved. Low volume road is the one, which is designed for lower design speeds and relatively

G. R. Harish (✉) · K. Madesh · N. Swathi · G. R. Vijayabhaskara Reddy · H. Deepak
Department of Civil Engineering, New Horizon College of Engineering Bangalore, Bengaluru, Karnataka, India
e-mail: gr.harish1@gmail.com

A. K. Agnihotri et al. (eds.), *Sustainable Engineering*, Lecture Notes in Civil Engineering 30, https://doi.org/10.1007/978-981-13-6717-5_27

has less usage. A well planned, designed and constructed road network is significant for the development of communities, efficient transportation of goods and providing services between communities. A poorly designed and constructed road network fails to meet the user needs and also involves high repair and maintenance costs. According to AASHTO, about 20% of the pavements fail due to insufficient structural strength. These days, good quality construction materials are depleting owing to the fact of rapid industrialization and urbanization and are not available everywhere. Under such circumstances, a sustainable option to overcome this problem is to develop an innovative pavement stabilization technique with a suitable reinforcement alternative, which can improve the overall strength of the structure, reduce operational cost and minimize maintenance requirements. Over five decades, geosynthetic reinforcement is being widely used for subgrade improvement and base reinforcement. Geosynthetics have been used in different civil engineering works viz., reinforced slopes, erosion control, retaining walls, subgrade improvement and reinforcement to base courses of roads. In the past, several researchers have studied the performance of geosynthetic reinforcement in the construction of roads through full-scale and laboratory model tests. The studies found that the provision of geosynthetics at the interface of subgrade and base course within a pavement significantly improves the pavement performance (Barksdale 1989; Al-Qadi et al. 1994, 2007; Perkins 1999; Hufenus et al. 2006; Mamatha and Dinesh 2017a, b, c). Most of the studies were carried out on reinforcing unpaved roads and found that the geosynthetic reinforcement is beneficial for pavements. The geosynthetic reinforcement results in reduced permanent deformation, and/or reduced aggregate thickness (Potter and Currer 1981; Lawson 1992; Webster and Watkins 1977; Giroud and Noiray 1981; Haliburton and Barron 1983; Love 1984; Austin and Coleman 1993; Al-Qadi et al. 1994; Fannin and Sigurdsson 1996; Hufenus et al. 2006; Madhavi Latha et al. 2010; Mamatha and Dinesh 2017a). The confinement offered by the geosynthetic (cellular form) reinforcement increases the modulus of the infill material resulting in an improved vertical stress distribution over the subgrade and consequent reduction in the subgrade deformation (Love 1984; Hass et al. 1988; Mamatha and Dinesh 2017b). Reduced shear stress transferred at the interface of subgrade and geosynthetic layer results in reduced permanent deformation.

The objective of the present study is to evaluate the performance of geosynthetic reinforced unpaved roads. Both planar and cellular form of reinforcement was considered. The relative advantages of different reinforcing materials placed at the interface of subgrade and sub-base course in terms of increased load carrying capacity and reduced rut depth were studied through a series of systematic field experiments.

2 Mechanism of Reinforcement

Early use of geosynthetics in roadways included separation, filtration and drainage in paved and unpaved roads and reinforcement in unpaved roads (Bender and Barenberg 1978). The two main benefits of the reinforcement are to improve the service life and/or to obtain equivalent performance with a reduced structural section. This improved performance of the pavement due to geosynthetics reinforcement has been attributed to three main mechanisms namely lateral restraint, increased bearing capacity and the tension membrane effect (Giroud and Noiray 1981; Perkins and Ismeik 1997a, b; Holtz 1998).

2.1 *Lateral Restraint*

In flexible pavements, the basic mechanism involved with reinforcement function is lateral restraint or confinement (Bender and Barenberg 1978) and the mechanism is depicted in Fig. 1. When the granular layer of the pavement is subjected to vehicular loading, the aggregate particles tend to displace in the lateral direction unless it is resisted by the subgrade and/or reinforcement. Interaction between the aggregate particle and the reinforcement material transfers the shear load from the granular layer to tensile load in the geosynthetic reinforcement. However, the lateral strain in the granular layer is limited by the stiffness of the geosynthetic reinforcement (Zornberg 2008).

2.2 *Increased Bearing Capacity*

The provision of geosynthetics increases the bearing capacity and this facilitates the development of an alternate failure surface with a higher bearing capacity. The mechanism involved in increased bearing capacity is depicted in Fig. 2. The geosynthetic reinforcement decreases the shear stress transferred to the subgrade and thereby provides vertical confinement outside the loaded area. With geosynthetic reinforcement, the expected mode of failure changes from punching failure to general shear failure.

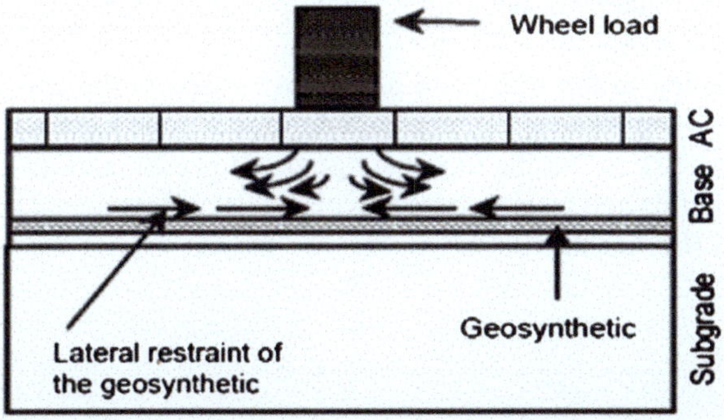

Fig. 1 Lateral restraint due to geosynthetics (Perkins and Ismeik 1997a, b)

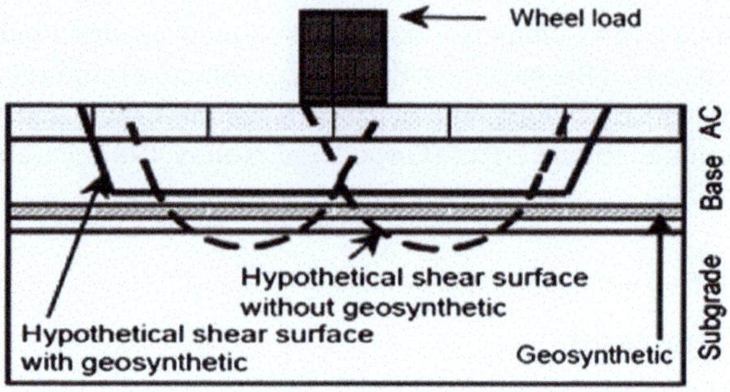

Fig. 2 Increased bearing capacity due to geosynthetics (Perkins and Ismeik 1997a, b)

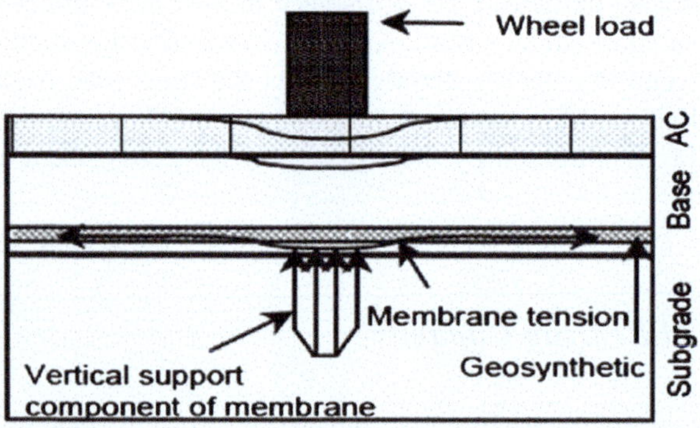

Fig. 3 Tension membrane effect due to geosynthetics (Perkins and Ismeik 1997a, b)

2.3 *Tension Membrane Effect*

The tension membrane effect involved with geosynthetic reinforcement is depicted in Fig. 3. In this mechanism, the reinforcement provides a vertical reaction component to the applied wheel load. The tension membrane effect is induced by vertical deformations leading to the formation of concave shape in the geosynthetic layer. The tension developed in the geosynthetic layer supports the wheel load and thereby reduces the vertical stress on the top of subgrade. However, significant rut depth is required to realize this and weak subgrade with a CBR of less than 3 is very effective in this viewpoint (Barksdale 1989).

3 Materials

The model pavement section was constructed in the campus of Vijaya Vittala Institute of Technology, Bangalore, Karnataka, India and the selected site was not subjected to any type of load in the past and therefore, it is ideal for simulating the condition of a newly constructed road.

3.1 Subgrade Soil

The natural soil available at the site was used as subgrade soil. The top 10 cm of the soil was excavated and tested for its engineering characteristics. All the tests were carried out as per relevant IS codes.

During testing, the in situ bulk unit weight of the soil was found to be 18.74 kN/m^3 and the natural water content was found to be 4.4%. The specific gravity of the soil is 2.65. The liquid limit and plastic limit of the soil was found to be 33% and 23%, respectively. The soil possesses a plasticity index of 10%. The grain size analysis revealed that the soil consists of 0% gravel, 74% sand, 21% silt and 5% clay. The soil is classified as A-2-4 as per HRB classification system. The maximum dry unit weight and optimum moisture content of the subgrade soil was found to be 15.7 kN/m^3 and 14%, respectively, under standard proctor condition. The CBR of the soil was found to be 7% and 4%, respectively, under unsoaked and soaked conditions.

3.2 Aggregates

Aggregates of different sizes i.e., 12.5 mm down, 40 mm down and stone dust were collected from a nearby quarry located in Byrathi Bande, Bangalore, Karnataka, India and is used for the preparation of sub-base course layer. Initially, the aggregates were tested for their engineering properties and the tests were conducted as per relevant IS codes. Table 1 shows the properties of aggregates along with MoRT&H specifications. Aggregates were found to satisfy the requirements specified by MoRT&H (2013). The selected fractions of aggregates were blended into a mix to obtain the sub-base course material confirming to grading II as per IRC:SP:72 (2007). Figure 4 shows the grain size distribution curve of the sub-base material with upper limit and lower limit for grading II as per IRC:SP:72 (2007). The blended mixture i.e., sub-base course material was found to possess a maximum dry unit weight of 23 kN/m^3 and an optimum moisture content of 4.5% under modified proctor condition. The sub-base material satisfies all the strength requirements of grading II as per IRC:SP:72 (2007).

Table 1 Properties of aggregates

Test	12.5 mm down	40 mm down	MORT&H specifications	Remarks
Specific gravity	2.6	2.26	2.5–3.2	Acceptable
Aggregate impact value (%)	14	15	Max 30%	Acceptable
Aggregate crushing (%)	21	20	Max 30%	Good
Water absorption (%)	0.4	0.4	Max 2%	Good
Los Angeles abrasion (%)	22	23	Max 40%	Acceptable
Combined indices (%)	27	24	Max 30%	Acceptable

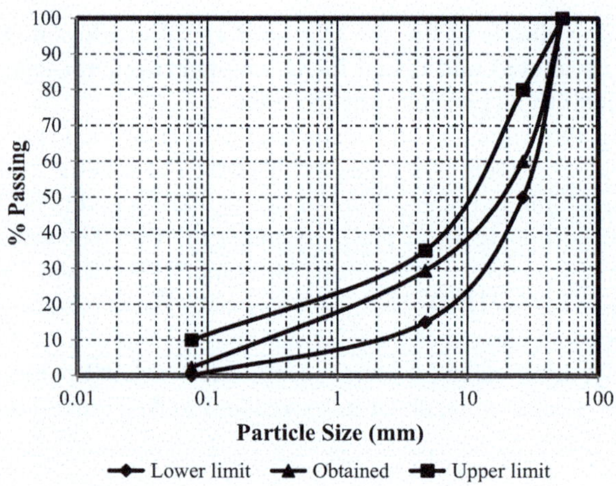

Fig. 4 Grain size distribution curve of sub-base course material

3.3 *Surface Layer*

For better ridability of vehicle, a levelled surface layer of 5 cm was prepared over prepared and levelled sub-base layer. The surface layer was prepared using the soil same as subgrade soil. In situ dry soil was used to prepare the surface layer. The soil was mixed with sufficient quantity of water to avoid dust and raveling when a vehicle passes over it.

3.4 *Reinforcing Materials*

Four different forms of geosynthetics namely non-woven geotextile, uniaxial geogrid, biaxial geogrid and geocell were considered for investigation.

A non-woven geotextile which is white in colour and has a thickness of 200 gsm was used.

The uniaxial geogrid and biaxial geogrid used in this work are made from high-density polyethylene. Uniaxial geogrid has rectangular openings of size 60 mm × 27 mm and has a tensile strength of 150 kN/m in longitudinal direction and 30 kN/m in transverse direction. Biaxial geogrid having square openings of size 18 mm × 18 mm and has a tensile strength of 30 kN/m in both longitudinal and transverse directions.

The geocell used in this work is made from high-density polyethylene with a cell depth of 100 mm. It had a weld spacing of 356 mm. The geocell possess minimum cell seam strength of 1420 N. The perforated cell wall consisted holes of 10 mm diameter.

4 Construction of Model Pavements

The model pavement section of size 1 m × 2 m was constructed in stages. The first stage involved the preparation of subgrade. The existing ground surface was excavated up to a depth of 10 cm and was compacted to the standard proctor condition. Over the prepared soil subgrade, a sub-base course layer of 15 cm was compacted to modified proctor condition. Over the prepared sub-base course layer, a surface layer of 5 cm was prepared to standard proctor condition using in situ soil. In the geosynthetic reinforced pavement sections, a geosynthetic layer was placed at the top of prepared soil subgrade and filled with sub-base course material. The compaction was done manually with a rammer. At the end of compaction of each layer, the unit weight and water content were examined at a minimum of three locations along the length of the model pavement section. A relative compaction of 98% was maintained throughout for all the layers as per MoRT&H (2013) guidelines.

5 Test Procedure

A two wheeler weighing 110 kg was driven by a person weighing 75 kg along the centre line of the prepared pavement section. The vehicle was passed in only one direction and the speed of the vehicle was maintained at an average of 20 kmph. When once the vehicle is passed over the prepared pavement section, the pavement gets deformed leading to the formation of ruts. Rut depth is the maximum measured perpendicular distance between the bottom surface of the straight edge and the contact area of the gauge with the pavement surface at a specific location (IRC:115 2014). The pavement section was divided into three sections in the longitudinal direction (section 1, section 2 and section 3) and five sections in the transverse direction. Arrangements were made to measure the rut depth at all the above-mentioned grid points. The rut depth was measured at the end of 10 passes and the test was stopped at a total of 100 passes.

6 Results and Discussions

Five field tests were conducted on unreinforced and geosynthetic reinforced test sections. The test results are presented and discussed in the following paragraphs. As discussed earlier, the pavement section was divided into three sections in the longitudinal direction and is designated as section 1, section 2 and section 3. Section 1 corresponds to the section at the entry, section 2 corresponds to centre and section 3 corresponds to the section at the exit with respect to the vehicular movement. During testing, the rut depth at various grid points was measured and recorded.

Figures 5, 6 and 7 shows the variation of rut depth with number of vehicular passes of unreinforced and geosynthetic reinforced pavement sections, respectively, at section 1, section 2 and section 3. At all the sections, the rut depth accumulation is linear with the number of passes in case unreinforced and uniaxial geogrid, geotextile and biaxial geogrid reinforced pavements. In case of geosynthetic reinforced sections, the magnitude of rut depth is reduced when compared with that of unreinforced pavement section. In case of geocell reinforced pavement section, the rut depth accumulation is significant up to 60 passes and beyond 60 passes, it stabilizes. At section 1, the accumulated rut depth was found to be 45 mm, 35 mm, 33 mm, 28 mm and 15 mm, respectively, in case of unreinforced, uniaxial geogrid, geotextile, biaxial geogrid and geocell reinforced pavements at end of 100 passes. At section 2, the accumulated rut depth was found to be 36 mm, 29 mm, 27 mm, 26 mm and 16 mm, respectively, in case of unreinforced, uniaxial geogrid, geotextile, biaxial geogrid and geocell reinforced pavements at the end of 100 passes. At section 3, the accumulated rut depth was found to be 34 mm, 27 mm, 25 mm, 24 mm and 14 mm respectively in case of unreinforced, uniaxial geogrid, geotextile, biaxial geogrid and geocell reinforced pavements at the end of 100 passes.

Figure 8 shows the surface profiles of unreinforced and geosynthetic reinforced pavement sections at section 2. It is observed that the provision of geosynthetics reduces heaving of the surface at the edges. The efficiency of reduction in heaving increases in the order of uniaxial geogrid, geotextile, biaxial geogrid and geocell.

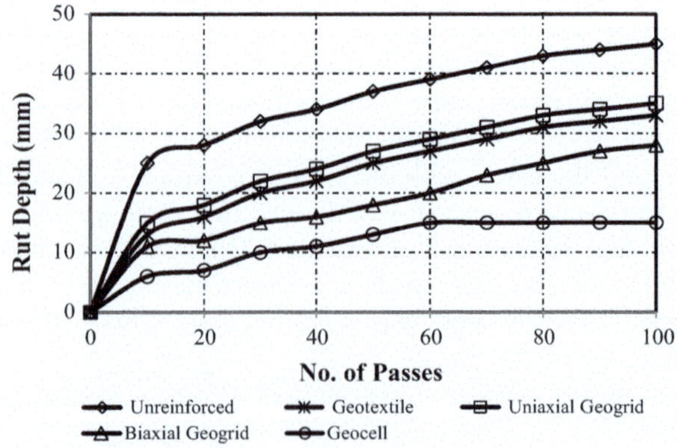

Fig. 5 Variation of rut depth with number of passes at section 1

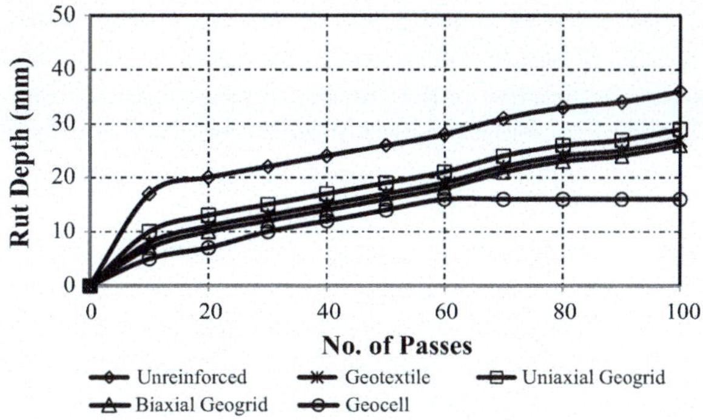

Fig. 6 Variation of rut depth with number of passes at section 2

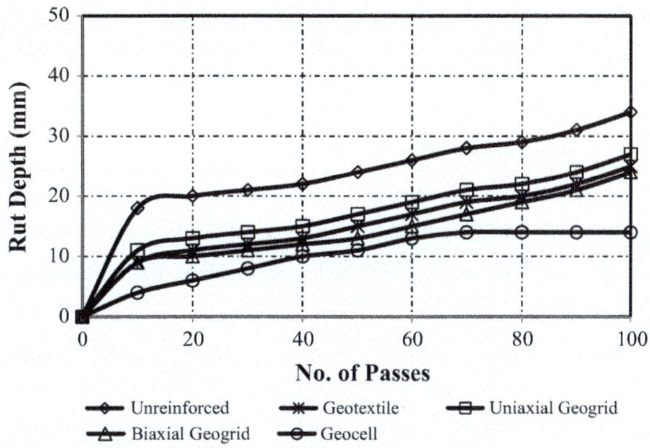

Fig. 7 Variation of rut depth with number of passes at section 3

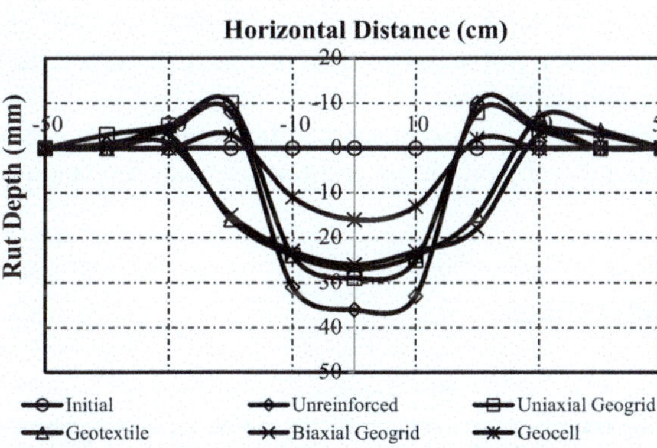

Fig. 8 Surface profiles of unreinforced and geosynthetic reinforced pavements (at section 2)

Geocells were found to be very effective in reducing heaving and thereby increases the riding quality of the pavements facilitating comfort to road users.

On comparing all the three sections, the accumulated rut depth is more at section 1 when compared to that at sections 2 and 3 irrespective of the test condition. The maximum rut depth at section 1 is attributed to the acceleration of the vehicle at the entry and at sections 2 and 3, the speed is uniform resulting in comparable rut depth. In case of unreinforced pavement section, the pavement exhibited maximum rut depth irrespective of the number of passes and sections. In case of geosynthetic reinforced pavements, the magnitude of rut depth was found to reduce and the reduction increases in the order of uniaxial geogrid, geotextile, biaxial geogrid and geocell irrespective of the number of passes and sections. The improvement in uniaxial geogrid and geotextile in attributed to partial separation and complete separation respectively. In case of biaxial geogrid, the improvement is attributed to particle interlocking and tension membrane mechanisms. In case of geocell, the improvement is attributed is confinement effect. Therefore, it is concluded that among various forms of geosynthetics, biaxial geogrid and geocells are feasible for designing sustainable pavements.

Among, biaxial geogrid and geocell, geocells are very effective and results in the improved level of service for the road users and sustainable pavements.

7 Conclusions

A total of five field tests were carried out to evaluate the influence of geosynthetics in as alternate materials for sustainable pavements. Based on the test results, the following results are drawn.

- Provision of geosynthetics at the interface of subgrade and granular layer reduces the rut depth and thereby improves the rut performance of the pavements.
- Among various forms of geosynthetics, biaxial geogrid and geocell are found to be suitable for pavement application.
- The rut depth was reduced in the order of 28% and 56%, respectively, with the provision of biaxial geogrid and geocell at the interface of subgrade and granular sub-base course.
- Provision of geosynthetics reduces the heaving resulting in increased ride quality and thereby providing comfort to road users leading to sustainable pavements.

References

Al-Qadi IL et al (1994) Laboratory evaluation of geosynthetic reinforced pavement sections. In: Transportation research record, 1439, pp 25–31. Transportation Research Board, National Research Council, Washington, DC

Al-Qadi IL et al (2007) Accelerated full scale testing of geogrid reinforced flexible pavements. In: 86th TRB meeting, Paper no. 07-2317. Transportation Research Board, National Research Council, Washington, DC

Austin DN, Coleman DM (1993) A field evaluation of geosynthetic reinforced haul roads over soft foundation soils. In: Proceedings of geosynthetics, industrial fabrics association international, pp 65–80. St. Paul, Minn

Barksdale et al (1989) Potential benefits of geosynthetics in flexible pavement systems. In: National co-operative highway research program report No. 315. Transportation Research Board, National Research Council, Washington, DC

Bender DA, Barenberg EJ (1978) Design and behaviour of soil–fabric–aggregate systems. In: Transportation research record, 671, Transportation Research Board, National Research Council, Washington, DC: 64–75

Fannin RJ, Sigurdsson O (1996) Field observations on stabilization of unpaved roads with geosynthetics. J Geotech Eng 122(7):544–553

Giroud JP, Noiray L (1981) Geotextile reinforced unpaved road design. Journal of Geotechnical Engineering 107(9):1233–1254

Haliburton TA, Barron JV (1983) Optimum method for design of fabric reinforced unsurfaced roads. In: Transportation research record, 916, pp 26–32. Transportation Research Board, National Research Council, Washington, DC

Hass R et al (1988) Geogrid reinforcement of granular bases in flexible pavements. In: Transportation research record, 1188, pp 19–27. Transportation Research Board, National Research Council, Washington, DC

Holtz RD et al (1998) Geosynthetic design and construction guidelines, Report no. FHWA-HI-98-038, Federal Highway Administration, Washington, DC

Hufenus R et al (2006) Full scale field tests on geosynthetic reinforced unpaved roads on soft subgrade. Geotext Geomembr 24(1):21–37

IRC-SP-72 (2007) Guidelines for the design of flexible pavements for low volume rural roads, The Indian Roads Congress, New Delhi

Lawson CR (1992) Some examples of separation geotextiles under road pavements. Proc Inst Civil Eng Transp 95(3):197–200

Love (1984) Model testing of geogrids in unpaved roads. In: Thesis submitted to University of Oxford, England

Madhavi Latha G et al (2010) Performance of geosynthetics in unpaved roads. Int J Geotech Eng 4:151–164

Mamatha KH, Dinesh SV (2017a) Performance evaluation of geocell reinforced pavements, submitted to Int J Geotech Eng. https://doi.org/10.1080/19386362.2017.1343988

Mamatha KH, Dinesh SV (2017b) Evaluation of strain modulus and deformation characteristics of geosynthetic reinforced soil—aggregate system under repetitive loading. Int J Geotech Eng. https://doi.org/10.1080/19386362.2017.1307309

Mamatha KH, Dinesh SV (2017c) Evaluation of flexural behaviour of geosynthetic reinforced unbound granular material beams. Road Mater Pavement Des. https://doi.org/10.1080/14680629.2017.1422790

MoRT&H (2013) Ministry of road transport and highways: specifications for roads and bridges, 5th edn. Indian Roads Congress, New Delhi

Perkins SW, Ismeik M (1997a) A synthesis and evaluation of geosynthetic-reinforced base course layers in flexible pavements: part I experimental work. Geosyn Int 4(6):549–604

Perkins SW, Ismeik M (1997b) A synthesis and evaluation of geosynthetic-reinforced base course layers in flexible pavements: part II analytical work. Geosyn Int 4(6):605–621

Perkins SW (1999) Mechanical response of geosynthetic reinforced flexible pavements. Geosyn Int 6(5):347–382

Potter JF, Currer EWH (1981) The effect of a fabric membrane on the structural behaviour of a granular road pavement. RRRL Report, Report no. LR996, Crowthorne, UK

Webster SL, Watkins JE (1977) Investigation of construction techniques for tactical bridge approach roads across soft ground. Report no. S-77-1, USAE Wateways Experiment Station

Zornberg JG et al (2008) Validating mechanisms in geosynthetic reinforced pavements. Report no. FHWA/TX-08/0-4829-1, Centre for Transportation Research, University of Texas, Austin

A Study on the Influence of Confining Pressure on the Behavior of Fiber-Reinforced Soil

Akash Priyadarshee, Arvind Kumar, Vaibhav Sharma and Vikas Kumar

Abstract The use of randomly distributed fiber as soil reinforcement is nowadays a popular ground improvement technique. The absence of weak plane is a major advantage of the fiber reinforcement over the other soil reinforcement techniques, where weak plane exit parallel to the reinforcement. Similar to the roots of plant, fiber binds soil particles through surface interaction between soil and fiber, which results in the improvement of the load carrying capacity of soil. In this study, a series of triaxial test were carried out to understand the effect of confining pressure on the strength and stiffness characteristics of the fiber-reinforced soil. The test results have shown that with increase in the confining pressure, peak stress and stiffness increases. But the performance of fiber reinforcement in strength improvement is more significant in case of lower confining pressure, because the reinforcing action of the fibers was suppressed by the increased confining pressure.

Keywords Confining pressure · Stiffness · Principal stress ratio · Strain softening · Strain hardening · Reinforced soil

1 Introduction

Due to ease of construction, the overall economy and simplicity of construction, soil reinforcement is most popular ground improvement technique. The availability of different materials and techniques as options is also a major advantage of the soil reinforcement. The use of reinforcements in the form of rod, sheets, membrane, etc., is traditional way, while randomly distributed fiber is one of the new technique.

Fiber has been used for a long time, but the utilization of these materials in construction is not frequent because of less understanding of potential of the fiber reinforcement. In fiber reinforcement technique, fiber is randomly mixed in the soil due to which isotropy can be achieved. The absence of weak plane in fiber–soil mixture is a major advantage of fiber reinforcement (Maher and Gray 1990). Gen-

A. Priyadarshee (✉) · A. Kumar · V. Sharma · V. Kumar
MIT, Muzaffarpur, India
e-mail: i.akashpriyadarshee@gmail.com

A. K. Agnihotri et al. (eds.), *Sustainable Engineering*, Lecture Notes in Civil Engineering 30, https://doi.org/10.1007/978-981-13-6717-5_28

erally, such plane exists parallel to the soil reinforcement interface in case of plane reinforcement.

Mechanism of the fiber-reinforced soil is based upon the interaction between soil particles and fiber. The resistance provided by the fibers due to surface friction and interlock leads to the load carrying capacity of the soil (Tang et al. 2007).

The behavior of fiber reinforcement is investigated by different case studies, model tests, and other laboratories tests. Different researchers through case studies like on embankment by Gregory and Chill (1998), on landfill liners by Refai (2000), on roads by Lindh and Erriksson (1990) and Santoni et al. (2001), etc. have shown the beneficial utilization of fiber reinforcement in different application. Gregory (2006) has suggested that fiber can be used effectively as secondary reinforcement in soil, with planar geogrid as main reinforcement. Through model test on strip footing supported by fiber-reinforced soil, improvement in the bearing capacity, and stiffness have reported by Al-Refai (1992) and Wasti and Butun (1996).

Stress–strain responses of the fiber-reinforced soil have been studied by different researchers. Gray and Ohashi (1983) have performed direct shear test on oriented fiber and found that strength improvement due to fiber-reinforced soil is proportional to the area ratio of fiber, stiffness of fibe,r and orientation of fiber. Gray and Al-Refai (1986), Maher and Gray (1990), Al-Refai (1991), Michalowski and Zhao (1996), Ranjan et al. (1994), Michalowski and Cermak (2003) through triaxial test have shown that performance of fiber reinforcement depends upon the fiber properties like length of the fiber, fiber content, stiffness of fiber and surface properties of soil and soil properties like soil type, size of particles, etc. Babu and Vasudevan (2008) have reported that increase in the diameter of fiber also improves the strength of fiber-reinforced soil. Ranjan et al. (1994), Santoni et al. (2001) have shown through the study that with increase in the length of fiber and fiber content strength of the fiber-reinforced soil increases up to a certain limit, but further increase in length and content of fiber does not contribute in strength improvement.

In the present study effect of confining pressure on the different properties of soil like stress–strain properties, strength properties, stiffness properties, and deformation properties have been studied through a series of triaxial test, the details of which are presented below.

2 Methodology

2.1 Test Materials

The soil used in this study was locally available sand. The properties of the soil are presented in Table 1. According to the Unified Soil Classification System (USCS), the soil is classified as poorly graded (SP).

Table 1 Properties of the soil

Properties	Value
Specific gravity	2.68
D_{50}	0.39
Uniformity coefficient (C_u)	2.61
Coefficient of curvature (Cc)	0.93
Maximum dry density (kN/m^3)	16.5
Minimum dry density (kN/m^3)	13.9

Table 2 Properties of glass fiber

Properties	Value
Specific gravity	2.57
Diameter (mm)	0.15
Tensile strength (GN/m^2)	2.61
Elongation at break (%)	1.8
Tensile Modulus (GN/m$^{2)}$	112.3

Glass fiber is used for reinforcement purpose in this study. Glass fibers were obtained from the local market and cut into different required length. The properties of glass fiber used in this study are presented in Table 2.

2.2 *Experimental Program*

In this study, triaxial tests were conducted on the unreinforced and fiber-reinforced soil specimen having a height of 200 mm and diameter of 100 mm. The tests on unreinforced soil specimens were also conducted to understand the beneficial effect of the reinforcement. For the study of the effect of confining pressure on the behavior of fiber-reinforced soil, relative density of soil, length of the fiber, and fiber content are taken as variable parameters. Two different relative densities (D_r) 30 and 80% which represent loose and dense condition, four lengths of fiber length (L) 10, 20, 40, and 80 mm, and fiber contents (f_c) of 1, 2, 3, 4, and 5% are used for the tests. All the tests were conducted on four different confining pressures (σ_3) 49, 98, 196, and 392 kPa. Strain-controlled triaxial test setup was used with a strain rate of 1.25 mm/min during the experiments.

2.3 *Sample Preparation*

To prepare the unreinforced soil sample, the required weight of soil was taken to maintain the particular relative density, and then it was poured in sampler and compacted in layers. In case of fiber-reinforced soil first mixture of soil–fiber was prepared. For this purpose fiber was taken as percent weight of soil mass. Fiber content (f_c) here is defined as the ratio of weight of fiber (w_f) to the weight of soil mass (w_s). To maintain the particular density required, the weight of soil and fiber was taken by assuming soil and fiber both as a part of solid part of soil. Mixing was done by hand mixing to achieve uniformity. Hand mixing is preferred by different researchers like Maher and Gray (1990), Al-Refai (1991). To avoid the segregation of fiber from soil, 3% of water was used to mixed soil and fiber. It was found that such amount of water does not affect the strength and stiffness of sand (Maher and gray 1990; Al-Refai 1991). After preparation of soil–fiber mixture, it was poured in the cylindrical sampler and compacted in layers to prepare soil sample.

3 Test Results and Discussion

The test results are presented here to understand the effect of confining pressure on the strength of the soil, stiffness of the soil and on the shear strength parameters. Careful observation of the soil specimen after the test was also done to understand the deformation behavior of the fiber-reinforced soil.

3.1 Effect on the Stress–Strain Behavior of Soil

Figures 1 and 2 represents the stress–strain behavior of fiber-reinforced soil for different lengths of fiber at low confining pressure (49 kPa) and high confining pressure (98 kPa) relatively, for the soil specimen with relative density of 80%, and fiber content of 4%. In the case of low confining pressure (49 kPa), stress–stain behavior of fiber-reinforced soil (Fig. 1) is showing stress drop after peak stress, which is similar to strain softening behavior. Stress drop is more in case of longer fiber, while in case of short fiber the behavior of fiber-reinforced soil is similar to the unreinforced sand, where stress drop is less. But in the case of high confining pressure (392 kPa), the soil is showing strain hardening behavior (Fig. 2) where material deforms at an approximately constant deviator stress after reaching the peak. Stress–strain behavior for all sizes of fiber is approaching toward asymptotic state. It can also be noted that improvement due to fiber reinforcement with respect to unreinforced soil is more in the case of low confining pressure. For example in case of fiber length of 40 mm improvement in peak stress is three times with respect to unreinforced soil at confining pressure of 49 kPa (Fig. 1), but in the case of high confining pressure of 392 kPa (Fig. 2), the improvement in the peak stress for fiber-reinforced soil is around 1.5 times of peak stress of unreinforced soil. It indicates that due to the increase in the confining pressure, reinforcing the action of fiber has suppressed. A similar type of behavior is found for the loose soil (Figs. 3 and 4).

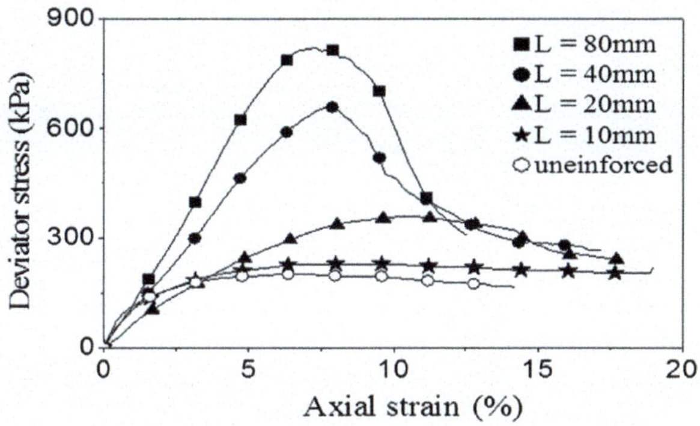

Fig. 1 Stress–strain behavior of fiber-reinforced soil ($D_r = 80\%$, $f_c = 4\%$, $\sigma_3 = 49$ kPa)

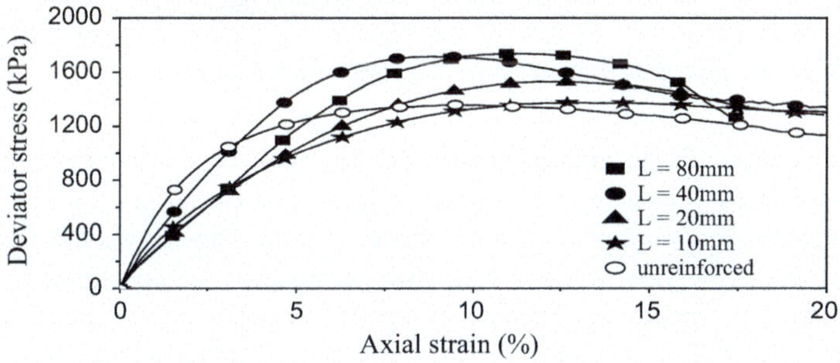

Fig. 2 Stress–strain behavior of fiber-reinforced soil ($D_r = 80\%$, $f_c = 4\%$, $\sigma_3 = 392$ kPa)

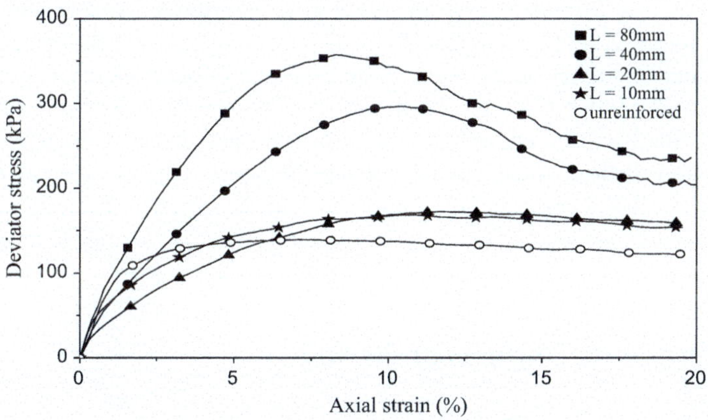

Fig. 3 Stress–strain behavior of fiber-reinforced soil ($D_r = 30\%$, $f_c = 2\%$, $\sigma_3 = 49$ kPa)

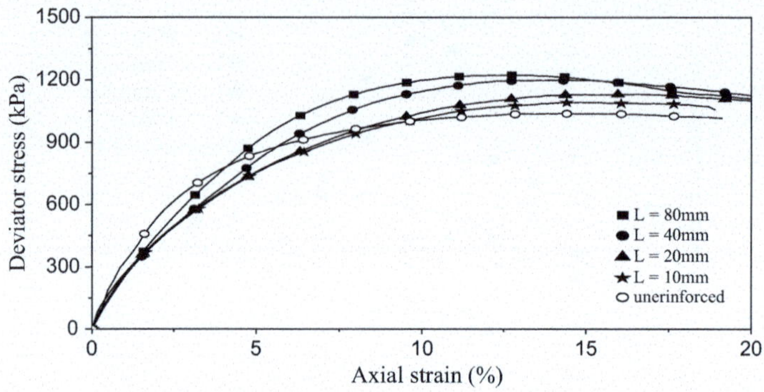

Fig. 4 Stress–strain behavior of fiber-reinforced soil ($D_r = 30\%$, $f_c = 2\%$, $\sigma_3 = 392$ kPa)

3.2 Effect on the Principal Stress Ratio at Failure

Principal stress ratio is the ratio of maximum principal stress (σ_1) of reinforced soil to the unreinforced soil at failure. Figure 5 shows the variation of principal stress ratio with confining pressure. It can be observed that at lower confining pressure, principal stress has higher value, while with increase in the confining pressure, it is decreasing. This behavior is similar for both loose ($D_r = 30\%$) and dense soil ($D_r = 80\%$). But in the case of dense soil, the principal stress ratio is higher than the loose soil. With increase in the relative density of soil, soil–fiber interaction increases. Increased interaction increases the strength improvement capacity of fiber reinforcement. Figure 5 also shows the impact of confining pressure on the principal stress ratio for different fiber content. At lower confining pressure, the principal stress ratio is greater for higher fiber content. If the amount of fiber is increasing, then more fibers are available for the interaction with soil, which increase the load carrying capacity of fiber-reinforced soil. It can further be noted that at higher confining pressure, the principal stress ratio is almost the same, which indicates that with increase in confining pressure, the impact of fiber parameters does not take place effectively.

3.3 Effect on the Stiffness of the Soil

To understand the effect of confining pressure on the stiffness of the fiber-reinforced soil, variation of stiffness modulus at particular strain for different confining pressures for dense sand (Dr = 80%) is presented in Fig. 6. Strain modulus at any strain is the ratio of deviator stress to the corresponding strain (Babu and Vasudevan 2008). It can be observed that for any strain, stiffness modulus is increasing with increase in the confining pressure. It can be further noted that at low strain stiffness, modulus is high and with increase in the strain, it is decreasing. The behavior of fiber-reinforced

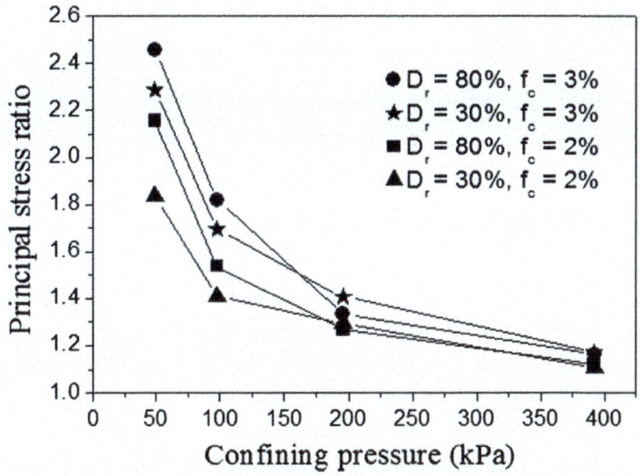

Fig. 5 Effect of confining pressure on the principal stress ratio

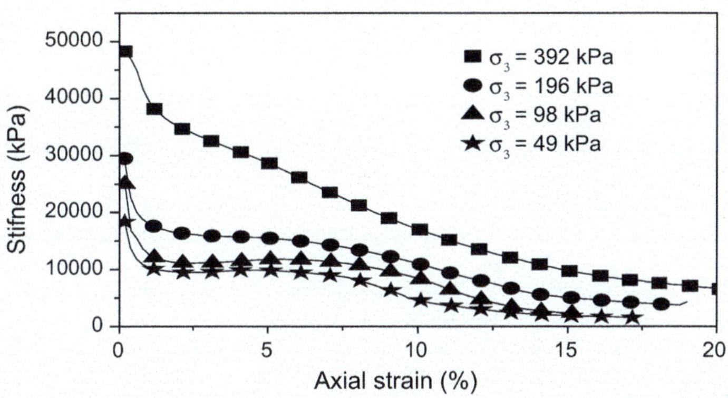

Fig. 6 Stiffness versus axial strain for fiber-reinforced soil ($D_r = 80\%$, $f_c = 4\%$, $L = 40$ mm)

soil mainly depends upon the surface interaction of soil and fiber. With increase in the confining pressure, the frictional resistance between soil particles and fibers increases, and these increases in the frictional resistance directly resists the deformation in the soil. Due to this resistance at a particular strain, the load carrying capacity has increased and soil has shown higher stiffness. A similar type of behavior is found for loose soil also.

4 Conclusions

In this study, the effect of confining pressure on the behavior of fiber-reinforced soil has investigated based upon the results of the triaxial test. The major conclusions drawn from this study are as follows:

- Fiber reinforcements have the potential to increase the load carrying capacity of the soil.
- Confining pressure has a significant impact on the load carrying capacity of soil. Peak stress increases with increase in the confining pressure of soil.
- At low confining pressure, the stress–strain behavior of fiber-reinforced soil is similar to strain softening behavior, while at high confining, its behavior is strain hardening type.
- With increase in the confining pressure, stiffness modulus for fiber-reinforced soil at a particular strain increases.

References

Al-Refai TO (1991) Behavior of granular soils reinforced with discrete randomly oriented inclusion. Geotext Geomembr. https://doi.org/10.1016/0266-1144(91)90009-1

Al-Refai TO (1992) Model tests on strip footing on reinforced sand. J King Saud Univ 4:155–169

Babu GLS, Vasudevan AK (2008). Strength and stiffness response of coir fiber reinforced tropical soil. J Mater Civil Eng ASCE. https://doi.org/10.1061/(asce)0899-1561(2008)20:9(571)

Gray DH, Al-Refai T (1986) Behavior of fabric-versus fiber reinforced sand. J Geotech Eng ASCE. https://doi.org/10.1061/(asce)0733-9410(1986)112:8(804)

Gray DH, Ohashi H (1983) Mechanics of fiber reinforcement of sand. J Geotech Eng ASCE. https://doi.org/10.1061/(asce)0733-9410(1983)109:3(335)

Gregory GH (2006) Shear strength, creep and stability of fiber-reinforced soil slope. PhD thesis, Oklahoma State University

Gregory GH, Chill DS (1998) Stabilization of earth slopes with fiber reinforcement. In: Proceedings of the sixth international conference on geosynthetics, pp 1073–1078. Atlanta, Georgia

Lindh E, Eriksson L (1990) Sand reinforced with fibers: a field experiment. In: Proceedings of the international reinforced soil conference on performance of reinforced soil structure, pp 471–474. Glasgow, UK

Maher MH, Gray DH (1990) Static response of sand reinforced with randomly distributed fibers. J Geotech Eng ASCE. https://doi.org/10.1061/(asce)0733-9410(1990)116:11(1661)

Michalowski RL, Cermak J (2003) Triaxial compression of sand reinforced with fibers. J Geotech Geoenviron Eng ASCE. https://doi.org/10.1061/(asce)1090-0241(2003)129:2(125)

Michalowski RL, Zhao A (1996) Failure of fiber-reinforced granular soils. J Geotech Eng ASCE. https://doi.org/10.1061/(asce)0733-9410(1996)122:3(226)

Ranjan G, Vasan RM, Charan HD (1994) Probablistic analysis of randomly distributed fiber reinforced soil. J Geotech Eng ASCE. https://doi.org/10.1061/(asce)0733-9410(1996)122:6(419)

Refai SM (2000) Impact of polypropylene fibers on desiccation cracking and hydraulic conductivity of compact clay liners. In: Dissertation submitted in partial fulfillment for requirements of doctoral degree, Wayne State University, Detroit, Michigan

Santoni RL, Tingle JS, Webster SL (2001) Engineering properties of sand-fiber mixture for road construction. J Geotech Geoenviron Eng ASCE. https://doi.org/10.1061/(asce)1090-0241(2001)127:3(258)

Tang C, Shi B, Gao W, Cai Y (2007) Strength and mechanical behavior of short propylene fiber reinforced and cement stabilized clayey soil. Geotext Geomembr. https://doi.org/10.1016/j.geotexmem.2006.11.002

Wasti Y, Butun MD (1996) Behavior of model footing on sand reinforced with discrete inclusion. Geotext Geomembr. https://doi.org/10.1016/s0266-1144(96)00044-1

Study of Degradation of Pharmaceutical Drug Cetirizine Using TiO_2 Photocatalysis

Steffi Talwar, Pooja Chaudhary, Vikas Kumar Sangal and Anoop Verma

Abstract Heterogeneous photocatalysis process has proven its points of interest over traditional strategies for wastewater treatment. In the present study, the heterogeneous photocatalysis treatment was applied for degradation of the pharmaceutical wastewater of cetirizine by using TiO_2 as photocatalyst. The experiments were run in slurry mode. The input parameters which were differed are pH, cetirizine concentration, and TiO_2 dose and time. Optimization was performed by using Response Surface Methodology (RSM). At the optimized conditions pH = 3.35 and TiO_2 dosage 2.32 g/l, % degradation was observed to be 98.38% for 420 min. Kinetic study showed that degradation of cetirizine follows first-order kinetics with rate constant value $k = 0.007\ min^{-1}$. The RSM results demonstrate that the predictions agreed with the experimental results.

Keywords Cetirizine · Photocatalysis · RSM · BBD · % degradation · TiO_2

S. Talwar
Department of Chemical Engineering, Thapar Institute of Engineering and Technology, Patiala, India
e-mail: steffi.talwar@gmail.com

P. Chaudhary
School of Chemistry and Biochemistry, Thapar Institute of Engineering and Technology, Patiala, India
e-mail: poojachaudhary527@gmail.com

V. K. Sangal (✉)
Department of Chemical Engineering, Malaviya National Institute of Technology, Jaipur, India
e-mail: vksangal@gmail.com; vksangal.chem@mnit.ac.in

A. Verma
School of Energy and Environment, Thapar Institute of Engineering and Technology, Patiala, India
e-mail: anoop.kumar@thapar.edu

A. K. Agnihotri et al. (eds.), *Sustainable Engineering*, Lecture Notes in Civil Engineering 30, https://doi.org/10.1007/978-981-13-6717-5_29

1 Introduction

The most widely recognized industries like textile, pharmaceutical, pesticides, and other organic chemicals manufacturing discharge wastewater which if not treated could pose problems to the environment (Ternes et al. 1999). Most commonly utilized pharmaceuticals are antibiotics, antidepressants, etc. which are not metabilized in the body and 90% of the medicine is excreted, which could enter in water bodies by mean of household or industrial discharges (Dinh et al. 2011). They are only found at trace level but this raises the issue of the environment (Gao et al. 2012). Conventionally used methods are not so much productive for completely removing contaminations present in wastewaters (Williams 2004). So, Advanced Oxidation Processes (AOPs) are of major interest nowadays to remove organic and inorganic toxins but also for effectively diminishing toxicity with complete mineralization.

Heterogeneous photocatalysis in presence of a semiconductor photocatalyst was used as an AOP for the treatment of wastewater (Achilleos et al. 2010; Kaur et al. 2016; Van Doorslaer et al. 2012; Yang et al. 2008). Photocatalysis by titanium dioxide has turned into an extremely dynamic field of research nowadays for the degradation of toxins present in wastewater. TiO_2 has proven to be the most efficient semiconductor photocatalyst (Fukahori and Fujiwara 2015; Venkata Subba Rao et al. 2003). The mechanism of TiO_2 photocatalysis is now well established where OH• radicals being nonselective can target any contaminant present in wastewater. To the best of our knowledge, there is no literature available on the photocatalytic degradation of cetirizine in presence of UV light. In the present study, we have focused on the photocatalytic degradation of the drug cetirizine in presence of suspended TiO_2 semiconductor and UV light in slurry pond reactor. To carry out parametric studies for photocatalytic degradation of the cetirizine, Box–Behnken Design (BBD) was used.

2 Methodology

2.1 Chemicals

Cetirizine 10 mg tablet (Okacet from Cipla) was brought from local market. TiO_2 P25 was obtained from Evonik Degussa Corporation, USA. 0.1 N hydrochloric acid (HCL) and 0.1 N sodium hydroxide (NaOH) were used to adjust the pH. For conducting experiments, a batch-type bench scale photocatalytic reactor system was fabricated. A borosil glass vessel of 500 ml capacity was used as the shallow pond reactor and UV rays were irradiated on the glass vessel. A magnetic stirrer was used for mixing, so that the TiO_2 remains suspended and the concentration of the pollutant within the reactor could be assumed to be constant at any time. The samples were taken after every 30 min for regular for analysis and analyzed using the double beam UV visible spectrophotometer (LABINDIA, T60 U) at 232 nm to find out % degradation (Eq. 1), where C_0 and C are initial and final concentrations, respectively.

$$\% \text{ degradation} = \frac{c_0 - c}{c_0} \times 100 \tag{1}$$

2.2 *Optimization by Response Surface Methodology (RSM)*

The optimization of analytical procedures has been carried out by using RSM. It is most relevant multivariate techniques used in analytical optimization which is Response Surface Methodology (RSM) (Sangal et al. 2014; Kaur et al. 2015). It can be well applied when a response or a set of responses of interest are influenced by several variables. However, to approximate a response function to experimental data that cannot be described by linear functions, experimental designs for quadratic response surfaces should be used. The BBD is an independent quadratic design that does not contain an embedded fractional factorial design. In BBD, the treated combinations are at the midpoints of edges of the process space and at the center. To design the experiments for wastewater treatment, the Statistical Design-Expert software version 6.06 (STAT-EASE Inc., Minneapolis, US) was used.

3 Results and Discussions

3.1 *Experimental Design and Analysis*

The numbers of experiments were executed in order to model the photocatalytic degradation of cetirizine and to find the optimum conditions (Table 1). BBD under Response Surface Methodology (RSM) was utilized for optimization of reaction parameters for the photocatalytic degradation of cetirizine. The four parameters, i.e., TiO_2 dose (0.1–0.5 g/l), pH (3–11), time (30–140 min), and conc cet (5–25 mg/l) were used as input variables and % degradation was taken as response. The response, i.e., % degradation of cetirizine was optimized according to the BBD matrix designed. The sequential F-test and other adequacy measures were exploited for selecting the best model. The quadratic model was suggested by sequential model sum of squares. Residual plot for % degradation of cetirizine has been shown in Fig. 1. Sequential model sum of squares showed that quadratic model was best-fit model for experimental data for % degradation of cetirizine. Cubic model was found to be aliased for degradation of cetirizine. The Prob > F value is 0.0001, which indicates that quadratic model is significant. The coefficient of determination is 0.95 for quadratic model.

The ANOVA (analysis of variance) for % degradation of cetirizine is shown in Table 2. This showed that model considered is significant for % degradation of cetirizine. For model terms to be significant, "prob > F" values should be less than 0.050 "prob > F" values larger than 0.10 indicates that model term is insignificant.

Table 1 Full factorial BBD matrix

Std	pH	TiO_2 dose (g/l)	Time (min)	Ctz Conc.	% degradation
1	3	1.5	30	15	27.37
2	11	0.5	225	15	12.67
3	7	1.5	225	15	41.28
4	3	2.5	225	15	99.33
5	7	0.5	225	25	20.25
6	7	1.5	225	15	41.28
7	7	1.5	225	15	41.28
8	7	0.5	225	5	17.02
9	11	1.5	225	25	6.78
10	3	0.5	225	15	37.29
11	11	1.5	30	15	1.7
12	7	1.5	30	5	4.41
13	3	1.5	225	5	75.49
14	7	2.5	225	25	38.95
15	7	1.5	30	25	1.96
16	11	1.5	420	15	31.21
17	11	2.5	225	15	6.78
18	7	2.5	225	5	24.87
19	7	1.5	225	15	41.28
20	11	1.5	225	5	21.58
21	7	0.5	30	15	9.79
22	7	0.5	420	15	20.147
23	7	1.5	420	5	61.81
24	7	2.5	420	15	60.59
25	3	1.5	420	15	92.71
26	3	1.5	225	25	48.7
27	7	1.5	420	25	56.25
28	7	1.5	225	15	22.32
29	7	2.5	30	15	2.3

From ANOVA, it is clear that the time and TiO_2 dose are highly significant terms for % degradation of cetirizine.

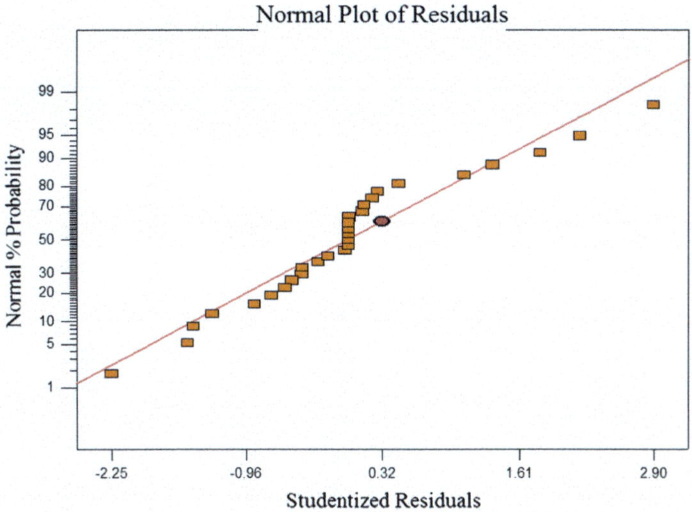

Fig. 1 Residual plot for photocatalysis of % degradation of cetirizine

3.2 *Effect of Selected Parameters on % Degradation and Optimization*

To study the effect of parameters, i.e., TiO_2 dose, pH, concentration of cetirizine, and time on % degradation (response), the three-dimensional response surface graph was considered from Fig. 2a–c.

From Fig. 2a, it is observed that with the increase in pH at all concentrations of TiO_2 dose, % degradation is decreasing. The % degradation is maximum at pH 3.35 with higher concentration of TiO_2. The 3D plot, Fig. 2b, shows the effect of pH and time on the % degradation. It has been clearly seen that at all pH values the % degradation is increased with increase in time but at lower range of pH with the increase in time the % degradation is maximum. In Fig. 2c, the effect of time and TiO_2 dose on % degradation is studied. With increase in time, the % degradation at every dose of TiO_2 also increases. But at the TiO_2 dose of 2.32 g/l the increase in time degradation obtained was maximum. Figure 2a–c concluded that photocatalytic degradation of a compound is done by the OH• attack which depends upon the pH and the dose of TiO_2 with increase in time. TiO_2 suspended in aqueous mode with the help of irradiation having energy greater than the bandgap energy of the semiconductor produces h^{+}_{TiO2} and e^{-}_{TiO2}. The migration of holes and electrons generated by the oxidants participates in the degradation of the cetirizine. From regression analysis, it was concluded that in acidic condition at the higher dose of TiO_2, % degradation of cetirizine was more.

In this study, % degradation of cetirizine was optimized in terms of maximization using BBD. The values of optimum parameters were found to be TiO_2 dose =

Table 2 Analysis of variance for the % degradation of cetirizine

Source	Sum of squares	DF	Mean square	F value	Prob > F	
Model	15153.06	14	1082.36	22.49	<0.0001	Significant
Ph	4440.17	1	4440.17	92.26	<0.0001	Highly significant
TiO_2 dose	1264.38	1	1264.8	26.27	0.0002	
Time	5072.81	1	5072.1	105.0	<0.001	Highly significant
Ctz Conc.	223.90	1	223.90	4.65	0.0489	
PH^2	24.26	1	24.26	0.50	0.4894	
TiO_2 dose^2	86.68	1	86.68	1.80	0.2010	
Time^2	943.95	1	943.95	19.61	0.0006	
Ctz Conc.^{2}	526.79	1	526.79	10.95	0.0052	
pH × TiO_2 dose	978.75	1	978.75	20.34	0.0005	
pH × Time	1114.22	1	1114.22	23.15	0.0003	
pH × Ctz Conc.	214.34	1	214.34	4.45	0.0533	
TiO_2Dose × Time	444.22	1	444.22	9.23	0.0089	
TiO_2Dose × Ctz Conc.	2.31	1	2.31	0.048	0.8297	
Time × Ctz Conc.	133.71	1	133.71	2.78	0.1178	
Residual	673.81	14	48.13			
Lack of fit	673.81	10	67.38			
Pure error	0.000	4	0.000			
Cor total	15826.7	28				

2.32 g/l, t = 420 min, and pH = 3.35 with a desirability of D = 0.99. At this optimum condition, the % degradation of cetirizine suggested by BBD was 98% for cetirizine concentration of 15 ppm.

Optimum condition for photocatalytic treatment of pharmaceutical drug was verified experimentally as % degradation came out to be 95.38%. Experiment was run for 420 min at pH 3.35 with TiO_2 dose 2.32 g/l. Hence, the predictions agreed well with the experimental and optimization results by RSM.

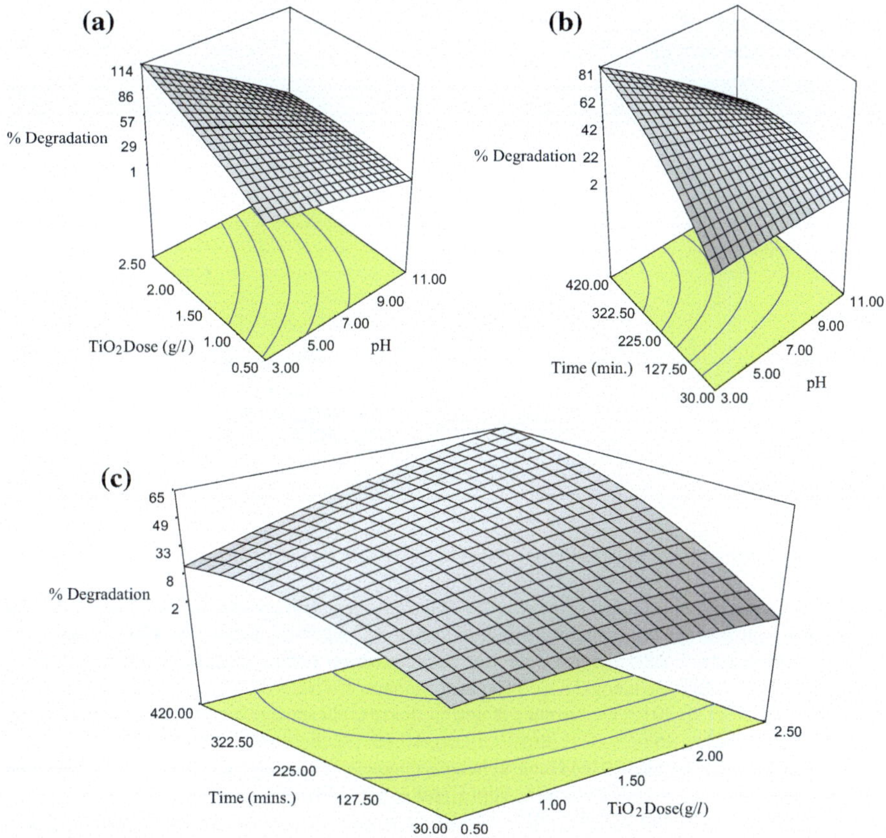

Fig. 2 **a** % degradation 3-D response surface graph between pH and TiO_2 dose and pH, **b** % degradation of cetirizine 3D response surface graph between pH and time, **c** % degradation of cetirizine 3-D response surface graph between TiO_2 dose and time

3.3 Kinetic Study

The kinetics study of cetirizine wastewater was studied for photocatalytic treatment process at optimum conditions. For the experimental results of photocatalytic treatment of cetirizine first-order model was fitted as shown in Fig. 3.

Photocatalytic degradation for the recalcitrant compound cetirizine wastewater with UV/TiO_2 was assessed. The optimization of the process parameters was successfully performed. It was examined that photocatalytic process was an efficient process for degradation of cetirizine at lower pH and higher TiO_2 dose with the increase in time. At optimum condition, the % degradation of cetirizine suggested by BBD was 98%. Experimentally, % degradation of cetirizine at optimum conditions was found to be 95.38% for slurry mode reactor. The kinetic study showed that the degradation followed first-order kinetics model.

Fig. 3 Kinetic study of photocatalytic treatment of cetirizine conclusion

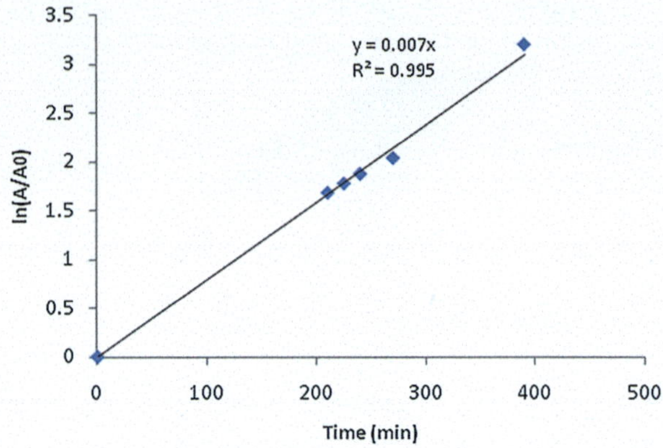

References

Achilleos A, Hapeshi E, Xekoukoulotakis NP et al (2010) Factors affecting diclofenac decomposition in water by UV-A/TiO_2 photocatalysis. Chem Eng J 161:53–59

Dinh QT, Alliot F, Moreau-Guigon E et al (2011) Measurement of trace levels of antibiotics in river water using on-line enrichment and triple-quadrupole LC-MS/MS. Talanta 85:1238–1245

Fukahori S, Fujiwara T (2015) Photocatalytic decomposition behavior and reaction pathway of sulfamethazine antibiotic using TiO_2. J Environ Manage 157:103–110

Gao L, Shi Y, Li W et al (2012) Occurrence of antibiotics in eight sewage treatment plants in Beijing, China. Chemosphere 86:665–671. https://doi.org/10.1016/j.chemosphere.2011.11.019

Kaur A, Umar A, Kansal SK (2016) General heterogeneous photocatalytic studies of analgesic and non-steroidal anti-inflammatory drugs. Appl Catal A 510:134–155

Kaur P, Sangal VK, Kushwaha JP (2015) Modeling and evaluation of electro-oxidation of dye wastewater using artificial neural networks RSC Adv 5:34663–34671

Sangal VK, Kumar V, Mishra IM (2014) Process parametric optimization of a divided wall distillation column. Chem Eng Commun 201:37–41

Ternes T, Stumpf M, Mueller J et al (1999) Behavior and occurrence of estrogens in municipal sewage treatment plants—I. investigations in Germany, Canada and Brazil. Sci Total Environ 225:81–90

Van Doorslaer X, Heynderickx PM, Demeestere K et al (2012) TiO_2 mediated heterogeneous photocatalytic degradation of moxifloxacin: operational variables and scavenger study. Appl Catal B Environ 111–112:150–156

Venkata Subba Rao K, Rachel A, Subrahmanyam M, Boule P (2003) Immobilization of TiO_2 on pumice stone for the photocatalytic degradation of dyes and dye industry pollutants. Appl Catal B Environ 46:77–85

Williams SAPM (2004) "Introduction" in advanced oxidation processes for water and wastewater treatment, pp 1–6. IWA Publis

Yang L, Yu LE, Ray MB (2008) Degradation of paracetamol in aqueous solutions by TiO_2 photocatalysis. Water Res 42(3480–3488):3

Liquefaction Potential of Sub-Surficial Soil Based on GRA of Smart City Jalandhar

Neha Khanna and Manish Khanna

Abstract Liquefaction and related phenomenon have been in charge of enormous measures of harm in recorded seismic tremor all over the world. The city Jalandhar is under zone IV. It sustains the vulnerability of city to seismic forces. A number of multistoried high-rise buildings are under progress and several high-rise buildings are proposed for next few years. From the study of past earthquakes, we have demonstrated that liquefaction-induced ground failure may pose catastrophic damage threats to rural and urban infrastructure including buildings. To fulfill the safety purpose it is imperative that the new structure in Jalandhar be analyzed for seismic forces. To fulfill the need of the objective, the area was analyzed with the help of simplified procedure given by Seed and Idriss (1971) of liquefaction potential evolution is used. With the use of SPT result, ground response analysis has been done. Peak ground acceleration has been calculated using software. The estimated peak ground acceleration will be used to check liquefaction of specific sites in Jalandhar. The bore log data has been obtained from 10 locations of Jalandhar. From this analysis, it is obtained that there are only two sites which are likely to be effected under liquefaction.

Keywords Ground response analysis · Liquefaction · Shear wave velocity · Peak ground acceleration

1 Introduction

Jalandhar is the oldest city in Punjab, and furthermore, one of the most seasoned in the nation, Jalandhar has seen fast urbanization and formed into an exceptionally industrialized focal point of business activity in the current years. Jalandhar was the capital of Punjab from India's autonomy (1947) until the point that Chandigarh was

---

N. Khanna (✉) · M. Khanna
Department of Civil Engineering, Geotechnical Engineering, DAVIET, Jalandhar, India
e-mail: nehakhanna13716@gmail.com

A. K. Agnihotri et al. (eds.), *Sustainable Engineering*, Lecture Notes in Civil Engineering 30, https://doi.org/10.1007/978-981-13-6717-5_30

built in 1953. Seismic tremor is a characteristic occasion which may cause huge death toll and property harm. One of the real difficulties confronting our nation is to lessen the helplessness to this wild and capricious danger by having a more prominent comprehension about its circumstances and end results. Historically in Punjab, there are very less effects of Earthquake, however, damage from earthquakes occurred during large events originating in the Himalayas was very high. The last earthquake that directly affected Punjab was in 1905 and that earthquake was very strong that it knocked down chimneys and parapets and even a few masonry walls. More than five roofs and high-rise building collapsed in the bazaar and the minaret of a mosque fell as well (http://moga.nic.in/earthquake_punjab.aspx).

"Liquefaction" is the process in which soil loses its strength because of dynamic loading, liquefaction is normally occurred in saturated and cohesionless soil. Soil starts to behave like a liquid until the shear stresses acting on the soil mass are as low as the reduced shear resistance. This phenomenon is called liquefaction. (Biswas and Naik 2010) such type of soil reduces the ability to strengthen the base of building and building starts to tilt and sink.

1.1 Geological Data of Jalandhar City

Based on old data, it provides complete information on the location of the city as well as its climatic conditions and the economic importance of the city. Jalandhar is spread over a huge area of 3,401 sq km and is situated between the fertile agricultural land of River Beas and Sutlej. Jalandhar is an important commercial hub of Punjab and it is situated at 31.3260°N latitude and 75.5762°E longitude at Jalandhar (Maps of India).

1.2 Seismicity

From the earthquake hazard map given in the above, it is seen that about 50% of the area of the state in the north, consisting of Kapurthala Jalandhar Patiala Ludhiana And Rup Nagar districts comes to MSK intensity VIII and it is also measured that about 45% could have intensity VII. An earthquake of M 5.5 occurred in Kapurthala district in 1952 and much larger earthquakes of M 7.0 to 8.0 have occurred in Himachal Pradesh at about 50–60 km from the State boundary, which could cause moderate to heavy damage in the districts of Gurdaspur, Amritsar, and Hoshiarpur.

2 Methodology

For the assessment of liquefaction of Jalandhar city, first of all, sites are selected in such a way that it covers the whole area of Jalandhar city. After the selection of sites standard penetration test has been done and data was collected. Collected data was used to calculated shear wave velocity and subgrade modulus by using the value of Vs and Gmax Peak Ground Acceleration and Peak Spectral Acceleration will be calculated. Calculated value will be used for the assessments of liquefaction.

2.1 Locations of Borehole

Total of 10 sites are selected for the assessments of liquefaction. Borehole was done until we have got a high number of blows value as well as stiff and hard strata below. List of the site with their locations are explained in Table 1, also the location of borehole are shown below in the graph of Jalandhar city Fig. 1.

2.2 Assessment of Liquefaction Potential

The assessments of soil liquefaction potential is done by using in situ tests such as standard penetration test (SPT), cone penetration test, shear wave velocity (Vs) measurement, etc., it is very difficult to get high quality of undisturbed soil samples

Table 1 List of sites

Site code	Location	Latitude	Longitude
J01	Inside link nagar MC garden, Lado Wali Road, Jalandhar	31.3144	75.60322
J02	New Sheetal Nagar, near DAV colony Maqsudan, Jalandhar	31.3493	75.55048
J03	Shanti Vihar, near Maqsudan chowk, behind army school, Jalandhar	31.3628	75.54149
J04	Near Shrama telecom, Guru Amar Das nagar, Jalandhar	31.3654	75.56676
J05	Opp of Panchvir Baba Dargah, Bulandpur, Jalandhar	31.3766	75.58831
J06	Karol Bagh, Jalandhar	31.3342	75.61986
J07	Near Wazir Singh Enclave, NH3, Jalandhar	31.3252	75.63357
J08	Jalandhar Cantt station near Phagwara National Highway, Jalandhar	31.3065	75.63085
J09	Near Flats At Jail Site, Jalandhar	31.2533	75.5778
J10	Near commercial building adjoining kidney hospital, Jalandhar	31.3043	75.5875

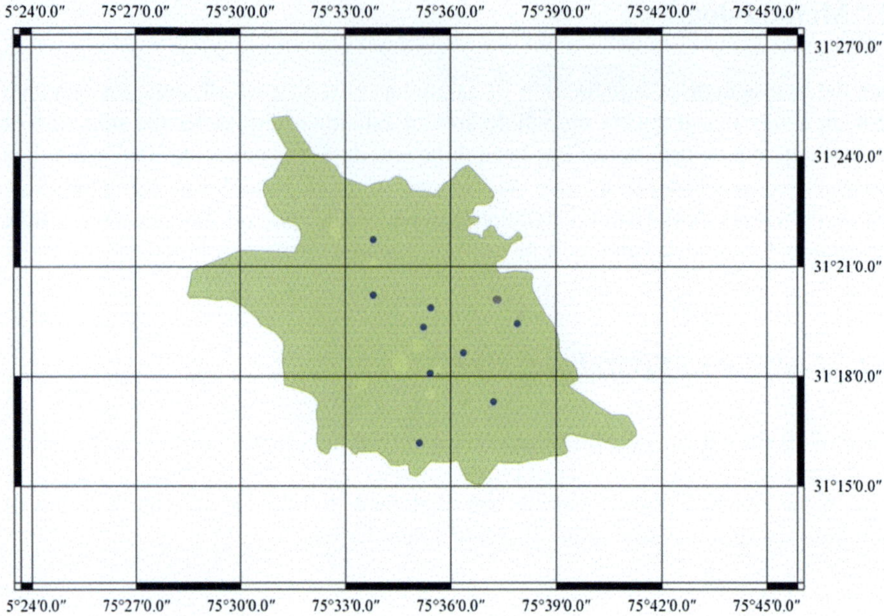

Fig. 1 Locations of borehole

for high-quality testing of soil (Sitaram and Vipin 2009). The use of in situ tests along with case histories—calibrated empirical relationships are generally resorted by the geotechnical engineers for the assessment of liquefaction potential of soils. The improved technique spearheaded by Seed and Idris (1971) generally relies on a limit bend, which shows an utmost state and isolates liquefaction cases from the non-liquefaction cases basing on field perceptions of soil in seismic tremors at the destinations where in situ information are accessible. The limit is generally drawn minimalistically with the end goal that all cases in which liquefaction have been watched lie above it. In this approach, the CSR is usually used as earthquake loading parameters and the cyclic resistance ratio (CRR) is represented by in situ test parameters that reflect the density and pore pressure generation properties of soil. Out of the various in situ methods as mentioned above SPT and CPT-based methods are widely used for liquefaction susceptibility analysis of soil. Evaluation of liquefaction potential of Jalandhar sites has been done using the Semiempirical procedure which was developed by Seed and Idriss in (1971).

2.3 Subsurface Ground Response Analysis

Ground response analysis is done for Jalandhar site for the estimation of peak ground acceleration and peak spectral acceleration of all the sites in Jalandhar, on the bases of

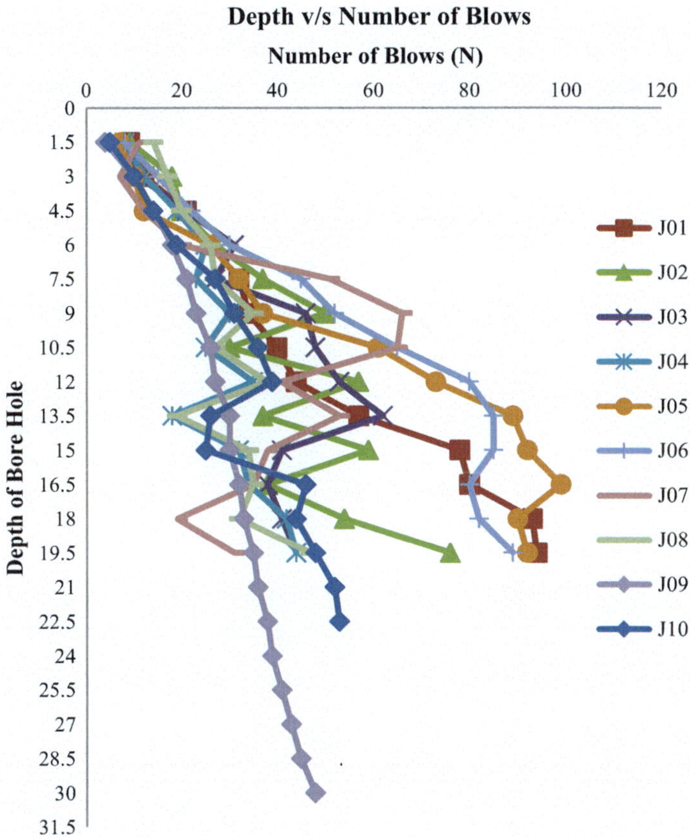

Fig. 2 Depth versus the number of blows

these, for the estimation of the same shear wave velocity have been calculated based on various co-relationships of shear wave velocity. A number of blows with respect to depth Fig. 2 and shear wave velocity with respect to depth Fig. 3 are explained below.

3 Results and Discussion

Peak ground acceleration (PGA) of all sites of Jalandhar

Sikkim (2011) input motion is applied for Peak Ground Acceleration using DEEPSOIL and results are observed as shown in Table 2. The variation in PGA of surface level ground motion due to the small difference in geotechnical condition is very

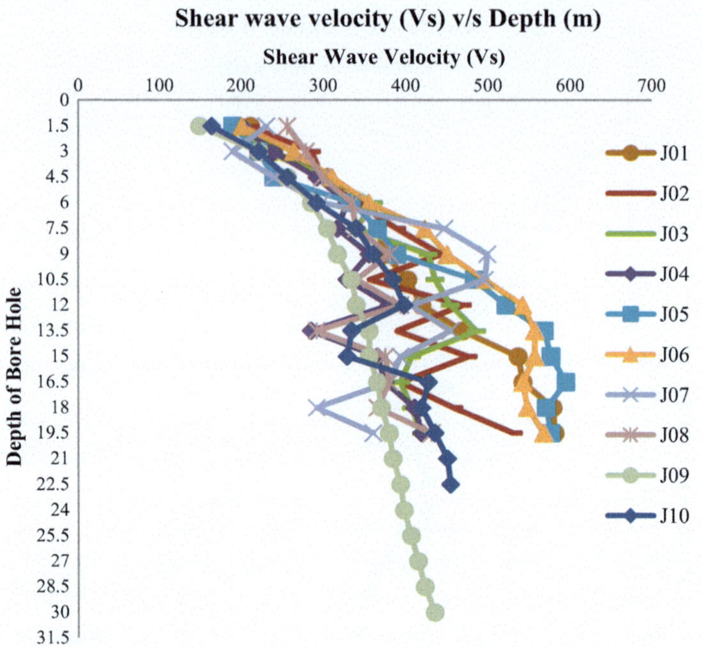

Fig. 3 Depth (m) versus shear wave velocity (Vs) for all sites in Jalandhar

Table 2 Peak ground acceleration of 10 sites of Jalandhar city

Site code	Average depth	PGA (Sikkim earthquake)	Site code	Average depth	PGA (Sikkim earthquake)
J01	19.5	0.088	J06	19.5	0.297
J02	19.5	0.222	J07	19.5	0.161
J03	19.5	0.128	J08	19.5	0.129
J04	19.5	0.174	J09	30	0.117
J05	19.5	0.313	J10	22.5	0.138

much dependent. So we can see that PGA value differs with respect to soil type and type of motion applied.

The spatial distribution of PGA at Jalandhar site is presented in the form of contour map in Fig. 4. Sikkim (2011) and input motion are applied for analysis. This value also gives us the characteristics knowledge of site.

Liquefaction potential study has been carried out for all the sites using PGA values calculated by ground response analysis of Jalandhar city. Calculated LP value shows us that there is very less susceptibility of liquefaction in sites of Jalandhar because of the type of soil available over there. The calculated liquefaction values are explained below in Table 3.

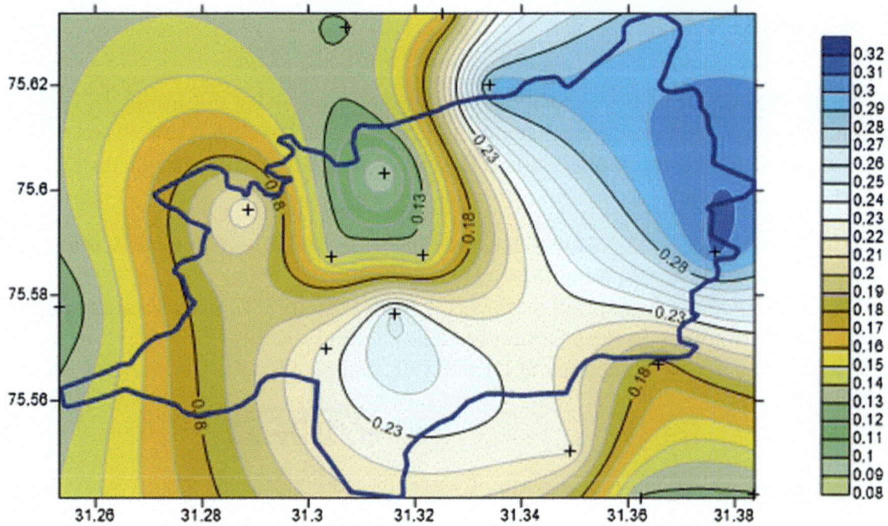

Fig. 4 Site using Sikkim (2011) earthquake motion

Table 3 Summary of liquefaction potential of all the sites

Location	Number of boreholes	FOS	Results
Inside link nagar MC garden, Lado Wali Road, Jalandhar	1	1.43	NL
New Sheetal Nagar, near DAV colony Maqsudan, Jalandhar	1	1.44	NL
Shanti Vihar, near Maqsudan chowk, behind army school, Jalandhar	1	1.30	NL
Near Sharma telecom, Guru Amar Das nagar, Jalandhar	1	1.32	NL
Opp of Panchvir Baba Dargah, Bulandpur, Jalandhar	1	1.19	NL
Karol Bagh, Jalandhar	1	1.29	NL
Near Wazir Singh Enclave, NH3, Jalandhar	1	1.73	NL
Jalandhar Cantt station near Phagwara National highway, Jalandhar	1	2.42	NL
Near flats at jail site, Jalandhar	1	0.86	L
Near commercial building adjoining kidney hospital, Jalandhar	1	0.97	L

4 Conclusions

Based on the ground response analysis and assessment carried out for all 10 sites of Jalandhar, it has been observed that various conclusions have been drawn out which can be useful for Geotech analyst and designers for their future design projects in Smart city Jalandhar

1. The values of J09, J10 shows that the FOS is below 1, which clearly indicates that these areas are liquefiable.
2. The liquefaction value of all sites varies from 0.86 to 2.42 which indicate there is a vast difference in liquefaction tendency within the city.
3. The PGA value produced using DEEPSOIL can be used for future references.

References

Biswas A, Naik AN (2010) Study on liquefaction of soil. Department of Civil engineering, NIT Rourkela

Ayothiraman R, Raghu Kanth STG, Sreelatha S (2012) Evaluation of liquefaction potential of Guwahati: gateway city to Northeastern India. Nat Hazards 63:449–460

IBC (2000) International building code, International code council

Idriss IM (1999) An update to the Seed-Idriss simplified procedure for evaluating liquefaction potential. In: Proceedings of TRB workshop on new approaches to liquefaction, January, Publication No. FHWA-RD-99-165, Federal Highway Administration

Idriss IM, Boulanger RW (2004) Semi-empirical procedures for evaluating liquefaction potential during earthquakes. In: Proceedings of the 11th international conference on soil dynamics & earthquake engineering (ICSDEE) and the 3rd international conference on earthquake geotechnical engineering, January 7–9, Berkeley, California, USA, pp 32–56

Idriss IM, Boulanger RW (2006) Semi-empirical procedures for evaluating liquefaction potential during Earthquakes. Soil Dyn Earthq Engrg 26:115–130

Inazaki T (2004) In situ measurement of the liquefaction potential of soils using a shear-wave vibrator. In: Proceedings of IWAM04, Mizunami, Japan

IS (1893–2002) Part 1- Indian standard criteria for earthquake resistance design of structures. BIS, New Delhi

Nath SK (2006) Seismic hazard and microzonation atlas of the Sikkim Himalaya

Sitaram GT, Vipin KS (2009) Seismic soil liquefaction based on insitu test data. IGS Gunter India, pp 857–864

Seed HB, Idriss IM (1971) Simplified procedure for evaluating soil liquefaction potential. J Soil Mech Found. Engrg, ASCE, 97(SM9):1249–1273

Studies on Transport of Reactive and Non-reactive Elements in Fractured Media

S. Rupali and Vishwas A. Sawant

Abstract In recent years, the focus has been shifted to groundwater quality because the consumptive lifestyle has been the reason for deterioration of groundwater. The contaminant transport in the subsurface has to be studied to know the extent of contaminant transport and also know the final magnitude of contaminant. Normally, these contaminants travel through a saturated or unsaturated porous media and sometimes fractured till clays or silts are encountered in the surroundings. The contaminant transport is very different when it has to travel through a fractured media as the hydraulic conductivity is high in the fractures even though the porosity is low, and hence makes it a very important topic of discussion. In the present study, a numerical model for contaminant transport in fractured porous media has been presented. Transport process in single fracture–matrix system is described by two coupled equations, one for fracture and other for matrix. Analysis is performed using meshfree method called element-free Galerkin method. A FORTRAN programme has been developed to obtain numerical solution, and it is validated with the results in the literature. Overall, the results predicted are found in good agreement. Parametric study has been performed to examine the transport of different reactive and non-reactive elements in the fractured media by varying their dispersivity, retardation factor and dispersion coefficient. Parameters of fracture like fracture thickness and hydraulic conductivity shall be varied to observe the transport of these elements in fractured porous media.

Keywords Contaminant transport · Fractured media · Meshfree methods · Element-free Galerkin method · Non-reactive elements · Reactive elements

1 Introduction

In recent years, the focus has drifted from water supply to groundwater quality. The present consumptive lifestyle has deteriorated the groundwater quality, and it

---

S. Rupali (✉)
NIT Jalandhar, Jalandhar, India
e-mail: satavalekarr@nitj.ac.in

V. A. Sawant
Department of Civil Engineering, IIT Roorkee, Roorkee 247667, India

A. K. Agnihotri et al. (eds.), *Sustainable Engineering*, Lecture Notes in Civil Engineering 30, https://doi.org/10.1007/978-981-13-6717-5_31

is evident from the contamination of groundwater reserves as well as rivers and lakes. Often the time taken to identify the contamination of groundwater is large, and it is difficult to apply remedial measures, and hence the focus is to reduce the emission of contaminants in the system. The contaminants are transported by advection, dispersion and retardation, but the behaviour of contaminants in fracture media is slightly different. In fractured zones, porosity is low but still the groundwater velocities are higher. Fractures, offer least hydraulic resistance and the contaminant will enter the system primarily through the fractures. Matrix diffusion from the fractures to the porous matrix constitutes in removing contaminant from the primary flow channels, and hence it acts as a dynamic storage mechanism within the fracture flow system. As a result, contaminant transport in fractured porous media can be different from contaminant transport in porous granular media in the conventional sense, where such a mechanism is not generally represented (Freeze and Cherry 1979). Solute transport through fractured porous media has been studied by many researchers in the past decade. Grisak and Pickens (1980) developed a finite element model for transport in fracture and matrix. The model was developed by considering two-dimensional transport by dividing the subregions into linear triangular elements and assigning each element of different properties. Tang et al. (1981) developed an analytical solution considering coupling of matrix flux in fracture and matrix. Wu et al. (2010) proposed a conceptual model of fracture-flow-enhanced matrix diffusion, which correlates with fracture flow velocity. The proposed model incorporated an additional matrix diffusion process, which was induced by rapid fluid flow along the fractures.

Sharma and Dixit (2014) conducted experimental study to investigate the behaviour of solute transport in fractured media. Sharma and Srivastava (2012) developed an implicit finite difference model to observe the reactive transport in fracture and matrix simultaneously. The model is useful for remediation of fractured porous media and disposal of hazardous waste material. In recent years, meshfree methods are getting attention as they aim to eliminate the structure of mesh and construct approximate solutions for the equations in terms of nodes (Liu and Gu 2005). Most common and successful meshfree method is the element-free Galerkin method (EFGM) and has been used for solving boundary value problems related to various field studies (Belytschko et al. 1994; Kumar and Dodagoudar 2008, 2009; Mategaonkar and Eldho 2012; Satavalekar and Sawant 2015).

In this paper, contaminant transports for different ions like Cl^-, Na^+, F^-, Mg^{2+}, Ca^{2+} and Cu^{2+} (non-reactive and reactive ions) have been observed through a fracture–matrix-saturated porous media, modelled using Element-Free Galerkin Method (EFGM). The governing equation employed is one-dimensional two-coupled contaminant transport through the fracture–matrix porous media. A detailed parametric study is conducted to examine the effect of velocity in fracture and matrix diffusion on the transport of contaminant in the media.

2 Methodology

The conceptual model is presented in Fig. 1. Transport process in single fracture—matrix system is described by two coupled equations, one for fracture and other for matrix. The coupling is incorporated through continuity of fluxes and concentrations along the interface. Groundwater velocity in fracture is assumed constant. The solute source is considered at the inlet of the fracture. The width of fracture is much smaller than the length, and the migration of contaminant is faster in fracture than matrix. The diffusive flux from fracture to matrix takes place perpendicular to the fracture.

Linear diffusive transport equilibrium equation for the concentration of solute in the solution along the matrix C_m (M/L^3) considering sorption, diffusion and first-order degradation for porous matrix (Tang et al. 1981) is written as

$$R_m \frac{\partial C_m}{\partial t} - D_{xm} \frac{\partial^2 C_m}{\partial x^2} - D_{zm} \frac{\partial^2 C_m}{\partial z^2} + \lambda C_m = 0 \tag{1}$$

in which, D_{xm} and D_{zm} are the diffusive flux in the matrix in the x- and z-directions (L^2/T), λ is the decay constant (T^{-1}) and R_m is the retardation factor in matrix. Retardation factor R_m is related to bulk density of soil matrix ρ_b (M/L^3), distribution coefficient K_m (L^3/M) and matrix porosity θ as $R_m = 1 + (k_m \rho_b / \theta)$

One-dimensional transport equation for the concentration of solute in the solution along the fracture C_f (M/L^3) including linear equilibrium sorption and first-order degradation for fracture (Tang et al. 1981) is written as

$$R_f \frac{\partial C_f}{\partial t} - D_f \frac{\partial^2 C_f}{\partial z^2} - \frac{\theta D_{xm}}{b} \frac{\partial C_m}{\partial x} + v_f \frac{\partial C_f}{\partial z} + \lambda C_f = 0 \tag{2}$$

where b is the half-aperture (L), λ is the decay constant (T^{-1}) and D_f is the diffusive flux in the fracture (L^2/T) given as $D_f = \alpha_f v_f + D_m$. R_f is the retardation factor in

Fig. 1 Domain of fracture–matrix media

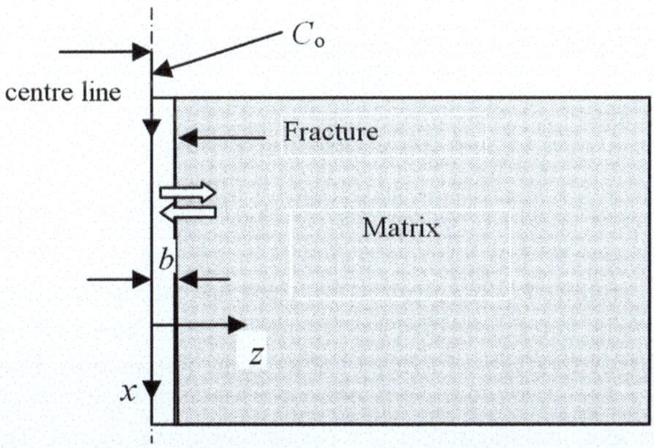

fracture related to distribution coefficient in fracture K_f (L^3/M) as $R_f = 1 + k_f / b$. α_f is the dispersivity in the direction of fracture axis.

Third term in Eq. 2, $\frac{\theta D_{xm}}{b} \frac{\partial C_m}{\partial x}$ is the coupling term, which provides the continuity of fluxes and concentrations along the interface of the boundary.

The following initial and boundary condition for fracture and porous matrix are used:

Initial condition

$$C_f(x, 0) = 0 \ ; \quad \left. \frac{\partial C_f}{\partial z} \right|_{(\infty, t)} = 0 \tag{3}$$

Constant-type boundary condition

$$\begin{aligned} &C_f(0, t) = C_0 \\ &C_m(x, z, 0) = 0 \ ; \ C_m(x, b, t) = C_f(z, t); \quad \left. \frac{\partial C_m}{\partial x} \right|_{(x, L, t)} = 0 \end{aligned} \tag{4}$$

where C_0 is the initial concentration injected at the inlet of the fracture (M/L^3), EFGM uses only set of nodes to model the boundary and generate discrete equations. It employs moving least squares (MLS) approximants formulated by Lancaster and Salkauskas (1981) to approximate the function C(x) with $C^h(x)$ in which C(x) is the contaminant concentration at x, where x is a position coordinate. EFGM does not satisfy the Kronecker delta criterion, and hence the Lagrangian multiplier technique (Dolbow and Belytschko 1998) is used to enforce the Dirichlet boundary condition. A detailed mathematical formulation and algorithm referred from Satavalekar (2015) and Satavalekar and Sawant (2015).

3 Model Verification

The model developed using element-free Galerkin method is verified with the results of Tang et al. (1981). Tang et al. (1981) analysed the one-dimensional two-coupled equation for fracture and matrix by developing a general analytical solution. Tang et al. (1981) considered one-dimensional flow in both fracture and matrix, and the coupling of fluxes was provided along the interface. The material properties are mentioned in Table 1 used in the present study. Sensitivity analysis is carried out with nodes varying from 11 × 11, 21 × 21 and 51 × 11. The results for the variation in normalized concentration with distance along fracture at last time step are presented in Fig. 2. It can be observed from Fig. 2 that the results by EFGM using 21 × 21 nodes are in good agreement with the results obtained by Tang et al. (1981). The domain of the fracture–matrix system is presented in Fig. 1. The fracture is a one-dimensional model, and the matrix is a two-dimensional model. The Dirichlet boundary is at the first node of the fracture.

Table 1 Material properties, Tang et al. (1981)

Parameters	Value
Velocity in fracture (m/d)	0.1
Width of fracture b (μm)	50
Dispersivity (m)	0.5
Diffusion coefficient D* (m^2/d) in matrix	0.0001382
Porosity	0.01
Tortuosity	0.1
$T_{1/2}$ (years)	12.35
Time (days)	100

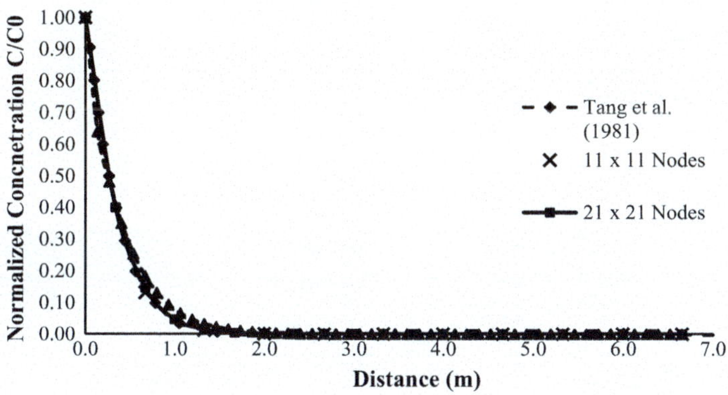

Fig. 2 Comparison of EFGM results with Tang et al. (1981)

4 Parametric Study

A parametric study is performed considering ions like Cl^-, Na^+, F^-, Mg^{2+}, Ca^{2+} and Cu^{2+}. The diffusion and distribution coefficients of these ions are presented in Table 2. These values have been referred from Shackelford and Daniel (1991), Chakraborty and Ghosh (2010), Sharma and Dixit (2014) and Tombul et al. (2005). A parametric study considering variation in fracture velocity and dispersivity has been performed. Parametric study provides a better idea of migration of contaminant when a particular parameter is changed. Material properties considered in the parametric study are reported in Table 2 except for the values of diffusion coefficient and time. The diffusion coefficient for different ions is considered from Table 3, and time of 100 days is considered. In the study, a particular parameter under consideration (whose effect is to be examined) is varied, while the other parameters are kept constant as reported in Table 1.

Table 2 Material properties, Grisak and Pickens (1980)

Parameters	Value
Length of the fracture (m)	0.76
Velocity in fracture (m/d)	0.6
Dispersivity α (m)	0.76
Width of fracture b (cm)	0.1
Diffusion coefficient D* (m^2/d) in matrix	0.0
Time (days)	3
Number of divisions in x-direction	20
Number of divisions in y-direction	20
Δx (m)	0.006
Δy (m)	0.038

Table 3 Material properties of different ions

Ions	Diffusion coefficient (m^2/day)	Retardation factor
Cl^-	0.000175	1.0
F^-	0.000127	1.2
Na^+	0.000115	1.45
K^+	0.000169	5.25
Mg^{2+}	0.0000609	13.5
Cu^{2+}	0.0001	1.8

4.1 Effect of Velocity

In this section, the effect of change of flow velocity in fracture is observed on the transport of different ions presented in Table 3 while maintaining other parameters like diffusion coefficient, dispersivity, distribution factor, aperture size same as reported in Tables 2 and 3. The fracture velocity is varied as 0.3 and 0.6 m/day, and the effect is presented in Fig. 3 for distance 0.76 m and time 100 days. From Fig. 3a, it can be observed that with a fracture velocity of 0.6 m/day, chloride ions (Cl^-) migrate faster and at the end of the fracture (0.76 m) will have a final normalized concentration of 0.95, whereas F^-, Na^+, K^+, Cu^{2+} and Mg^{2+} have final normalized concentration of 0.92, 0.9, 0.75, 0.9 and 0.78 as these ions are reactive in nature and concentration reduces due to adsorption of these in the matrix. From Fig. 3b, it can be observed that with a flow velocity of 0.3 m/day, chloride ions (Cl^-) migrate faster and at the end of the fracture (0.76 m) will have a final normalized concentration of 0.86, whereas F^-, Na^+, K^+, Cu^{2+} and Mg^{2+} have final normalized concentration of 0.83, 0.8, 0.53, 0.77 and 0.53. By reduction of velocity by half, the concentrations of Cl^-, F^-, Na^+, K^+, Cu^{2+} and Mg^{2+} are reduced by 9, 11, 12, 28, 15 and 31%, respectively. It is observed that when the flow velocity of fracture is lower, the transport of contaminant is lesser and it varies for different ions. Non-reactive ions having zero

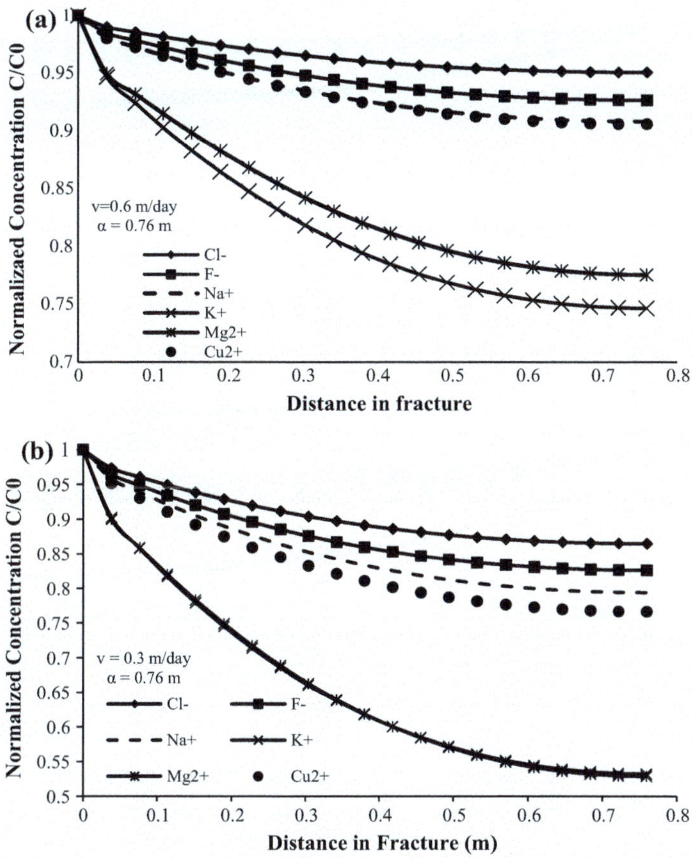

Fig. 3 Effect of velocity on concentration

distribution coefficient observe faster transport as compared to ions having tendency to retard due to adsorption. Similar results were also reported by Grisak and Pickens (1980, 1984).

4.2 *Effect of Fracture Aperture*

In this section, the effect of fracture aperture is observed on the transport of different ions presented in Table 3 while maintaining other parameters like diffusion coefficient, distribution factor, aperture size same as reported in Tables 2 and 3. The fracture aperture is varied from 0.0000005 to 0.0005 m and is presented in Fig. 4. It is observed from Fig. 4a that when fracture aperture is increased to 0.0005 the final normalized concentrations for different ions (Cl^-, F^-, Na^+, K^+, Cu^{2+} and Mg^{2+}) are

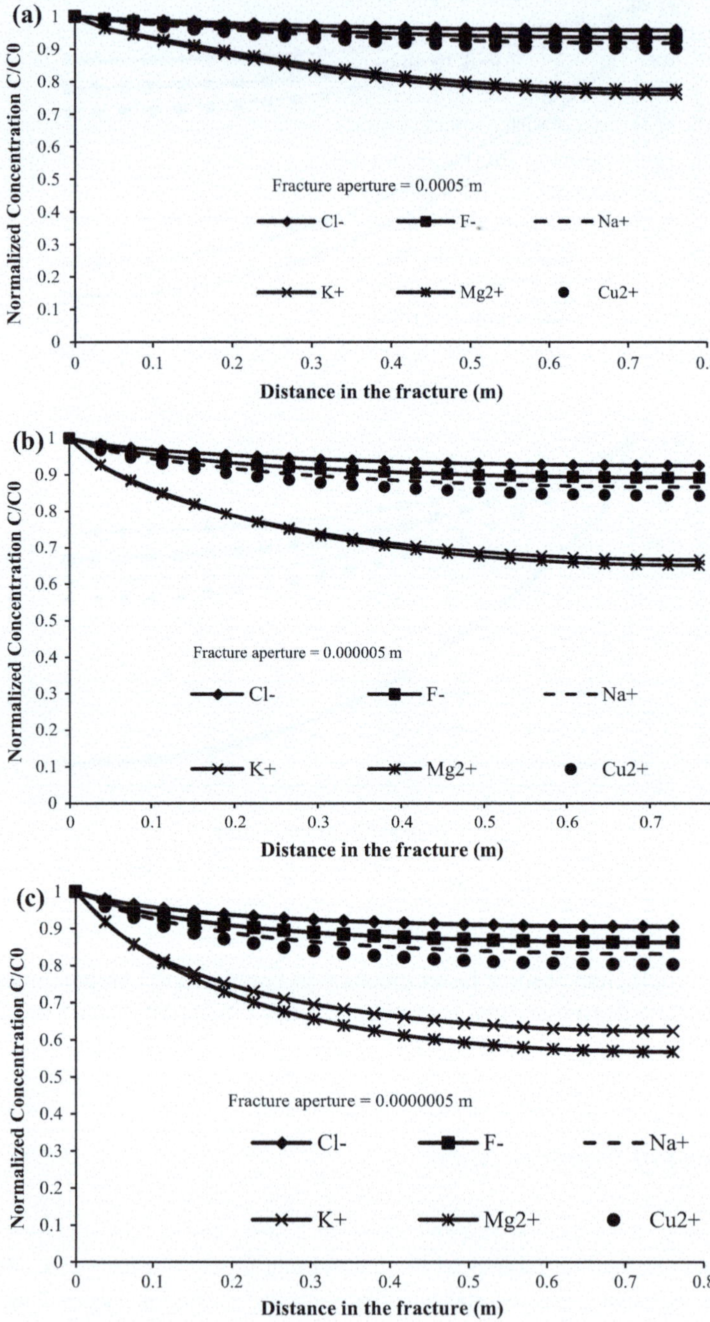

Fig. 4 Effect of fracture aperture on concentration

increased by 0.77, 0.96, 1.11, 2.12, 0.36 and 0.05%. From Fig. 4b, it is observed that when aperture is reduced to 0.000005 m, and the normalized concentrations for different ions (Cl^-, F^-, Na^+, K^+, Cu^{2+} and Mg^{2+}) are reduced by 2.72, 3.8, 4.6, 10.8, 7 and 16%; and when the aperture is further reduced to 0.0000005 m as presented in Fig. 4c, the final normalized concentrations for different ions (Cl^-, F^-, Na^+, K^+, Cu^{2+} and Mg^{2+}) are reduced by 4.7, 6.8, 8.3, 16.5, 11.4 and 26.8%. This indicates that concentration of reactive ions reduces when fracture aperture is reduced. Similar results were reported by Grisak and Pickens (1980) and Satavalekar (2015).

5 Conclusions

The one-dimensional two-coupled formulation for the contaminant transport through fracture and matrix can be modelled using EFGM. It is validated with the example from Tang et al. (1981). The size of the support domain (D_{max}) should be maintained by 1.25 for better results. In the present study, transport of different ions like Cl^-, F^-, Na^+, K^+, Cu^{2+} and Mg^{2+} has been considered in the fracture, and the behaviour of the same has been presented by performing parametric study. The parameters considered are fracture flow velocity, dispersivity, diffusion coefficient and distribution coefficient, and the following conclusion has been drawn.

- It can be observed that the flow velocity in fracture is responsible for the rate of contaminant transported in the fracture media. It is also observed that the transport is also governed by diffusion coefficient and distribution coefficient of different ions. A non-reactive ion-like Cl^- will move faster than Mg2+ ion.
- Fracture aperture plays a similar role like fracture velocity in reducing the concentration in the fracture. The larger the aperture, the larger is the concentration transported in the fracture and vice versa. It is observed that with smaller aperture sizes, reactive ions diffuse more in the matrix and less in the fracture.

References

Belytschko T, Lu YY, Gu L (1994) Element-free Galerkin methods. Int J Num Meth Eng 37:229–256

Chakraborty R, Ghosh A (2010) Finite difference method for computation of 1D pollutant migration through saturated homogeneous soil media. Int J Geomech ASCE 11:12–22

Dolbow J, Belytschko T (1998) An Introduction to programming the meshless element free Galerkin method. Arch Comp Meth Eng 3:207–241

Freeze RA, Cherry JA (1979) Groundwater. Prentice Hall Inc. New Jersey

Grisak GE, Pickens JF (1980) Solute transport through fractured media 1. The effect of matrix diffusion. Water Resour Res 16:719–730

Grisak GE, Pickens JF (1984) Contaminant transport through fractured geologic media. Can Water Resour J 9:108–116

Kumar RP, Dodagoudar GR (2008) Meshfree modelling of one-dimensional contaminant transport. Geotechnique 58:523–526

Kumar RP, Dodagoudar GR (2009) Element free galerkin method for two dimensional contaminant transport modelling through saturated porous media. Int J Geotech Engg 3:11–20

Lancaster P, Salkauskas K (1981) Surfaces generated by moving least-squares methods. Math Comp 37:141–158

Liu GR, Gu YT (2005) An introduction to meshfree methods and their programming. Springer, Berlin

Mategaonkar M, Eldho TI (2012) Two dimensional contaminant transport modelling using meshfree point collocation method. Eng Anal Bound Elem 36:551–561

Satavalekar RS (2015) Meshfree modelling of contaminant migration. PhD thesis, IIT Roorkee

Satavalekar RS, Sawant VA (2015) Two dimensional contaminant transport through layered soil using element free galerkin method. Ind Geotech J 46:192–205

Shackelford CH, Daniel DE (1991) Diffusion in saturated soil. I: background. ASCE J Geotech Eng 117:467–484

Sharma PK, Dixit U (2014) Contaminant transport through fractured-porous media: an experimental study. J Hydro Res 8:223–233

Sharma PK, Srivastava R (2012) Concentration profiles and spatial moments for reactive transport through porous media. J Haz Tox Radio Waste 16:125–133

Tang DH, Frind EO, Sudicky EA (1981) Contaminant transport in fractured porous media: analytical solution for a single fracture. Water Resour Res 17:555–564

Tombul M, Koparal AS, Ogutveren UB (2005) Transport modelling of copper contamination in groundwater caused by a wood preservation plant. Water Qual Res J 4:469–475

Wu Y, Ye M, Sudicky EA (2010) Fracture-flow-enhanced matrix diffusion in solute transport through fractured porous media. Transp Por Med 81:21–34

Energy-Efficient Green Building with Sustainable Engineering of Natural Resources

Aesha Bhatt, Soham Desai and Shuchi Gulabani

Abstract The concept of green building is about achieving energy efficiency and cost efficiency. The green buildings are an indication of significant sustainable development and represent the responsibility of users toward the environment. Using natural resources like rainwater, daylight, proper air ventilation, solar power and wind energy, etc. Buildings can be made both, energy and cost efficient. Thus, focusing on the daylight factor, air ventilation and use of clay wall within the construction of the building can help develop a green building. The application of these factors is easy and also helps the environment. Higher the ventilation, the better IAQ is, and having a better IAQ meets the parameters that define green buildings. Using more of the daylight through openings will reduce the use of artificial lights and power consumption. This project has the study of daylight factor, proper ventilation, and the application of clay wall in the modern building. Calculating daylight factor and proper ventilation will help increase the natural lighting, and with the help of clay wall, the temperature inside the building gets reduced by natural cooling. Also, the concept of rainwater harvesting is used to keep the clay wall humid at several time intervals.

Keywords Green building · Sustainable engineering · Energy efficiency · Natural resources · Clay wall

1 Introduction

The concept of green building is wide and important from the environmental perspective. Conventionally constructed buildings use a lot of natural resources which lead to problems like deforestation and soil erosion. These buildings when constructed also do not ensure the longevity of the building, the construction materials may lead to the depletion of the strength and life of a building ultimately leading to untimely

A. Bhatt (✉) · S. Desai · S. Gulabani
Environmental Science and Engineering Department, MEFGI, Rajkot, Gujarat, India
e-mail: aesha11.bhatt@gmail.com

A. K. Agnihotri et al. (eds.), *Sustainable Engineering*, Lecture Notes in Civil Engineering 30, https://doi.org/10.1007/978-981-13-6717-5_32

demolition (Sinha et al. 2013). Thus, if these resources are used in an efficient manner then the outcomes can be better, such that there is minimal harm to the environment (Mehta and Porwal 2013). Also, proper use can enhance the value of the building and raise the quality of living. With the energy consumption reduced in a few sectors, internal energy usage is focused all around. To achieve a few aspects of green buildings two considerations have been made in this project namely, clay wall and daylight factor (van derhoeven 2013).

Clay Wall: The first is the clay wall concept that is applied to the building. Use of natural clay has been in the construction since historical ages, the benefits of using clay being to not only providing sound structures but also temperature control (Zuo and Zhao 2013). The clay huts that prevail in the rural areas and desert regions have a significant temperature difference from the outer atmosphere. In modern days, the high-rise buildings cannot be constructed just with the clay, thus the application of clay walls can be a possible solution. After the construction of a normal wall, a separate sheet of clay can be put up and can be kept wet with the use of the rainwater collected through the conventional rainwater harvesting system. The clay wall will be of a size that is a little less in both length and height of the original room walls of the building.

India is a country with a hot temperate climate and would need a lot of artificial cooling in its summer days. Similarly, other hot countries like Africa, UAE can also use such concepts that can keep the inner building temperatures lower (Green home Guide). The use of cooling systems use up a lot of the electricity and is not efficient, thus we must have a solution to this. Getting inspired by the ancient Indian building of hot climate areas, we have brought this concept for our modern buildings. The people living in arid regions of Rajasthan live in houses made of white clay. The white clay has the capacity to lower the heat by nearly 10°. The other clay, known as brown clay, were used to make vessels in the early ages to store food, water, and other goods at a lower temperature than the room temperature. Using these clay walls in our building we want to provide cooling sans electricity. The vessel used to store food stays a little wet which, as per our plans, harvested rainwater can act as a coolant. The wall will be set up with the same working like that of an indoor waterfall fountain so that it is continuously kept wet. Water will continuously be passed through the wall dampening it, and the air which comes in contact with the wall cools down, dropping the temperature of the room in turn. The walls will be set up on the built walls a little distance away so that after a little time if the walls need to be changed they can easily be replaced. The application figures and tables are added further.

Daylight factor: Daylight is a natural source of light which meets all requirements of good lighting while enhancing user efficiency and productivity. The clear sky conditions can help in a satisfactory indoor illumination (Littlefair 1985). The daylight factor can be measured and this is an important factor. The concept is the passive design approach (Zhai et al. 2015). The passive design approach is the calculation of sunlight entering the building, the shading, and also the use of properly constructed ventilated buildings (Rodriguez-Ubinas et al. 2014). This approach can also be a support to the clay wall concept applied to the building as the proper ventilation leads to better indoor air quality and more of the use of the passive sunlight can be used (TAMU Mays 2011). There should be a better daylight factor; daylight factor is the quantification of available daylight inside a building with respect to the outer open area illumination. Also, daylight area is a term to be considered as the measurement is with reference to the total area used for living. The illumination and sun slight can also help in ventilation (Yu and Kim 2010) and this keeps the passive cycle of air flow going on in the building. This also improves the IAQ of the building with a flow of air naturally (Chuck and Kim 2011). This reduces energy consumption at the max rates of 10% as described in various journals (CSIR 2001). Thus the target can be to reduce the energy consumptions to this level. The cost reduction can be in the form of the cost required for artificial illumination (Leaman and Bordass 2007).

2 Methodology

Study of the efficiency of clay walls was done on two models prepared of different materials in conjunction with clay. One wall composed of soil and water and the other wall made of soil with husks and water. The walls can be given an additional treatment of heating and cooking the clay at high temperatures to increase durability and wall life.

The process for constructing a lab scale model is simple; a mixture of clay and water was prepared and allowed to set. Sprinkling of water at regular intervals to prevent cracks from forming and to increase durability during setting process was done (Durmus-Pedini and Ashuri 2010). The water used for sprinkling on the walls is from the collection of water done through the water harvesting.

Two walls have been constructed as for the project study, the composition of Wall 1 being soil and water; and the composition of Wall 2 being soil, rice/wheat husks, and water. The time required for construction and watering intervals are mentioned in Table 1.

Supervision time is the time interval after which if watering is stopped cracks may start developing on the surface of plates and which can result in the breakdown of the whole plate. Thus, care has to be taken for watering the surfaces of the wall with light sprinkles. Care has to be taken for the watering the surfaces of the wall with light sprinkles.

Table 1 Process times

	Wall 1 (h)	Wall 2 (h)
Construction	2	2
Watering interval	0.5	0.5
Without supervision	6	4

Design for use, reuse and sustainability of clay wall

- **Design for Use**

The two clay walls that have been made have a different life period. The wall has an extra support of the husks in wall 2 and the wall 1 has more cooling.

Black soil or sticky red soil can be used. 1 part of soil needs 0.3 part of water. When the soil has an addition of husks in it the soil part is added with husks parts is 0.05 in the whole of 1 part of soil.

- **Reliability**

Wall 1: temperature drop by 4 °C
Wall 2: temperature drop by 5 °C

The temperature drop in the walls depends on the composition and the cooling capacity.

- **Maintenance**

1. Watering every half hour manual or automatic
2. Frequent and more amount of watering

The above-mentioned points are keys to the sustainable use of the clay wall. After this, the wall does not need any type of other maintenance. This is why the clay walls can be put into use easily.

- **Reuse design**

Collection of the used clay wall sand can be done and moulding it into a new one with the addition of certain amount of new soil 1. The wall when breaks down can be collected and again moulded into a new plate. This way it can be reused. The soil can be easily moulded again with addition of half amount of new soil and water addition. This helps in increasing the quality of the green building.

- **Design for sustainability**

Proper framing of the wall for sustainability. When a proper framing is given to the wall in balanced form (Meyer and Sardjono) and also strength is increased and the wall can be run longer.

Manufacturing of the wall should be stronger. The manufacturing of the wall if strong then the wall does not need maintenance for long. The properly made wall takes a good time interval even to form cracks.

- **Water usage**

The water used for this watering of the wall is 2% of the lot at every 12 h gap. Only sprinkling of water is needed on the wall to keep it wet so that no cracks are developed. Overwatering may also wash away the soil from its formation.

- **Temperature change**

The temperature change with the application of clay wall is a difference of 2 °C overall. The measurement has been done by placing the wall into a well-ventilated box such that air passes through the wall and the temperature difference inside the box has been measured on an average of 3–4 °C.

The components compared over a period of time in between both of the walls are as follows (Table 2 and Figs. 1, 2, 3 and 4):

Table 2 Components

	Wall 1	Wall 2
Composition	Black soil 4 kg Water 790 ml	Black soil 2.5 kg, Straws (husks) 120 gm Water 520 ml
Temperature difference caused	3	4
Durability	Long	Short
Measurements	25 × 30 1.4 in.	25 × 30 0.8 in.
Time to water plate	Half hour	Half hour
Break down without watering	12–13 days	7–8 days
Drying	Slower	Fast
Life of plate with watering	Long	Comparatively shorter

Fig. 1 Wall 2 made of soil and rice husks

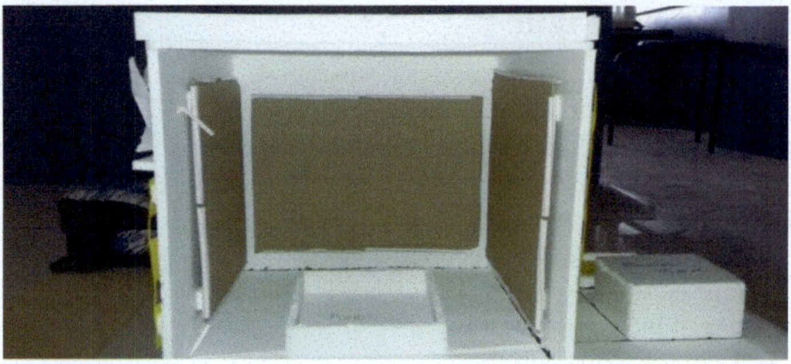

Fig. 2 Position of clay wall inside the room

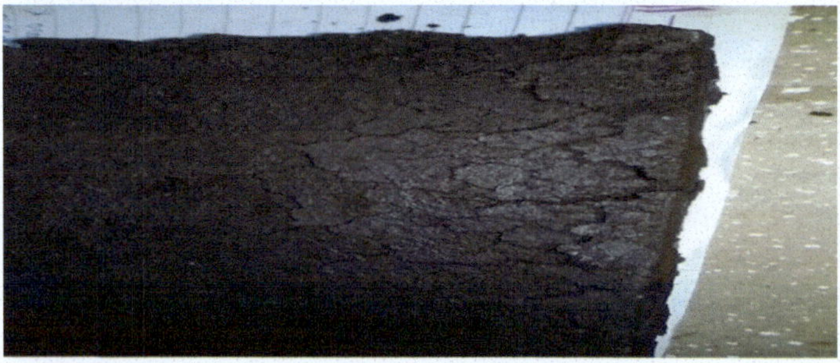

Fig. 3 Initial development of cracks in wall after an interval of 4 months without maintenance but periodic watering

Fig. 4 Dried wall which is almost cracked and no longer in use after a period of 3 months and no watering

3 Results and Discussion

The outcome of this project is that the clay wall concept and its possible application. As the increasing energy demands and consumptions, this can be an alternate energy-efficient concept applied in homes along with the passive daylight energy designing for proper illumination inside the building and also providing proper ventilation to the wall so that the outcomes are maximized and the sustainable solution of natural clay wall can be used. The watering of the wall can be done by the water collected by the rainwater harvesting systems.

4 Conclusions

The use of clay wall can lower the temperature of the surrounding environment by 2° minimum; thus this sustainable application of clay wall is useful. Also is cost-efficient and a use of natural resource for cooling helps in reducing energy consumption than that of the usage of air conditioners. The ventilation systems can help increase IAQ and the clay wall can reduce the use of cooling materials and daylight factor also helps the increment of illumination of the inner building.

References

Chuck WF, Kim JT (2011) Building environmental assessment schemes for rating of IAQ in sustainable buildings. Indoor Built Environ 20(1):5–15

CSIR (2001) An overview of green building rating tools. Built Environ 21:376–386

Durmus-Pedini A, Ashuri B (2010) An overview of the Benefits and risk factors of going green in existing buildings. IJFM 1:1

Green Home Guide-comprehensive resources for green remodeling also includes green landscaping and asks the expert advice. greenhomeguide.com

Leaman A, Bordass B (2007) Are users more tolerant of green buildings? Build Res Inform 35:662–673

Littlefair PJ (1985) The luminous efficacy of daylight: a review. PhD, Building Research Establishment, Garston, Watford, Herts 17(4):162–182

Mahlia TMI, Taufiq BN, Ismail, Masjuki HH (2007) Correlation between thermal conductivity and the thickness of selected insulation materials for building wall 39:182–187

Mehta HS, Porwal V (2013) Green building construction for sustainable future. Civil Environ Res 3(6):1

Meyer DA, Sardjono W Wall frame structure. US5325651A 1:1

Rodriguez-Ubinas E, Montero C, Porteros M, Vega S, Navarro I, Castillo-Cagigal M, Matallanas E, Gutiérrez A (2014) Passive design strategies and performance of net plus energy houses. Energy Build 83:23–35

Sinha A, Gupta R, Kutnar A (2013) Sustainable development and green buildings wood industry. Drvna Industrija (Citeseer) 64(1):45–53

TAMU mays business school corporative study (2011) Quantifying the hidden benefits of high performance building. ISSP Insight 1:1

van derhoeven M (2013), Energy efficient building envelopes. OECD/IEA, International Energy Agency 15(1):1
Yu CWF, Kim JT (2010) Building pathology, investigation of sick buildings-VOC emissions. Indoor Belt 19(1):30–39
Zhai XQ, Song ZP, Wang RZ (2015) Comprehensive review on passive design approaches in green building rating tools 15(2):25−46. 1
Zuo J, Zhao Z-Y (2013) Green building research current status and future agenda: a review. Renew Sustain Energy Rev 30:271–281

Study on the Influence of Attached Mortar Content on the Properties of Recycled Concrete Aggregate

Kanish Kapoor and Amad Ur Rehman Bohroo

Abstract The evergrowing construction industry is one of the largest consumers of natural resources. The increasing demand for the infrastructure, leads to the usage of raw materials like aggregates, is at its peak. Due to which the natural nonrenewable resources, which are being used for producing these materials, are starting to get extinct as these resources have been continuously and tremendously used. Depletion of these natural resources is a matter which needs sincere attention. To counteract this problem, one of the possible solutions is to use recycled aggregates as the substitute for the natural aggregates. The only barrier in using recycled aggregate is that the strength of recycled concrete aggregates is less than that of virgin aggregates because of the adhered mortar present on the recycled concrete aggregates. This reduction of the strength of recycled concrete aggregates depends on the amount of adhered mortar present on the aggregates. The possible solutions to improve the quality of recycled concrete aggregates are by strengthening the adhered mortar and the other by removing the adhered mortar also known as "beneficiation". Both these ways include a number of methods, which are used to improve the quality of recycled concrete aggregates either by strengthening or by removing adhered mortar. This paper reviews the methods to improve the quality of recycled concrete aggregates and also points out the improvement done by each method.

Keywords Beneficiation · Recycled concrete aggregates · Recycled aggregate concrete · Hydrochloric acid · Construction and demolition waste · Compressive strength

---

K. Kapoor
Dr B R Ambedkar National Institute of Technology, Jalandhar, Punjab, India

A. U. R. Bohroo (✉)
Lovely Professional University, Phagwara, Punjab, India
e-mail: amadrehman157@gmail.com

A. K. Agnihotri et al. (eds.), *Sustainable Engineering*, Lecture Notes in Civil Engineering 30, https://doi.org/10.1007/978-981-13-6717-5_33

1 Introduction

Concrete is the second most used construction material after water in the world. The evergrowing construction industry forces more use of construction materials like concrete and hence, the quantity of concrete used is increased day by day. Since concrete is a composite of cement, sand, and coarse aggregates, so the use of these materials is also increased. Aggregate form a major part in concrete about 60–70% (Kosmatha et al. 2002) and more use of concrete means more use of aggregates. Due to which the natural resources from which these aggregates are obtained are continuously and tremendously used. Depletion of these nonrenewable natural resources is a matter which needs major attention.

To counteract this problem, a lot of researches have been done in which the coarse aggregates have been replaced by a lot of substitutes, which include plastics, broken glass, etc., but no satisfactory results have been found. In some cases, the results of the goods were not up to satisfactory limits so that the alternate material could replace aggregate in major quantities. Recycled aggregates derived from construction and demolition waste (CDW) has also been the topic of interest for researchers for the replacement of aggregates. A lot of researchers have used the recycled aggregates in their researches but the use of recycled aggregate leads to the decrease in strength of new concrete. The strength can be reduced up to 40% (Tam and Tam 2005). That is why, the use of recycled aggregates is restricted by several standards.

Construction and demolition waste (CDW) is produced for the demolition of old buildings and structures whose life period has ended or from those structures that are damaged due to earthquake or any other disasters. A lot of concrete is wasted in ready-mix plants after testing on a daily basis, which is then added to this waste. On construction sites, some volume of concrete is wasted every day. In India, 48 million tons of solid waste is produced every year and out of which CDW is 25% (Ghosh and Aich 2011). This amount of solid waste which is produced every year is not completely recycled. Only negligible amount is used for researches, etc., and the rest is dumped in the dump yards. The increase in the amount of solid waste increases the area of land, which is used as dump yards. So, the lands which would otherwise be used for useful purposes are turning into dump yards. Thus, if somehow recycled aggregates are used in the concrete, two purposes can be solved: first, the pressure on nonrenewable natural resources will be decreased and second, the waste will be reused and the useful lands will be saved from becoming dump yards. But the only problem is the decrease in strength in new concrete when recycled aggregates are used.

The main reason behind the decrease in strength of recycled aggregates is the adhered mortar. Recycled aggregates produced from CDW have a layer of porous old mortar, which surrounds the actual aggregate (Katz 2003). The volume of this adhered mortar varies from 25 to 60% (Corinaldesi 2010). The strength is also decreased because there is an additional interfacial transition zone (ITZ) in recycled aggregate concrete. The presence of another interfacial transition zone (ITZ) in the case of RAC: one ITZ is between aggregate and old mortar (adhered mortar) and other ITZ is between old mortar and new mortar, reduces the strength of recycled concrete aggregates. So, there are two planes of weakness in the recycled aggregate concrete (Guneyisi et al. 2014). The only way to increase the number of recycled aggregates used in concrete is to treat the aggregates before use. The researchers have found that the quality of recycled aggregates can be improved by either strengthening the adhered mortar or by removing adhered mortar also known as beneficiation. Various physical, chemical, biological, etc., methods are used to improve the properties of recycled concrete aggregates (Silva et al. 2015).

2 Methods to Treat Recycled Aggregates

Recycled concrete aggregates (RCA) are the waste from the demolition of old concrete, brick masonry, or rubble masonry structures. Nowadays, with the development in construction, a lot of concrete is used. Use of heavy amount of concrete puts danger on resources. Hence, the use of recycled concrete aggregates comes into play. The methods that have been used to improve the quality of recycled concrete aggregates are divided into the following two broad processes: by strengthening the adhered mortar or by removing the adhered mortar. Sometimes, a combination of both the processes is used to get effective results. The following are the methods which are used to improve the quality of aggregates using abovementioned techniques.

2.1 Presaturation

It is a process in which the recycled concrete aggregates are immersed in water and are kept there for few minutes before these are taken out and used in concrete. Sometimes, these aggregates are again soaked to improve the properties. Thus, the aggregates are soaked one or two times before their use. This is generally done to improve the fresh properties of the concrete as the soaking of aggregates in water generally improves the workability of the concrete as the water-to-cement ratio is not altered. According to Garcia-Gonzalez et al. (2014), recycled aggregates were soaked for short intervals before these were used in concrete. It was seen that there was an insignificant reduction of compressive strength (11 and 13% for 3 and 5 min soaking periods). It was observed that the recycled aggregates reached saturation after 10 days and 56.1% of saturation was observed in 60 min.

2.2 Mechanical Grinding

It is the mechanical process of removing the adhered mortar from the recycled aggregates. In this process, the aggregates are rotated in a rotating drum where crushing and ball milling action is involved. The aggregates are positioned inside the milling drum and the drum is allowed to rotate at a certain speed for a particular period. The aggregates strike with each other and with the walls of drum and some portion of adhered mortar is removed from these aggregates. Sometimes, steel balls are also placed in the drum and revolved along with the recycled aggregates to increase the quantity of adhered mortar removed. The quantity of adhered mortar depends upon the speed of the rotation and the time of rotation. This method is very simple and famous but it is not frequently adopted because the crushing action sometimes develops the micro-cracks on the surface of actual aggregates present inside the layer of adhered mortar of recycled aggregates. This generally happens if the amount of adhered mortar is less and the surface of actual aggregate is exposed. The researcher (Pepe et al. 2014) used the mill drum of 300 mm diameter and 500 mm height. It was filled with recycled aggregates up to 33% of volume of drum and the drum was allowed to rotate at 60 rpm. After that, the fine particles and dust from aggregates were removed by washing, and then the aggregates were allowed to dry. The results showed a decrease in the water absorption with an increase in the duration from 2 to 15 min. Recycled aggregates before treatment had 30% adhered mortar. But after mechanical grinding, the decrease of adhered mortar was up to 15%.

2.3 Heat Treatment

This method involves high temperature for the treatment of recycled aggregates. High temperature and high capacity ovens are required for this process. Recycled aggregates are placed in the high-temperature ovens at a certain high temperature for a specific duration, and then the aggregates are taken out and were allowed to cool down. The properties of aggregates are improved in terms of water absorption and weathering, etc. According to the researchers (Al-Bayati et al. 2016; Fathifazl et al. 2009), treatment of recycled aggregates was done at four different temperatures, 250, 300, 500, and 750 °C for 1 h in a conventional oven. The results after the treatment showed that there was an improvement in the water absorption, specific gravity, porosity, and freeze–thaw resistance. It was concluded from the research that the recommended temperature is between 300 and 350 °C. Temperature range 400–600 °C leads to the expansion and development of internal stresses. Temperature from 600 to 800 °C leads to visible micro-cracking which deteriorates the aggregates.

2.4 Two-Stage Mixing

Two-stage mixing approach (TSMA) method improves the properties of recycled aggregates. In this method, the mixing water is divided into two parts: First part is mixed with cement and a slurry is made. This cement slurry is added to the fine and coarse aggregates, and then the mix is properly mixed. This will allow the pores and fissures of recycled aggregates to be filled with the slurry. After that, the remaining water is added and mixed to complete the hydration. According to the researcher (Tam et al. 2007a, b), mixing water was divided into two parts: First part was mixed with cement, sand, and recycled aggregates and mixed well. The thin cement slurry filled the aggregate pores and cracks. Then, the second part of water was added to complete the hydration. The results indicated that the strength was achieved up to 21.19% for 20% recycled aggregates replacement for 28 days.

2.5 *Three-Step Method*

In this test, a part of residual mortar is removed by a process and then the quality of adhered mortar of recycled aggregates is improved in another step. This test (Gao et al. 2013) is a combination of three steps involving rough crushing of the concrete aggregates in the first step, thermal treatment of the crushed concrete aggregates to separate the adhered mortar from the aggregates in the second step, and then chemical treatment with salicylic acid of remaining attached mortar from the aggregates in the final step. The first step, crushing, involves the crushing of RCA manually by hammer to a size of 50 mm or more considering the maximum size of aggregates normally used is 20 mm. This is done to avoid destroying of aggregates. The second step of the three-step process, i.e., physical treatment, involves two variants: a soundness test (ST) consisting of cycles of freezing (−17 °C) and heating (+60 °C) immersed in 26% Na_2SO_4 solution and the second thermal test involved liquid nitrogen microwave heating cycles. Both methods showed similar efficiencies, i.e., a direct recovery rate of 84% of clean aggregates of size 20 mm (52% were recovered out of 62% of 20 mm aggregates initially present in the concrete). The final step, chemical treatment, involves the removal of adhered mortar that is left after the first two steps. The chemical treatment using salicylic was done in the third step with an efficiency of 67−69% (Gao et al. 2013).

2.6 *Mineral Admixture Addition Approach*

This method improves the quality of adhered mortar. In this method, the surface of the aggregates is coated with low water to cement ratio paste or the surface impregnation of recycled aggregates is done using silica fume or any other mineral admixture. The surface coating and impregnation of mineral admixture help in the penetration of admixture in the cracks and fissures of loose layers of recycled aggregates. Interfacial transition zone (ITZ) is improved by silica fume impregnation. This is achieved when cracks are filled with the silica fume and the pozzolanic reaction of silica fume with calcium hydroxide $Ca(OH)_2$ produces secondary calcium silicate hydrate (CSH) gel, which strengthens recycled concrete aggregates. Silica fume treatment is stronger in early stage but gets slower with more addition as the cracks get filled and no calcium hydroxide is left to react. The other pozzolanic substances like fly ash, GGBS, etc., show the similar effects. This method improves the strength and durability of the recycled concrete aggregates (Wang et al. 2013).

2.7 *Self-healing*

This is the method which improves the quality of adhered mortar attached to the recycled aggregates. It is a time-consuming process. In this process, recycled aggregates are immersed in water for one month. This period of 1 month gives enough time to the unhydrated cement particles to react again with water to form calcium silicate hydrate which will add to the strength of the concrete (Elhakam et al. 2012).

2.8 *Polymer Emulsion*

In this method, silicon-based emulsions composed of silane or siloxane or both of them are used to impregnate recycled concrete aggregates. The impregnation can be simple impregnation or double impregnation. In the case of single impregnation, the recycled concrete aggregates are impregnated with polymer solution for 5 min then dried at room temperature maintained at 20 °C and relative humidity of about 50% for 24 h and then in oven at temperature of 50 ± 5 °C until the difference in the mass after 24 h is less than 0.1%. In case of double impregnation the recycled concrete aggregates are impregnated with polymer solution for 3 min then dried at room temperature maintained at 20 °C and about relative humidity of 50% for 20 h; then these recycled concrete aggregates are again impregnated with polymer solution for 5 min followed by drying fir 24 h at room temperature maintained at 20 °C and in oven at 50 ± 5 °C until the difference in the mass is less than 0.1%. These processes of impregnation improves the pore structure of recycled concrete aggregates and there is a significant reduction in the water absorption as well (Spaeth and Tegguer 2013).

2.9 *Calcium Carbonate Bio-deposition*

This method is used to improve the properties of attached mortar attached to recycled aggregates. This is a natural process as all the substances required for this method are naturally available in the environment. A bacteria named Sporosarcina pasteurii is used in this process. The bio-deposition is based on the ability of the bacteria to precipitate the calcium carbonate on the outer surface of the cell wall due to the occurrence of negative zeta potential of adequate strength. Sporosarcina pasteurii cell can attract calcium ions (Ca_2^+), which react with carbonate ions ($CaCO_3^{2-}$) originating from urea ($CO(NH_2)_2$) hydrolysis. Simultaneously, ammonia ions (NH_4^+) increase the pH of the surrounding medium which makes the environment suited for bacteria and thus calcite precipitation efficiency is increased. The results showed a reduction in the water absorption and results were even better when finer fragments of aggregates derived from the inferior quality of concrete were used (Grabiec et al. 2012).

2.10 *Carbonation*

Carbonation is the process in which the calcium hydroxide (which is one of the main cement hydration products) reacts with the carbon dioxide in the atmosphere. This method improves the quality of recycled concrete aggregates as it increases the solid volume by depositing calcium carbonate on the surface of aggregates. The results showed that the CO_2 curing process can densify the adhered mortar. After CO_2 curing, there was a reduction in the water absorption and porosity of the aggregates. The carbonation was easily done for RCA with smaller particle size. The moisture content played an important role in carbonation process as the dry matrix could not provide the sufficient water for the carbonation reactions and the pores in the water saturated matrix were filled with a water blocking CO_2 penetration. The carbonation reaction was fast for the first 2 h but slowed down sharply after that (Silva et al. 2015).

2.11 *Presoaking in Acid*

Soaking of the RCA in acid solutions of low molarities has also been used to remove the adhered mortar from the RCA. Different kinds of acids have been used for this method which includes hydrochloric acid (HCl), sulphuric acid (H_2SO_4), and phosphoric acid (H_3PO_4). Some other types of acids have also been used. RCA is soaked in the acid solution of low molarities for a few hours. Before soaking in acid, these aggregates are washed with distilled water to remove any stray particles on them. These aggregates are again washed once the soaking is done. The duration of soaking depends upon the type and concentration of the acid used for soaking. The acid helps in removing some of the adhered mortar from the aggregates. Good results can be achieved with this technique. The care should be taken while choosing the type and concentration of acid as it should not harm the actual aggregate and should only remove the adhered mortar. Sometimes, mechanical grinding is also done after soaking to get good results but the same care should be taken here with the actual aggregates that are surrounded by adhered mortar. Researcher (Tam et al. 2007a, b), soaked the RCA for 24 h at 20 °C, and then RCA was washed to remove any acid on them. The acids used were hydrochloric acid, sulphuric acid, and phosphoric acid with a molarity of 0.1 M. The results show that the water absorption was reduced and the mechanical properties of concrete were improved. Researcher (Ismail and Ramli 2014) soaked the aggregates in 0.5 M HCl for 24 h and then washed with distilled water. Then, the RCA was impregnated with calcium meta silicate ($CaSiO_3$) solution for 24 h. The impregnation was done to fill the pores and fissures of RCA. These meta-silicate particles present on recycled concrete aggregates will dissolve in concrete mix and will act as filler which will densify the microstructure. The result showed improved microstructure, bond strength, and less water absorption. RCA was soaked in acidic solution of hydrochloric acid (37%) and acetic acid (99.7%) at a low concentration of 0.1 M for 24 h at 20 °C (Al-Bayati et al. 2016). According to the researcher, the RCA was soaked in HCl solution for 24 h at 20 °C and then washed with distilled water. The researcher (Ismail and Ramli 2013) studied the effect at different concentrations of acid on the RCA. Three types of molarity 0.1, 0.5, and 0.8 M were used in the study. The immersion time was 1, 3, and 7 days. The result showed linear correlation between the amount of mortar loss and increase in concentration. However, the duration of treatment did not have any significant effect on the loss of adhered mortar.

3 Method Adopted (Chemical Beneficiation)

A method was adopted to see the amount of adhered mortar removed from the recycled aggregates and test was conducted on it. The more the quantity of adhered mortar removed, the more is the possibility of recycled aggregates to be used in the concrete. The test method and results are described below:

3.1 Test Method

Acid soaking method was adopted and hydrochloric acid (HCl) was used to make the solution. To remove the adhered mortar, recycled were allowed to soak in the acid of different molarities and for different time periods. Recycled concrete aggregates were soaked in HCl with molarities of 0.1, 0.3, 0.5, and 0.7 M for a soaking period of 1, 3, and 7 days. Three samples of aggregates were taken for each molarity and were soaked in acid for the said three durations. The aggregates were weighed in their natural dried form pre- and post-soaking treatment to calculate the percentage of adhered mortar removed. Distilled water was used for making acid solution thereby the acid remained unreacted until the recycled aggregates were soaked. The aggregates were stirred after every 24 h and at the end of the soaking period, aggregates were washed with distilled water to remove the remains of acid and chunks of particles. The recycled aggregates used were recycled concrete aggregates (RCA). The RCA was taken from a ready-mix plant near the author's host institute and were crushed to a size of 20 mm. The acid used was hydrochloric acid (36.46%) of specific gravity 1.189.

Table 1 Percentage of adhered mortar removed with respect to total sample of aggregates

S No.	Molarity of HCl (M)	Adhered mortar remove (1 day) (%)	Adhered mortar removed (3 days) (%)	Adhered mortar removed (7 days) (%)
1	0.1	0.7	1.1	2.1
2	0.3	1.4	1.9	2.7
3	0.5	2	2.5	2.9
4	0.7	2.6	2.9	3.3

Complete adhered mortar was also removed to calculate the total mortar content of the RCA, a solution with rthe atio of 200 ml of acid (36.46%) in 600 ml of distilled water (≈4 M) is prepared and a sample of RCA is immersed in this solution (IS: 1199-1959). The acid solution was changed after every 24 h. This method is used to find mortar content. The weight of aggregates before and after treatment is calculated in the natural dried state.

3.2 *Results*

The amount of adhered mortar was calculated with respect to the total weight of the sample taken and presented in Table 1. To determine the amount of attached mortar with respect to the total mortar content, the total mortar content of recycled concrete aggregates was determined. The test results showed that the adhered mortar removal is directly proportional to the increase in molarity of acid and increase in soaking duration. Complete adhered mortar removal was done to calculate the adhered mortar removed with respect to total mortar content (Table 2). The researches showed that more adhered mortar removal leads to the increasing strength of the recycled aggregate concrete. That means the strength of recycles aggregate concrete increases with the increase in the soaking duration and molarity of acid. For each soaking duration, the increase in the molarity increased the percentage of adhered mortar removal. The rate of adhered mortar removal is decreased. The trend shows that the percentage of adhered mortar removed increases with increase in the molarity of acid (Fig. 1).

4 Conclusions

- It can be concluded that the use of alternative materials will reduce the impact on nonrenewable natural resources. Untreated recycled aggregates decrease the strength and durability properties but if recycled aggregates are treated, the properties can be improved. The attached mortar is the main factor for decreased

Table 2 Percentage of adhered mortar removed with respect to total mortar content

S No.	Molarity of HCl (M)	Adhered mortar removed (1 day) (%)	Adhered mortar removed (3 days) (%)	Adhered mortar removed (7 days) (%)
1	0.1	1.6	2.51	4.79
2	0.3	3.2	4.34	6.16
3	0.5	4.57	5.71	6.62
4	0.7	5.94	6.62	7.53

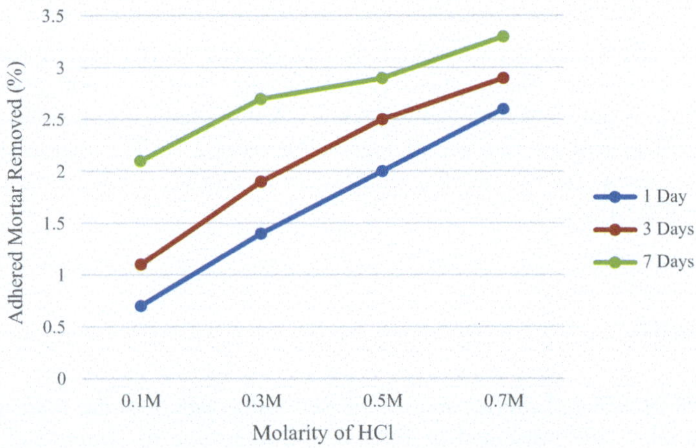

Fig. 1 Percentage of adhered mortar removal with increase in molarity

strength. This problem can be solved by either strengthening adhered mortar or by removing it.

- The amount of adhered mortar removed is directly proportional to the increase in molarity and increase in the soaking period.
- The rate of adhered mortar removal is decreased with increase in the soaking duration. It is clear from the results that for 0.7 M, the percentage of mortar removal is 5.94% (1 day), 6.62% (3 days), and 7.53% (7 days). For an increase in duration of 2 days (for 1–3 days), the percentage removal is increased by 0.68% and for further increase in the duration of 4 days (from 3 to 7 days), the percentage removal is increased by 0.91%. Likewise, trends are observed at 0.1, 0.3, and 0.5 M.
- Research show that the strength of the treated RCA is be more than untreated RCA.

References

Al-Bayati H et al (2016) Evaluation of various treatment methods for enhancing the physical and morphological properties of coarse recycled concrete aggregate. Constr Build Mater 112:284–298
Corinaldesi V (2010) Mechanical and elastic behaviour of concretes made of recycled concrete coarse aggregates. Constr Build Mater 24(9):1616–1620
Elhakam AA et al (2012) Influence of self-healing, mixing method and adding silica fume on mechanical properties of recycled aggregates concrete. Constr Build Mater 35:421–427
Fathifazl G et al (2009) New mixture proportioning method for concrete made with coarse recycled concrete aggregate. J Mater Civil Eng 21(10):601–611
Gao XX et al (2013) A three-step method for the recovery of aggregates from concrete. Constr Build Mater 45:262–269
Garcia-Gonzalez J et al (2014) Pre-saturation technique of the recycled aggregates: solution to the water absorption drawback in the recycled concrete manufacture. Materials 7:6224–6236
Ghosh S, Aich A (2011) Rebuilding C&D waste recycling efforts in India. Waste Manag World 12:5
Grabiec A et al (2012) Modification of recycled concrete aggregate by calcium carbonate bio deposition. Constr Build Mater 34:145–150
Guneyisi E et al (2014) Effect of surface treatment methods on the properties of self-compacting concrete with recycled aggregates. Constr Build Mater 64:172–183
Ismail S, Ramli M (2013) Engineering properties of treated recycled concrete aggregate (RCA) for structural applications. Constr Build Mater 44:464–476
Ismail S, Ramli M (2014) Mechanical strength and drying shrinkage properties of concrete containing treated coarse recycled concrete aggregates. Constr Build Mater 68:726–739
IS: 1199-1959: Methods of sampling and analysis of concrete, Bureau of Indian Standards
Katz A (2003) Properties of concrete made with recycled aggregate from partially hydrated old concrete. Cem Concr Res 33(5):703–711
Kosmatka SH et al (2002) Design and control of concrete mixtures, 7th edn. Cement Association of Canada, Ottawa
Pepe M et al (2014) Alternative processing procedures for recycled aggregates in structural concrete. Constr Build Mater 69:124–132
Silva RV et al (2015) Carbonation behaviour of recycled aggregate concrete. Cem Concr Compos 62:22–32
Spaeth V, Tegguer AD (2013) Improvement of recycled concrete aggregate properties by polymer treatments. Int J Sustain Built Environ 2:143–152
Tam VWY, Tam CM (2005) Recycled Aggregates from concrete waste for higher grade of concrete construction. PhD thesis, Department of building and construction, City University of Hong Kong
Tam V et al (2007a) Optimization on proportion for recycled aggregate in concrete using two-stage mixing approach. Constr Build Mater 21:1928–1939
Tam V et al (2007b) Removal of cement mortar remains from recycled aggregate using pre-soaking approaches. Resour Conserv Recycl 50:82–101
Wang H et al (2013) Improving performance of recycled aggregate concrete with superfine pozzolanic powders. J Cent South Univ 20:3715–3722

Properties of Concrete Made with Plastic Aggregates

Kanish Kapoor, Bhupesh Kumar Gupta and Satwinder Singh

Abstract Owing to the advantages of plastics like cheapness, endurance, lightness, hygiene, and design adaptability, its use is increasing day by day. The negative side of the picture is that it is nonbiodegradable product and they cannot be disposed to open environment, because they affect the environment, soil, and subsoil. They release toxic gasses when exposed to open environment and can block sewer lines, drainages, and other waterways. Plastics are also harmful to human health; they may contain or lead to harmful acids. Thus, recycling of plastic is essential in today's condition. Incorporating the waste plastics in concrete by converting them into aggregates is one of the possible solutions to consume them. In the present study, work of various authors on natural fine aggregates partially replaced with recycled plastic fine aggregates in concrete is analyzed. The various mechanical and durability properties of concrete made with incorporation of plastic fine aggregates are presented in the paper.

Keywords Plastic aggregates · Compressive strength · Flexural strength · Shrinkage · Carbonation · Water absorption

1 Introduction

In blessing situation, plastic waste turns into an essential dialog. All through a decade ago, the great populace increment worldwide along the edge of the need of people to receive enhanced states of living light-transmitting diode to a significant increment of the use of polymers (essentially plastics). Materials appear to be entwining with the extraordinary society wherever we tend to live. Machines produced from plastic region unit getting utilized at each water or compound snugness, quality, and flexibility can be formed effectively into various shapes and so on. Plastics are utilized as

K. Kapoor
Dr B R Ambedkar National Institute of Technology, Jalandhar, Punjab, India

B. K. Gupta (✉) · S. Singh
Lovely Professional University, Phagwara, Punjab, India
e-mail: bgsunny847@gmail.com

A. K. Agnihotri et al. (eds.), *Sustainable Engineering*, Lecture Notes in Civil Engineering 30, https://doi.org/10.1007/978-981-13-6717-5_34

a part of majority share for packaging, pipes and plumbing, toys, bottles, polythene packs, electronic division, and so forth. India utilizes around 43% of plastic in packaging (Seetharaman 2017). Plastics are acquired after polymerization of monomer units to shape long chain of hydrocarbons either by buildup or expansion responses. Unadulterated plastics have less poisonous quality since they are insoluble and synthetically latent in nature, yet because of expansion of specific added substances in plastic to enhance some of its properties or granting diverse hues, and hence plastic wind up plainly poisonous in nature. Overabundance utilization of plastic outcomes in the age of substantial measure of plastic waste stacks which makes terrible situation and furthermore open dumping pull in ailments. These waste plastics if enter in sewer line can cause blockage, and the entire sewer line will move toward becoming breakdown and proficiency of sewer will diminish. It contains various bonds (twofold or triple bond) between carbon particles which do not break effectively under ordinary conditions and typically enjoys long time to reprieve bond because of which plastic takes quite a long while up to 1000 years to corrupt (Thosar and Husain 2017). Plastic being nonbiodegradable and artificially inert discharges poisonous mixes into the air, water, and soil in specific situations which causes long-haul natural contamination. Traditionally, recycling or incinerating plastic at 850 °C is the only possible way to dispose solid waste. But if solid waste is incinerated, it releases flue gasses into the atmosphere which can even cause cancer and finally end product of incineration will be solid and highly toxic which must be buried in earth pits safely. Leaching can occur which can degrade soil underneath. The above facts bring a must focus on well-mannered unseating of plastic waste.

Current development industry is an enormous industry, and it is as yet developing step by step, as the improvement is going on all around and in quick developing country like India, it is developing at a quicker rate. Subsequently, use of natural materials is at its pinnacle, because of which the regular non-sustainable assets which are being utilized for delivering these materials are beginning to get wiped out as these assets have been consistently and colossally motivation behind our life as they have differed valuable properties in setting of simple life. Plastic has made its lasting spot in our day-by-day life because of its efficiency, simple accessibility, cleanliness, lightweight, solidness, utilization. Consumption of characteristic assets is an issue which needs earnest consideration. Inside the pursuit for different choices to natural materials, the utilization of plastics as a substitute may serve two needs: one it energizes the waste and second it may save the natural materials. Plastic waste can be utilized as a part of concrete as partial substitution of common fine aggregates by plastic aggregates prepared from Polyethylene Terephthalate (PET) plastic containers; it can be shelter for environment and developing construction industry. Utilizing this strategy, we can discover option and safe technique to arrange plastic waste as it will diminish cost of reusing, transportation cost, and essentially helpful land which will end up plainly fruitless in the wake of dumping plastic waste. Moreover, mistreatment of already consumed plastic waste in concrete will not solely be its safe disposal methodology; however, it might improve the concrete properties like durability, defiance to chemicals, shrinkage, and short- and long-term creep.

2 Lightweight Plastic Aggregates' Production

A waste organization configuration arranges the improvement practices toward a normally welcoming technique by diminishing the measure of waste materials and their discard in landfills. The normal and money-related central focuses happen when waste materials are diverted from landfills fuse—a. safeguarding of raw materials; b. diminish in the cost of waste disposal; and c. beneficial usage of the materials. Waste materials must be kept clean and in confined groups with a particular true objective to be utilized or recycled in an effective way. The procedure of creating lightweight aggregates contrasts from researcher to researcher. They are made from the waste PET bottles by coordinate cutting in the scope of 5–15 mm and afterward interrupted in the blender. The blender at that point pivots with a speed of 30–50 rpm for 20 s with an internal temperature of 250 ± 10 °C and after that Granulated Blast-Furnace Slag (GBFS) is acquainted with a blender to harden the surface of aggregates (Choi et al. 2005). Another procedure is to granulate the waste PET bottles, having a thickness of 1–1.5 mm, in a cutting edge plant to the extent of 0.1–5 mm. At that point, the acquired particles were isolated, through sieves and along these lines reassembled in like manner to the granulometric curve of the fine aggregates that they would substitute (Frigione 2010). The plastic aggregates can likewise be acquired after mechanical granulating of PET waste took after by cleaning and detachment by physico-compound techniques. Plastic aggregates can likewise deliver from crushed plastic stones. PET bottles are squashed into little division and washed to expel the remote particles. At that point, it is warmed at a specific temperature with the goal that the essential brittleness is acquired. After expulsion, the liquid plastic is chilled off and gathered in rocks of 100 mm estimate roughly (Mathew et al. 2016).

3 Influence of Plastic on Fresh Concrete Properties

Workability

With the expansion of plastic particles, workability of concrete augmentations as Waste Plastic Aggregate (WPA) has spherical and smooth shape. The smooth surface WPA poses less contact which drives an increase in slump value. Likewise, less water osmosis point of confinement of WPA prompts more prominent availability of water in the transition zone, which is another factor for the advancement of workability (Choi et al. 2005). Some different investigations likewise say diminishment or no impact in workability with augmentation of WPA. This lessening is credited to the way that a few particles are angular, and others have nonuniform shapes bringing about less fluidity (Ismail and AL-Hashmi 2008; Frigione 2010). Past writers likewise found that WPA particles have more specific surface area contrasted with the characteristic sand due to their soldier of fortune shape. Consequently, there would be more friction between the particles prompting less workability in the blends (Rahmani et al. 2013). Table 1 depicts some results of slump value from previous authors.

Table 1 Slump test result from various authors

Author	Replacement level (%)	Range of slump value (mm)
Guendouz et al. (2016)	0–40	70–120
Amalu et al. (2016)	0–25	70–78
Harini and Ramana (2015)	0–20	120–150
Safi et al. (2013)	0–50	240–380

Table 2 Fresh density result from various authors

Author	Replacement level (%)	Density range (kg/m^3)
Guendouz et al. (2016)	0–40	2090.0–1900.0
Akçaözoğlu et al. (2013)	0–60	2240.0–1590.0
Rahmani et al. (2013)	0–15	2413.96–2337.26
Batayneh et al. (2007)	0–20	2320.0–2495.0

Fresh Density

Fresh density of WPA Concrete mixture (WPAC) is essentially lessened as the density of the WPA being lower than the sand by 69.7% (Ismail and AL-Hashmi 2008). The specific weight of WPA (1.27 g/cm^3) is essentially lower than that of natural aggregate (2.45 and 2.57 g/cm^3) which prompts diminishment in fresh density of WPAC (Akçaözoğlu et al. 2013). Table 2 enlists some fresh density range from previous studies.

4 Influence of Plastic on Hardened Concrete Properties

Dry Density

The inclusion of WPA into solid blend demonstrates a diminishing pattern for dry density, which is like fresh density. The dry densities at each curing age tend to diminish with expanding the waste plastic proportion in each solid blend, yet the dry densities tend to increment with time for each solid blend at all curing ages (Ismail and AL-Hashmi 2008). The lessening of the mass of the blend happened, as the thickness of natural aggregate is higher than the density of WPA. At higher rate substitution, more diminishment in mass of cement was seen by scientists. Table 3 highlights some dry density range from previous studies.

Table 3 Dry density result from various authors

Author	Replacement level (%)	Density range (kg/m^3)
Thosar and Husain (2017)	0–60	2574.80–2354.43
Osei (2014)	0–100	2483–1433
Akçaözoğlu et al. (2013)	0–60	2230–1530
Rahmani et al. (2013)	0–15	2221.83–2147.68

Table 4 Compressive test result from various authors

Author	Replacement level (%)	28 days compressive strength (MPa)
Gaur et al. (2017)	0–25	32.51–24.8
Thosar and Husain (2017)	0–60	26.47–25.83
Jibrael and Peter (2016)	0–5	46.6–38.95
Guendouz et al. (2016)	0–20	20–26
Yongo et al. (2016)	0–20	46.87–42.51
Arivalagan (2016)	5–20	29.3–13.33
Amalu et al. (2016)	0–25	24.44–17.55
Ismail Al-Hadithi and Fawzi Ahmed Alani (2015)	0–7.5	69–50.35
Harini and Ramana (2015)	0–20	47–40

5 Compressive Strength

Diminishment in strength because of decrease in bonding with presentation of plastic particles was observed. The event of higher content of bleeding water, generally situated around the aggregates of WPAC, created a weaker bond between the concrete lattice and WPA (Frigione 2010). The abatement in adhesive strength between the surface of the WPA and the concrete paste, and additionally the particles' size of the WPA increment brings about decrement of compressive strength of WPAC. Moreover, hydrophobic property of plastic confines the water fundamental for bond hydration from entering through the structure of the concrete amid the curing time frame prompting the loss (Ismail and AL-Hashmi 2008). At low substitution levels, interlocking between the WPA on the cracked surfaces is incremented because of the special shape of the WPA and their flexibility. Be that as it may, when the substitution level expands, on account of the powerless attachment between the surface and the WPA, they act as obstruction and keep the concrete glue from holding fast to natural aggregates. Subsequently, the contact is not critical and the concrete strength reductions step by step (Rahmani et al. 2013). Table 4 shows a comparison of compressive strength results from some previous authors.

Table 5 Flexural test result from various authors

Author	Replacement level (%)	28-day flexural strength (MPa)
Jibrael and Peter (2016)	0–5	3.8–2.9
Guendouz et al. (2016)	0–20	3.5–5
Yongo et al. (2016)	0–20	4–3.5
Amalu et al. (2016)	0–25	48.32–45.58
Ismail Al-Hadithi and Fawzi Ahmed Alani (2015)	0–7.5	9.11–10.17
Harini and Ramana (2015)	0–20	4.6–3.75
Reis and Carneiro (2012)	0–20	7.82–4.85

6 Flexural Strength

Decrement in adhesive quality between the surface of waste plastic particles and the bond glue and also the hydrophobic nature of WPA which may restrict the hydration of cement watches out for the decrement of flexural strength of WPAC (Ismail and AL-Hashmi 2008). With the augmentation of WPA, flexural strength has an expanding pattern to begin with; however, it drops sooner or later. For instance, at 5% replacement level, WPAC indicates 8.02% expansion in flexural strength, where the same for 15% replacement level is 6.25% lessening in the flexural strength. This event informs interlocking between the two cracked surfaces as a result of the special shape of WPA and their flexibility which keep the entire beam failure (Rahmani et al. 2013). Moreover, WPA can defer split start and delay the break engendering interim, thereby expanding structural strength (Ismail Al-Hadithi and Fawzi Ahmed Alani 2015). Table 5 enlists a comparison between flexural strength results obtained from previous studies.

7 Split Tensile Strength

The negative impact of the smooth surface on the bond quality is featured because of the expansion surface territory of WPA contrast with sand prompts loss in split tensile strength as the compressive strength does (Frigione 2010; Rahmani et al. 2013). Table 6 depicts some split tensile strength range obtained by previous authors.

8 Modulus of Elasticity

Decrease in modulus of elasticity of WPAC due to the moderately bringing down modulus of elasticity of WPA is watched. Moreover, the week joint between the

Table 6 Split tensile test result from various authors

Author	Replacement level (%)	28-day split tensile strength (MPa)
Thosar and Husain (2017)	0–60	11.133–7.233
Yongo et al. (2016)	0–20	3.48–2.81
Arivalagan (2016)	5–20	2.97–2.26
Ismail Al-Hadithi and Fawzi Ahmed Alani (2015)	0–7.5	3.87–3.36
Harini and Ramana (2015)	0–20	3–2.8
Saikia and Brito (2013)	0–15	3.49–2.28

Table 7 Modulus of elasticity result from various authors

Author	Replacement level (%)	28-day elastic modulus range (GPa)
Ismail Al-Hadithi and Fawzi Ahmed Alani (2015)	0–7.5	73.08–40.53
Rahmani et al. (2013)	0–15	34–25.9
Choi et al. (2005)	0–75	23.5–18.5

texture and PWA is another purpose behind this wonder (Rahmani et al. 2013). Table 7 shows the variation of modulus of elasticity with the increment of plastic content obtained from some previous studies.

9 Shrinkage

Reduction in shrinkage with the inclusion of PWA is watched, which is because of the lessening of drying limit. Shrinkage happens because of the dissipation of water from the vessels and may likewise because of the gain in stiffness with hydration time (Akçaözoğlu et al. 2010; Da Silva et al. 2014). Different examinations likewise demonstrate a slight increment of the drying shrinkage of WPAC. This occasion is expressed by the compactness of concrete. Inclusion of PWA makes concrete compact, which does not permit the exchange of water inside the specimens. Additionally, it likewise identifies with the abatement in elastic modulus (Frigione 2010).

10 Carbonation Depth

Carbonation depth relies upon penetrated CO_2. PWA and sand aggregates utilized together did not join with each other adequately which makes solid more permeable,

Table 8 Comparison of UPV results at different replacement levels (Akçaözoğlu et al. 2013)

Replacement level (%)	Ultrasonic wave velocities (km/s)	
	28 days	90 days
0	4.36	4.42
30	3.46	3.4
40	3.18	3.10
50	2.89	2.94
60	2.44	2.32

and thus allows more penetration of phenolphthalein solution. Along these lines, a moderately higher carbonation depth is watched. The maximum carbonation depth observed for mortar containing only WPA at 28 days curing age is 1.7 mm, while that for mortar containing both WPA and sand together is 2.5 mm (Akçaözoğlu et al. 2010).

11 Ultrasonic Pulse Velocity

The Ultrasonic Pulse Velocity (UPV) decreases with increment in w/c ratio. The Presence of Extra water in the pores offer ascent to diminishment in the UPV (Rahmani et al. 2013). Moreover, the low unit weights of WPA and more permeable structure of WPAC bring about a diminishment in the UPV of specimens (Akçaözoğlu et al. 2013; Safi et al. 2013). Table 8 shows a comparison of UPV at different substitution levels of PWA.

12 Thermal Conductivity

PWA poses lower Thermal Conductivity (TC) coefficient (0.15 W/m K) than natural aggregate (2 W/m K); due to this property, PWA slows down the heat propagation inside the mix (Akçaözoğlu et al. 2013). This causes reduction in TC of PWAC. Moreover, the presence of more pores in WPAC reduces thermal conductivity. Table 9 enlists a comparative result for TC at different replacement levels.

13 Conclusion

Utilization of plastic as fine aggregate impacts tremendously the concrete properties. Generally, it lessens the strength of concrete while supplanted by a higher amount, while an insignificant diminishment happens at bringing down substitution. Up to a

Table 9 Comparison of TC results at different replacement levels (Akçaözoğlu et al. 2013)

Replacement level (%)	TC (W/m K)	Reduction rate (%)
0	0.9353	0
30	0.6118	34.6
40	0.5723	38.8
50	0.5348	42.8
60	0.3924	58.0

substitution level of 20% usage of plastic in concrete is by all accounts fitting with lesser diminishment in strengths. In the interim, the issue in bonding strength can be enhanced by utilizing some supplementary cementing materials as metakaolin, granulated blast-furnace slag, and so on in concrete or treating WPA with the same. So the overall strength of concrete can be enhanced, which permits a higher level of substitution. Also, consideration of plastic into concrete tends to increase the porosity of concrete, which helps in enhancing durability properties like UPV and TC. Diminishment in shrinkage is likewise a recipient impact which prompts decrease in shrinkage cracks. This investigation gives fundamental data that apart from the reduction in strength plastic may enhance the durability properties of concrete. In this manner, it is attainable to utilize plastic in concrete with taking vital care of the bonding. At last, it is presumed that use of plastic in concrete in aggregate form usefully serves environment by giving an elective method for disposal and also gives an elective decision against the natural aggregate.

References

Akçaözoğlu S, Atiş CD, Akçaözoğlu K (2010) An investigation on the use of shredded waste PET bottles as aggregate in lightweight concrete. Waste Manag 30:285–290. https://doi.org/10.1016/j.wasman.2009.09.033

Akçaözoğlu S, Akçaözoğlu K, Atiş CD (2013) Thermal conductivity, compressive strength and ultrasonic wave velocity of cementitious composite containing waste PET lightweight aggregate (WPLA). Compos B Eng 45:721–726. https://doi.org/10.1016/j.compositesb.2012.09.012

Amalu RG, Ashraf A, Hashim M, Rejith KU, Vijitha VR (2016) Use of waste plastic as fine aggregate substitute in concrete. Int J Sci & Eng Res 7:172–177

Arivalagan S (2016) Experimental investigation on partial replacement of waste plastic in concrete. Int J Eng Sci & Res Technol 5:443–449. https://doi.org/10.5281/zenodo.167190

Batayneh M, Marie I, Asi I (2007) Use of selected waste materials in concrete mixes. Waste Manag 27:1870–1876. https://doi.org/10.1016/j.wasman.2006.07.026

Choi YW, Moon DJ, Chung JS, Cho SK (2005) Effects of waste PET bottles aggregate on the properties of concrete. Cem Concr Res 35:776–781. https://doi.org/10.1016/j.cemconres.2004.05.014

Da Silva AM, De Brito J, Veiga R (2014) Incorporation of fine plastic aggregates in rendering mortars. Constr Build Mater 71:226–236. https://doi.org/10.1016/j.conbuildmat.2014.08.026

Frigione M (2010) Recycling of PET bottles as fine aggregate in concrete. Waste Manag 30:1101–1106. https://doi.org/10.1016/j.wasman.2010.01.030

Gaur K, Jyotsana J, Arya AK, Singh NK (2017) Use of plastic as partial replacement of fine aggregate in fibre reinforced concrete. IOSR J Mech Civ Eng 14:71–74. https://doi.org/10.9790/1684-1403037174

Guendouz M, Debieb F, Boukendakdji O et al (2016) Use of plastic waste in sand concrete. J Mater Environ Sci 7:382–389

Harini B, Ramana KV (2015) Use of recycled plastic waste as partial replacement for fine aggregate in concrete. Int J Innov Res Sci, Eng Technol 4:8596–8603. https://doi.org/10.15680/IJIRSET.2015.0409106

Ismail Al-Hadithi A, Fawzi Ahmed Alani M (2015) Mechanical properties of high performance concrete containing waste plastic as aggregate. J Eng 21:100–115

Ismail ZZ, AL-Hashmi EA (2008) Use of waste plastic in concrete mixture as aggregate replacement. Waste Manag 28:2041–2047. https://doi.org/10.1016/j.wasman.2007.08.023

Jibrael MA, Peter F (2016) Strength and behavior of concrete contains waste plastic. J Ecosyst & Ecography 6:2–5. https://doi.org/10.4172/2157-7625.1000186

Mathew P, Ambika KP, Prakash P et al (2016) Comparative study on waste plastic incorporated concrete blocks with ordinary concrete blocks. Int Res J Eng Technol 3:1894–1896

Osei DY (2014) Experimental investigation on recycled plastics as aggregate in concrete. Int J Struct Civ Eng Res 3:168–174

Rahmani E, Dehestani M, Beygi MHA et al (2013) On the mechanical properties of concrete containing waste PET particles. Constr Build Mater 47:1302–1308. https://doi.org/10.1016/j.conbuildmat.2013.06.041

Reis JML, Carneiro EP (2012) Evaluation of PET waste aggregates in polymer mortars. Constr Build Mater 27:107–111. https://doi.org/10.1016/j.conbuildmat.2011.08.020

Safi B, Saidi M, Aboutaleb D, Maallem M (2013) The use of plastic waste as fine aggregate in the self-compacting mortars: Effect on physical and mechanical properties. Constr Build Mater 43:436–442. https://doi.org/10.1016/j.conbuildmat.2013.02.049

Saikia N, de Brito J (2013) Waste polyethylene terephthalate as an aggregate in concrete. Mater Res 16:341–350. https://doi.org/10.1590/S1516-14392013005000017

Seetharaman G (2017) The Economic Times. ET Bureau

Thosar PC, Husain M (2017) Reuse of plastic waste as replacement of sand in concrete. Int J Innov Res Sci, Eng Technol 6:789–794

Yongo E, Manyala JO, Kito K, Matsushita Y, Outa NO, Njiru JM (2016) Diet of Silver Cyprinid, Rastrineobola argentea in Lake Victoria, Kenya. Int J Adv Res 4:144–149. https://doi.org/10.21474/IJAR01

Properties of Permeable Concrete Made by Recycled Aggregates

Kanish Kapoor, Firdous Ahmad Dar and Danish Ahmad Rather

Abstract Day-to-day problem arises due to logging of water on pavements due to improper drainage systems and impermeable surface at the time of heavy rainfall. One of the beneficial methods of removing these constraints is using pervious concrete either as pavement or on the sides of pavement. A brief literature study to gain high-strength and high-permeable concrete is carried to improve application of pervious concrete in construction. The goal of using recycled coarse aggregates is to reuse hardened concrete for economical construction and decrease environmental issues with its different applications. Pervious concrete which is not conventionally used for construction can be used because of its feasible strength and permeability which can help in decreasing issues of drainage systems and environmental threats.

Keywords Pervious concrete · Recycled coarse aggregates · Permeability · Void ratio · Compressive strength · Infiltration · Waterlogging

1 Introduction

Pervious concrete is a mixture of cement or binding materials, coarse aggregates, admixtures, and water with no fine aggregates or less fine aggregates. It is also called "no fines" and "zero slumps" concrete that makes it different from conventional concrete (Biji et al. 2016). The varying sizes of aggregates are bonded together by a

K. Kapoor
Dr. B. R. Ambedkar National Institute of Technology, Jalandhar, Punjab, India

F. A. Dar (✉) · D. A. Rather
Lovely Professional University, Phagwara, Punjab, India
e-mail: darfirdous4@gmail.com

D. A. Rather
e-mail: drshabeer123@gmail.com

A. K. Agnihotri et al. (eds.), *Sustainable Engineering*, Lecture Notes in Civil Engineering 30, https://doi.org/10.1007/978-981-13-6717-5_35

paste formed by cement and water at its points of contact. Cement plays the main role of binding material by surrounding around the aggregates resulting in voids during drying of concrete in the absence of fine aggregates. Usually, the void ratio varies between 3 and 5% in conventional concrete, but depending upon the application of pervious concrete can have void content from 15 to 30%, the compressive strength of 2.8 to 28 MPa, and a percolation rate of 3 to 8 gal/ft^2/min (Mehta et al. 2016).

The increase in percentage of voids is due to lack of filler material in concrete, i.e., fine aggregates. Due to high percentage of voids, pervious concrete allows rapid percolation of water as compared to conventional concrete. The property of rapid percolation helps in groundwater recharge by decreasing runoff water during high-intensity rains and also decreases water logging due to improper drainage. Water-logging due to impervious pavement sometimes results in flash floods during high-intensity rains in monsoon, and using pervious concrete is the beneficial method to remove these constraints (Kumar 2015). Pervious concrete has lower density than conventional concrete. It has also lower cost because of lower cement content. It has also low drying shrinkage, no segregation, and high capillary movement than conventional concrete (Patil and Murnal 2014). Pervious concrete has lower compressive strength and higher permeability as compared to conventional concrete. Pervious concrete pavement decreases the use of costly stormwater management practices in rural areas, and it also gives efficient use of land by eliminating the need of retention ponds and other water management practices which uses a lot of lands (Shah et al. 2013). In pervious concrete, natural aggregates cannot be replaced fully by recycled coarse aggregates until it satisfies the criteria of compressive strength and hydraulic conductivity. Recycled aggregates consist of not only natural aggregates but also a small amount of adhered low-density hydrated cement paste (Berry et al. 2012). Pervious concrete has a lot of applications but its primary use is in pavements which are in residential roads, low traffic volume pavements, alleys and driveways, parking lots and sidewalks, slope stabilization, etc. (Ajamu et al. 2012). Impervious pavement absorbs heat in the day and releases it in the night, and Urban Heat Index (UHI) leads to thermal discomfort. Pervious concrete has a property that it does not cause any thermal discomfort because pores present in the concrete does not allow concrete to absorb any heat but transfers heat to the below soil (Huang et al. 2010). A lot of research has not been done usage of pervious concrete in areas having a cold climate with wet freezing characteristics. Every region has its unique properties and uses local materials in its concrete mix designs so it will depend on these materials whether a pervious concrete with good compressive strength, hydraulic conductivity, and durability can be produced (McCain and Dewoolkar 2009).

2 Methodology

2.1 *Materials*

Ordinary Portland cement (C 43 grade) fulfilling the requirements of IS 8112:1989 was used in the study. Crushed angular stones are used as coarse aggregates; also recycled coarse aggregates were used as the replacement of coarse aggregates. Coarse aggregates passing 12.5 mm sieve and retained on 4.75 mm sieve satisfying IS 383-1970 size aggregates were used. The bulk specific gravity of coarse aggregate was 2.56 as per IS 2386 (Part III)-1963, and their water absorption value was 0.56% (Tables 1 and 2).

2.2 *Specimens*

Cylinders of 12 in. diameter and height 12 in. were casted in accordance with ASTM C192/C192 M for hydraulic conductivity test to check the infiltration rate and also for compressive strength cubes with dimension 150 × 150 mm were casted.

2.3 *Mix Proportion*

See Table 3.

Table 1 Physical properties of ordinary Portland cement 43 grade

Properties	Obtained value (OPC)	IS code recommendation
Specific gravity	3.14	3.10–3.15
Initial setting time (min)	38 min	Minimum 30 min
Final setting time (min)	190 min	Maximum 600 min
Consistency (%)	29.4	30–35

Table 2 Physical properties of coarse aggregates

Experiment	Obtained value	Permissible values
Los angle abrasion value test (%)	15.2	30 and 45
Impact test (%)	12.5	30
Specific gravity test	2.58	2.1–3.2

Table 3 Mix proportion

Mix	A/C ratio	W/C ratio	Max. size of coarse aggregates (mm)	Fine aggregate used (%)	RCA used (%)
M1	4:1	0.35	10 mm (12.5–4.75 mm)	10% natural sand (2.36 mm–90 µm)	0
M2	4:1	0.35	10 mm (12.5–4.75 mm)	10% RCA sand (2.36 mm–90 µm)	25
M3	4:1	0.37	10 mm (12.5–4.75 mm)	10% RCA sand (2.36 mm–90 µm)	50
M4	4:1	0.38	8 mm (12.5–4.75 mm)	10% RCA sand (2.36 mm–90 µm)	75
M5	4:1	0.40	8 mm (12.5–4.75 mm)	10% RCA sand (2.36 mm–90 µm)	100

3 Results and Discussions

3.1 Compressive Strength

The aim of the study is to check the compressive strength of cubes having dimension 150 × 150 mm in accordance with ASTM C39/C39 M, standard test method for compressive strength of cylindrical concrete specimens. The compressive strength of cubes was checked at 7, 14, and 28 days of curing. It was found that compressive strength of porous concrete increases with increase in age as in normal concrete and also the compressive strength of porous concrete increases with decrease in size of coarse aggregates. The compressive strength of porous concrete made with 25% of RCA comes out almost the same as the porous concrete made with natural aggregates. Further increase in RCA percentage replacement decreases the compressive strength as compared to porous concrete made with natural aggregates (Table 4 and Fig. 1).

3.2 Hydraulic Conductivity

Permeability test of porous concrete was done by using single-ring infiltrometer method. In this method, infiltrometer ring with diameter 12 ± 1/2 mm is placed on the sample and plumbers putty is applied on sides to avoid any flow through sides. Pour 8 lbs. of water into the ring for pre-wetting of sample maintaining a head of ½ in. inside the ring. Determine water quantity for test 40 lbs., if time taken to pass

Table 4 Compressive strength at 7, 14, and 28 days of curing for different mixes

Mix	Age	Water/cement	Compressive strength (MPa)
	7		17.22
M1	14	0.35	23.85
	28		26.5
	7		16.26
M2	14	0.35	22.52
	28		25.02
	7		14.69
M3	14	0.37	20.34
	28		22.6
	7		12.74
M4	14	0.38	17.64
	28		19.6
	7		10.01
M5	14	0.40	13.86
	28		15.4

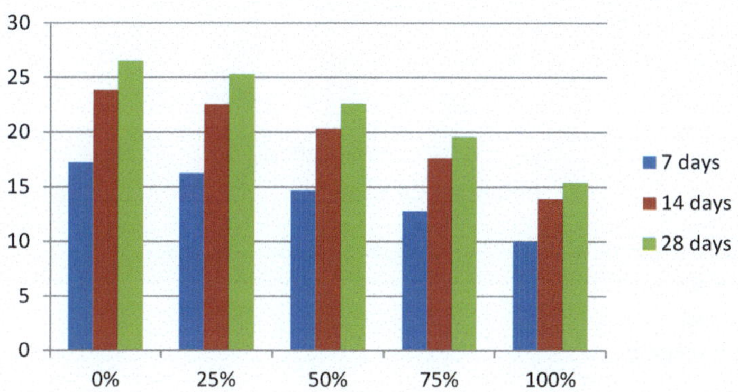

Fig. 1 Compressive strength of porous concrete at 7, 14, and 28 days of curing

through sample is less than 30 s. 8 lbs. and if time taken to pass through sample is 30 s or more.

Start within 2 min of pre-wetting and start pouring the determined water quantity in infiltrometer ring maintaining the head of ½ in. within ring. Record the time of infiltration from startup to no water present on the surface of sample. Using formula $I = KM/D^2 * t$ (in./hr)

$K = 126{,}870$ (in.^{3} s)/(lbh), where K is the coefficient of permeability

M = mass of water, D = Internal diameter of the sample, t = time of infiltration (Table 5 and Fig. 2).

Table 5 Permeability result for different mixes

Mix	Time (sec)	Infiltration rate (cm/s)
M1	80	0.30
M2	65	0.37
M3	52	0.47
M4	40	0.61
M5	31	0.79

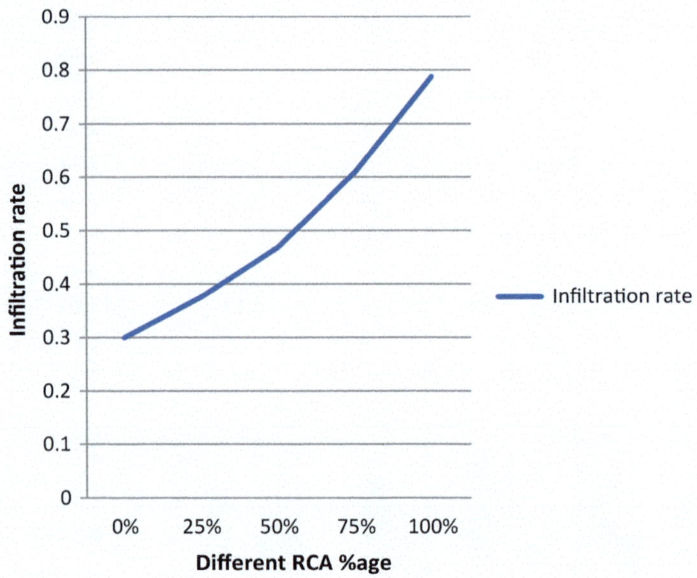

Fig. 2 Infiltration rate for various RCA replacements for different mixes

4 Conclusions

1. Compressive strength of porous concrete made by replacing 25, 50, 75, and 100% of natural coarse aggregates by recycled coarse aggregates comes out to be 94.41, 85.28, 73.96, and 58.11% of porous concrete made with natural aggregates.
2. Replacing up to 25% of natural coarse aggregates by recycled coarse aggregates give feasible results for both compressive strength and infiltration and may be used for high load traffic pavements.
3. Decreasing the size of recycled aggregates for replacement above 50% helps in improving compressive strength and maintaining infiltration rate within feasible limits.

References

Ajamu SO, Jimoh AA, Oluremi JR (2012) Evaluation of structural performance of pervious concrete in construction. Int J Eng Technol

Berry BM, Suozzo MJ, Anderson IA, Dewoolkar MM (2012) Properties of pervious concrete incorporating recycled concrete aggregate. TRB 2012 Annual Meeting

Biji UI, Rajeshwari M, Dharaani KK, Karthikai Raja K (2016) Studies on applicability of pervious concrete for pavements. Int J Emerg Technol Adv Eng

Huang B, Wu H, Shu X, Burdette EG (2010) Laboratory evaluation of permeability and strength of polymer–modified pervious concrete

Kumar SR (2015) Characteristic study on pervious concrete. Int J Civ Eng Technol (IJCIET)

McCain GN, Dewoolkar MM (2009) Strength and permeability characteristics of porous concrete pavements. TRB Annual Meeting

Mehta S, Shah Y, Surti V, Shah R (2016) Study on compressive strength and permeability of pervious concrete. Int J Eng Res & Technol (IJERT)

Patil PK, Murnal SM (2014) Study on the properties of pervious concrete. Int J Eng Res & Technol (IJERT)

Shah DS, Pitroda J, Bhavsar JJ (2013) Pervious concrete: new era for rural road pavement. Int J Eng Trends Technol (IJETT)

Experimental Study on Infiltration Properties of Pervious Concrete

Kanish Kapoor, Tanveer Ahmad and Gowhar Afzal

Abstract The problem of run-off water and water logging is a severe problem in the cities after heavy rainfall. The cities are converting into concrete jungles and the groundwater recharging is decreasing day by day. One of the possible solutions to these problems is the use of pervious concrete either as main pavement or as sidewalks. Pervious is also known as no-fines, gap-graded, zero-slump concrete having 18–35% of void content; thus, it allows water to pass through it. Pervious concrete also reduces the use of extra drainage system in roads and has become the environment-friendly material from other concrete materials as it allows the rainwater to seep into the ground, decreases flooding possibilities especially in metro cities, recharges the groundwater level and enhances freeze and thaw resistance than conventional concrete. The only barrier in using the pervious concrete is its clogging problem, which comes into picture during its age. The objective of the present study is to calculate the infiltration rate of pervious concrete at different clogging percentages.

Keywords Pervious concrete · Porosity · Durability · Compressive strength · Infiltration · Urban heat island

1 Introduction

Pervious concrete also called as no-fines concrete, porous or permeable concrete is composed of cement, water, coarse aggregates with little or no fine aggregates. Development of interconnected pores takes place in this unique concrete by using proper quantity of cementitious materials and water which too occurs due to estab-

K. Kapoor
Dr B R Ambedkar National Institute of Technology, Jalandhar, Punjab, India

T. Ahmad · G. Afzal (✉)
Lovely Professional University, Phagwara, Punjab, India
e-mail: gowharafzal841@gmail.com

T. Ahmad
e-mail: tanveerulislam79@gmail.com

A. K. Agnihotri et al. (eds.), *Sustainable Engineering*, Lecture Notes in Civil Engineering 30, https://doi.org/10.1007/978-981-13-6717-5_36

lishment of slim layer on the paste (Montes 2006). Pervious concrete is used since 30 years ago in different countries of the world, viz., United States and Japan. In view of the fact that the environmental advantages of pervious concrete, for instance, controlling the storm water run-off, recharging the groundwater supplies, diminishing of water and soil pollution, it is widely used in United States (Kajio et al. 1998; Tennis et al. 2004). Meanwhile, it has the capability of decreasing the urban heat island effects and can also act as noise barriers in roads (Tennis et al. 2004; Cackler et al. 2006). In porous concrete, little to no fine aggregates are used, and the accurate amount of cement paste is employed to produce an arrangement of high porosity and interconnected voids that can percolate water expeditiously. Usually, the void content is the range of 15–25%, and the permeability is generally 2–6 mm/s (Schaefer et al. 2006; Tennis et al. 2004). Although for high porosity pervious concrete, comparatively low strength is usually linked with it. The low strength of pervious concrete bounds its implementation not only in heavy traffic highways but also effects durability and stability of the structures because of which pervious concrete is vulnerable to frost damage and low defiance to aggressive chemicals so that is why pervious concrete with less strength can be used for sidewalks, parking lots, recreation areas and moulding for typical pavements (Wang 1997; Nader et al. 1995). Pervious concrete can be applied for pavement shoulders and local roads by using smaller sized aggregates or by inclusion of fine aggregates so as to increase the strength. Strength and abrasion resistance of pervious concrete can be improved to large extent by using properly selected ingredients, viz., coarse and fine aggregates mixed with organic intensifiers and by regulating the mix proportion (Yang and Jiang 2003). Past studies make it clear that gradation, particle size of aggregates and mass ratio of aggregate to cement are the main elements which affect porosity, permeability and compressive strength of pervious concrete while water–cement ratio has a little influence on the pervious concrete (Jiang et al. 2005). The quantity of aggregates per unit volume of concrete can be increased by using smaller sized aggregates which obviously results from increase in specific surface of aggregates, the binding area and hence enhancement in strength of concrete. Yang and Jiang proved that strength of pervious concrete can be improved remarkably by using silica fume and superplasticizers in it (Yang and Jiang 2003). The fibres in pervious concrete increase the void content, considerably increase the permeability and more significantly enhance the splitting tensile strength of pervious concrete (Kevern 2008; Wang 1997). Another method to increase the strength is to use chemical additives like polymer Kevern; it was shown that workability, strength, permeability and freeze–thaw resistance can be enhanced by the addition of polymer (styrene butadiene rubber, SBR) (Kevern 2008) (Fig. 1).

Fig. 1 Pervious concrete cylinder and cube

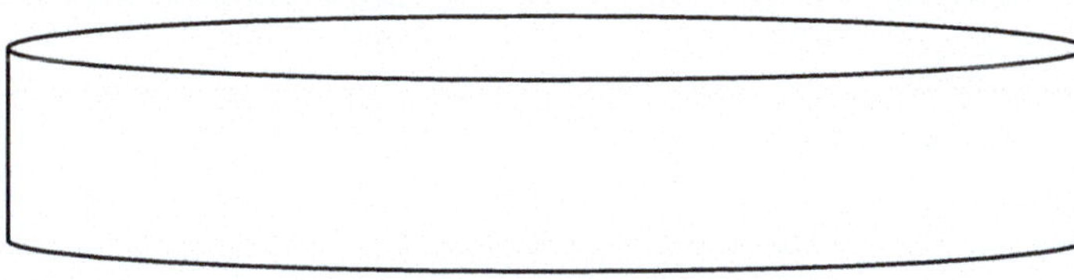

Fig. 2 Ring of dia. 12 in.

2 Experimental Programmes

2.1 Single Ring Infiltration Test

The testing of the sample starts by placing the ring at the top of the sample and then pre-wetting of the sample using 8 lb's of water. Then for main testing, 40 lb's water was used and the time taken by the water to pass through the sample was noted during the whole process. The head of water was maintained between 10 and 15 mm, respectively, on the top surface of the sample.

According to ASTM C 1701, the acceptable infiltration ranges from (3 to 8 gal/ft^2/mint), which is calculated by using the equation $I = KM/D^2 * t$.

Where

- I infiltration rate, mm/h (in./h);
- M mass of infiltrated water kg (lb);
- D inside diameter of infiltration ring mm (in.);
- t time required for the measured amount of water to infiltrate the concrete in seconds; and
- K constant (126870 inch-pound) (Fig. 2).

Table 1 Aggregate testing

S. No.	Tests performed	Results of 20 mm	Results of 10 mm	Permissible values
01	Abrasion value test	12.2	16.4	30% 50%
02	Specific gravity test	2.54	2.64	2.1–3.2
03	Soundness of aggregates	2.1–3.2	8.2	12%
04	Impact value	13.9	14.6	30% 45%
05	Water absorption test	0.59	0.58	5%
06	Combined elongation and flakiness	2.64	15.6	40%
07	Crushing value test	13.6	13.3	30%

2.2 *Aggregate Testing*

Ordinary Portland cement was selected in the experiments. Different gradations of coarse aggregates were observed studied under ASTM specifications whose properties are enlisted in Table 1. The samples used in these experiments are cylindrical in shape having diameter and height of 12 in.

3 Results and Discussion

3.1 *Discussion on Permeability and Compressive Strength*

The variation in the size of aggregates and in the water–cement ratio was only to maintain the specified value of infiltration as Mix 1 was found to be having zero permeability because the ratio of cement with respect to aggregate was very high, and the size of aggregates was also between (4.75 and 20 mm) the smaller sized aggregates where also responsible for making the mix impermeable, in Mix 2, in order to get specified value of infiltration, the amount of cement was reduced and also the size of aggregates were reduced and the size was adopted between (10 and 20 mm) so that a higher amount of voids can be created but after checking the infiltration the value was much higher as the amount of cement was reduced also the size range was increased. In Mix 3, we increase the water–cement ratio in order to see the effect on the infiltration while increasing the water–cement ratio, the infiltration was reduced a little bit if comparing with Mix 2. Now the aim was to maintain the infiltration without changing the mix proportion, and the only way was to use some amount of sand, so a proper replacement of coarse aggregates by sand of adequate size and amount was done in Mix 4 (replacement 10% of coarse aggregates with sand) but the size and range of aggregates where same as that of Mixs 3 and 2 again the value of infiltration was reduced comparing with Mixs 2 and 3 but it was again too high if compared with drainage limits. Then, the size of aggregates in Mix 5 was reduced in a range of (6.3–16 mm) but this time no sand was used. In Mix 5, the maximum size of the aggregates was further reduced and that in turn increased the impermeability. In Mix 5 and Mix 6 the impermeability was increased to a higher value and this was done by maintaining the aggregate size between (4.75 and 12 mm) and by doing an adequate replacement of coarse aggregates by sand. The size was the key to maintain the infiltration without replacement but to get the desired value of both strength and infiltration replacement of coarse aggregates by sand was done. The values of strength for mix were very effective but at the same time, the permeability was zero. It was observed from the Mixes 2 to 5 that the compressive strength was too low but both properties were in the specified range. In Mix 6, as the size of aggregates were reduced, the infiltration was reduced. It was also observed that in Mix 6 a well graded range of aggregates with 15 mm as maximum size aggregate and sand of 10% below 2.36 mm size have provided the desired infiltration under drainage limits (Tables 2, 3 and Figs. 3, 4).

It was observed that the sample in which cement and coarse aggregates ratio was 1.5:3 with water cement ratio of 0.35 has zero infiltration because of the presence of large quantity of smaller sized aggregates. The samples in which water–cement ratio was 1:4 with difference in maximum sized coarse aggregates were observed

Table 2 Mix design

Mix. No.	Mix proportion	Maximum sized coarse aggregates (mm)	Max quantity sized coarse aggregates (mm)	Replacement of coarse aggregate with sand	W/c ratio
M1	1.5:3	20 (from 4.75 to 20 mm)	10–18	0%	0.35
M2	1:4	20 (from 10 to 20 mm)	20	0%	0.32
M3	1:4	20 (from 10 to 20 mm)	20	0%	0.35
M4	1:4	20 (from 10 to 20 mm)	20	10% 2.36 mm passing	0.35
M5	1:4	16 (from 6.3 to 16 mm)	10	0%	0.33
M6	1:4	15 (from 4.75 to 12 mm)	10	10%	0.33

Table 3 Mix with infiltration

Mix. No.	Mix proportion	Time in seconds	Infiltration (in./h)
M1	1.5:3	0	0
M2	1:4	10	3524.166
M3	1:4	11	3203.787
M4	1:4	15	2349.444
M5	1:4	17	2073.039
M6	1:4	69	510.748

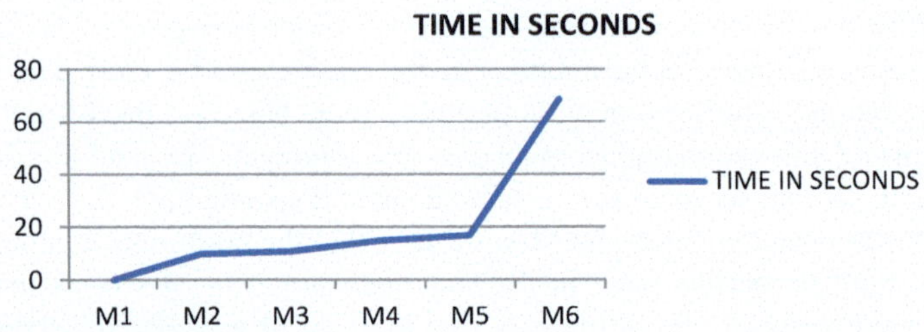

Fig. 3 Time of infiltration

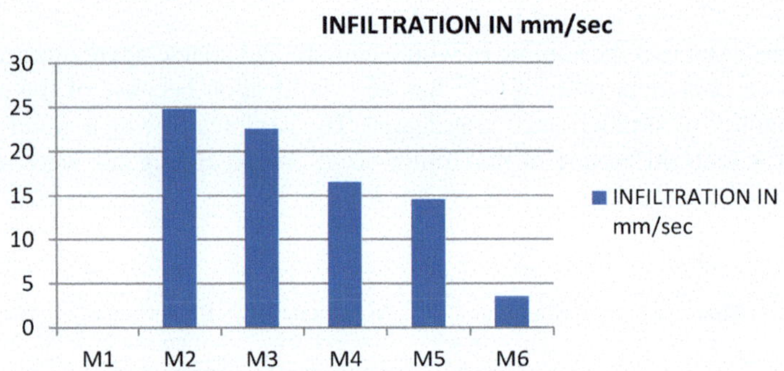

Fig. 4 Infiltration rate

highly porous. The sample in which coarse aggregates were replaced by 10% fine aggregates and maximum sized coarse aggregates used were 15 mm was observed to have desirable infiltration.

4 Conclusion

A laboratory evaluation was conducted in which infiltration of porous concrete was studied at different water–cement ratios and at varying mix proportions with different sizes of aggregates. In order to make effective porous concrete, appropriate mix proportion that includes water–cement ratio, size of coarse aggregates needs to be selected wisely or as prescribed in the present study.

- Partial replacement of coarse aggregates by fine aggregates plays a vital role in developing a pervious concrete having adequate compressive strength and desired infiltration which can be used for pavements, residential roads and in many civil engineering applications.
- Mix 6 can be used for traffic loading in various special cases because of its desired permeability and strength.

References

Cackler ET et al (2006) Evaluation of US and European concrete pavement noise reduction methods. National Concrete Pavement Technology Center, Concrete Pavement Surface Characteristics Project Report, and Iowa State University

Huang B et al (1999) Fundamentals of permeability in asphalt mixtures. J Assoc Asphalt Paving Technol 68:479–500

Jiang Z et al (2005) Effects of some factors on properties of porous pervious concrete. J Build Mater 8:513–519

Kajio S et al (1998) Properties of porous concrete with high strength. In: Proceedings 8th international symposium on concrete roads. Lisbon, pp 171–177

Kardon JB (1997) Polymer-modified concrete: review. ASCE J Mater Civil Eng 9(2):85–91

Kevern JT (2008) Advancement of pervious concrete durability. Ph.D. dissertation, Iowa State University, Ames, IA

Montes F (2006) Pervious concrete: characterization of fundamental properties and simulation of microstructure. Ph.D. Dissertation, University of South Carolina

Nader G, Shivaji D (1995) Development of no-fines concrete pavement applications. ASCE J Transport Eng:283–288

Schaefer VR et al (2006) Mix design development for pervious concrete in cold weather climates. Final report, National Concrete Pavement Technology Center, Iowa State University, and Ames, IA

Storm Water Technology Fact Sheet (1999) Porous pavement. EPA 832-F-99-023. Office of Water, Washington, DC

Tennis PD et al (2004) Pervious concrete pavements EB302 Portland cement Association Skokie Illinois and National Ready Mixed Concrete Association. Silver Spring, Maryland

Wang W (1997) Study of pervious concrete strength. Sci Technol Build Mater China 6(3):25–28

Yang J, Jiang G (2003) Experimental study on properties of pervious concrete pavement materials. Cem Concr Res 33:381–386

Youngs A (2005) Pervious concrete it's for real. Presentation at pervious concrete and parking area design workshop, Omaha

To Study the Strength Characteristics of Bricks Made with Fly Ash

Harsimranjit Singh, Jagbir Singh and Kanish Kapoor

Abstract The conventional bricks are generally made up of clay as its principal constituent. The urbanization and industrialization lead to the use of conventional bricks at a very high rate. Moreover, the rapid increase in population and to meet its needs, new houses, and high-rise buildings are constructed in large numbers, which requires a large number of bricks for its construction. To decrease the extensive use of the clay and finding its substitute is the need of the hour. In the present study, the Fly Ash is used as the substitute of clay and fly ash bricks were made by using fly ash, sand, lime, and Plaster of Paris (POP) in different mix proportions. The fly ash brick specimens of size 230 × 110 × 70 mm were manufactured and various quality parameters like compressive strength at an age of 7 and 28 days, water absorption and dry density were tested. The economical mix design of the fly ash bricks for maximum compressive strength and its rate analysis was also done. It was observed from the results that the compressive strength decreases with increase in fly ash content and increases with an increase in sand and lime content in fly ash bricks. Water absorption value increases with an increase in fly ash content and decreases with increase sand and lime content in fly ash bricks.

Keywords Fly ash · Coarse sand · Lime · Compression strength · Water absorption

1 Introduction

A vast number of bricks are required for construction like new houses, elevated structures for expanding populace at a high rate. The urbanization prompts the utilization of regular bricks at a high rate (Akhtar et al. 2011). The ordinary bricks are for the most part comprised of mud as its standard constituent thus it is nearly fatigue. To

H. Singh (✉) · J. Singh
Lovely Professional University, Phagwara, Punjab, India
e-mail: harsimranjit.22231@lpu.co.in

K. Kapoor
Dr B R Ambedkar National Institute of Technology, Jalandhar, Punjab, India

A. K. Agnihotri et al. (eds.), *Sustainable Engineering*, Lecture Notes in Civil Engineering 30, https://doi.org/10.1007/978-981-13-6717-5_37

Table 1 Chemical composition of fly ash

Sr. No.	Constituents	Percentage (%)
1	Silica	57–60
2	Alumina	24–25
3	Iron oxide	7–8
4	Lime	4
5	Magnesium	2
6	Loss on ignition	6

conquer this issue some option is required. So to beat this issue or to lessen this issue up to some degree, fly powder can be used in set up of earth in assembling the bricks named fly ash bricks (Mishra 2015). This will prompt the reuse of fly ash cinder and spare the best layer of soil, which is extremely ripe utilized as dirt in customary bricks. Also, the cost of fly ash bricks will be less when contrasted with ordinary bricks (Samander et al. 2013). According to Government standards, we can utilize at least 25% of fly powder in bricks, where fly ash debris is being created (Naganathan et al. 2015).

Fly ash is a finely divided residue resulting from the combustion of powdered coal, transported by flue gases and collected by electrostatic precipitators (Gawatre and Vairagade 2014). It contains a small amount of unburnt carbon, which is acidic in nature. Its main constituents are silica, aluminum oxide, and ferrous oxide. Chemical composition of fly ash is shown in Table 1.

With the increase in the limit of warm power age in India the production of fly powder is all time high, which is roughly 50 million tons for every year. There are more than 70 warm power plants are in production and in future this number may increase to satisfy the requests of expanding populace (Banu et al. 2013).

To overcome this problem and to empower the usage of fly ash debris in 2003, the Government of India made it mandatory to utilize fly ash cinder with soil on weight to weight premise. It ought to be utilized for assembling of block tiles or squares inside a span of 100 km from coal-based warm power plants. From decades ago, endeavors are being made to discover a satisfactory technique for the transfer and appropriate use of fly fiery remains.

Endeavor has been made by assembling the fly ash bricks utilizing fly ash cinder as a rule constitute and blending with Coarse sand, Lime, and Plaster of Paris (POP) (Kumar and Rinku 2012).

2 Methodology

To accomplish the goal, with the assistance of writing audit and gathered distinctive fly ash brick tests in the wake of testing, we choose one proportion for fly ash debris, coarse sand, lime, and POP. Three classes of bricks were made. In the first class, fly ash variety was done and different parameters were kept consistent. In the second class, the coarse sand variety was done and different parameters were kept steady. Also in the third classification, hydrated lime was kept consistent and different parameters were kept steady.

Blend extents fly ash debris and all other crude materials were taken from assembling the unit. Each material in the required rate by weight as appeared in configuration blend were brought close to the blend. Initially the lime and sand were blended for 2–3 min. After that fly ash was blended in and required amount of water was included and blended for 2–3 min to acquired homogeneous blend and consistency in shading. The varying quantitites of fly ash, coarse sand and lime in the mixes are presented in Tables 2, 3 and 4 respectively.

Table 2 Variation of fly ash

Sample No.	Fly ash (Kg)	Coarse sand (Kg)	Lime (Kg)	POP (kg)
1	40	20	20	5
2	50	20	20	5
3	60	20	20	5
4	70	20	20	5

Table 3 Variation of coarse sand

Sample No.	Fly ash (Kg)	Coarse sand (Kg)	Lime (Kg)	POP (kg)
5	50	10	20	5
6	50	20	20	5
7	50	30	20	5
8	50	40	20	5

Table 4 Variation of lime

Sample No.	Fly ash (Kg)	Coarse sand (Kg)	Lime (Kg)	POP (kg)
9	50	20	10	5
10	50	20	15	5
11	50	20	20	5
12	50	20	25	5
13	50	20	30	5

3 Result and Discussions

Category 1 (Fly Ash Variation)
See Fig. 1.

Compressive Strength (N/mm^2)
See Table 5.

Category 2 (Coarse Sand Variation)
It was observed from the conducted experiment, with an increase in the sand content in fly ash bricks the compressive strength increases. It was observed from the conducted experiments, with an increase in coarse sand content in fly ash bricks average water absorption value decreases and average dry density increases (Fig. 2 and Table 6).

Category 3 (Hydrated Lime Variation)
It was observed from the conducted experiments, with an increase in hydrated lime content in fly ash bricks average water absorption value decreases up to 25% and average dry density increases (Fig. 3 and Table 7).

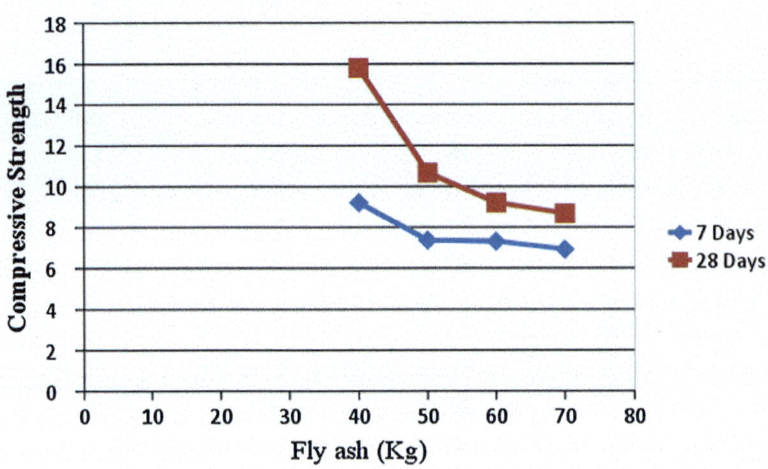

Fig. 1 Compressive strength

Table 5 Water absorption and dry density

Sample No.	Avg. water absorption (%)	Avg. dry density (Kg/m^3)
1	12.97	1562.39
2	16.64	1482.77
3	16.74	1470.35
4	16.86	1409.93

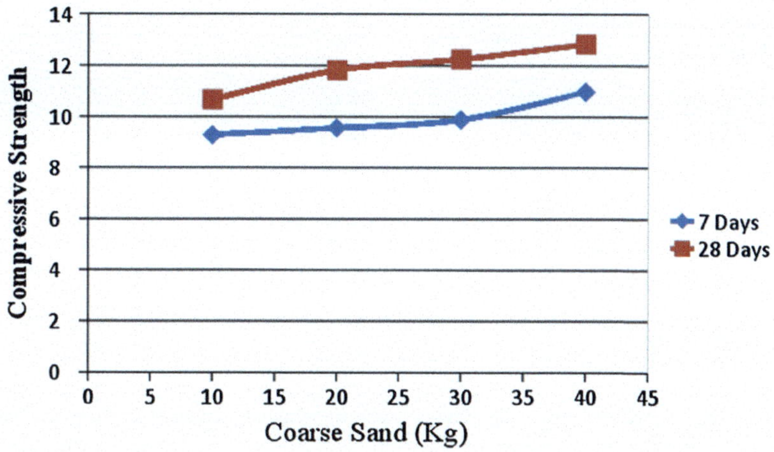

Fig. 2 Compressive strength

Table 6 Water absorption and dry density

Sample No.	Avg. water absorption (%)	Avg. dry density (Kg/m^3)
5	15.75	1465.83
6	14.08	1470.92
7	13.76	1513.83
8	12.01	1560.70

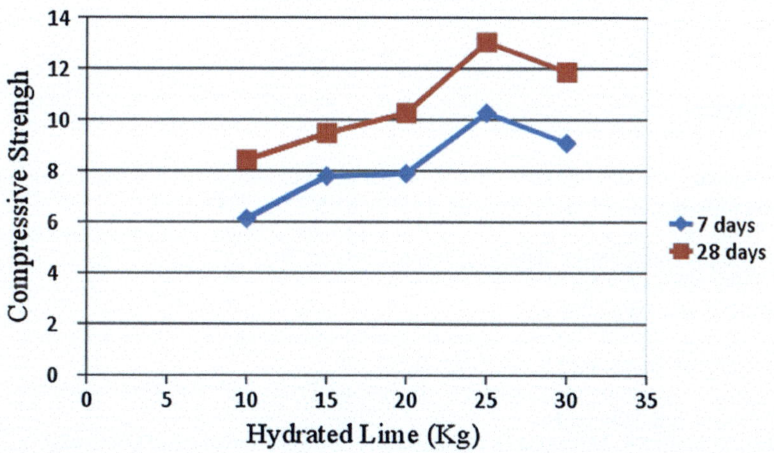

Fig. 3 Compressive strength

Table 7 Water absorption and dry density

Sample No.	Avg. water absorption (%)	Avg. dry density (Kg/m^3)
9	18.63	1415.01
10	18.54	1424.61
11	15.02	1439.29
12	12.02	1559.00
13	14.84	1570.21

4 Conclusion

1. The compressive strength decreases with increase in fly ash content in fly ash bricks. The compressive strength varies from 8.69 to 15.81 N/mm^2 and cost varies from Rs. 1.28 to Rs. 1.99 with a variation of fly ash from 40 to 70 kg keeping other parameters constant.
2. The compressive strength increases with increase in coarse sand content in fly ash bricks. The compressive strength varies from 10.67 to 12.84 N/mm^2 and cost varies from Rs. 1.47 to Rs. 2.03 with a variation of sand from 10 to 40 kg keeping other parameters constant.
3. The compressive strength increases with increase in hydrated lime content in fly ash bricks optimum at 25%. The compressive strength varies from 8.43 to 13.04 N/mm^2 and cost varies from Rs. 1.28 to Rs. 1.90 with a variation of lime from 10 to 30 kg keeping other parameters constant.

References

Akhtar JN, Alam J, Akhtar MN (2011) Bricks with total replacement of clay by fly ash mixed with different materials. Int J Eng Sci Technol (IJEST) 3(10):7338–7347

Banu T, Billah M, Gulshan F, Kurny ASW (2013) Experimental studies on fly ash-sand-lime bricks with gypsum addition. Am J Mater Eng Technol 1(3):35–40

Gawatre DW, Vairagade LN (2014) Strength characteristics of different types of bricks. Int J Sci Res (IJSR) 3(10):2247–2252

Kumar PA, Rinku P (2012) Comparative study of compressive strength of bricks made with various materials to clay bricks. Int J Sci Res Publ 2(7):7–10

Mishra V (2015) Evolution of properties of fly ash-lime bricks with soil. IJSTE—International J Sci Technol & Eng 1(9):12–18

Naganathan S, Yousef A, Mohamed O, Mustapha KN (2015) Performance of bricks made using fly ash and bottom ash. Constr Build Mater 96:576–580

Samander S, Dwivedi AK, Agarwal S (2013) Effect of silica fume on fly ash cement bricks—an experimental study. IOSR J Mech Civ Eng (IOSR-JMCE) 6(4):14–18

Nondestructive Analysis of Self-compacting Concrete Made with Recycled Concrete Aggregates

Kanish Kapoor and S. P. Singh

Abstract The paper reports the nondestructive analysis of Self-compacting Concrete (SCC) prepared with Coarse Recycled Concrete Aggregates (CRA) as a fractional or complete replacement of Coarse Natural Aggregates (CNA). The SCC mixtures were made with 0, 50, and 100% of CRA. Ordinary Portland Cement (PC) was partly substituted with Fly Ash (FA) up to 30% by weight, which was further replaced with 10% either by Silica Fume (SF). The effect CRA on fresh properties of SCCs was evaluated using J-ring test, L-box test, slump flow test, and V-funnel test. For the nondestructive analysis of concrete Ultrasonic Pulse Velocity (UPV) tests were conducted on the various SCC mixtures at the maturing ages of 28, 56, and 120 days.

Keywords Slump flow · Recycled aggregates · Ultrasonic pulse velocity · Fly ash · Silica fume

1 Introduction

Self-Compacting Concrete (SCC) is a concrete that can easily flow through the dense reinforcement without any application of external force (ACI Committee 237, 2007). The use of SCC in the construction of new infrastructure is increasing day-by-day. In the modern era, it is one of the highly appreciated types of concrete that become popular throughout the world. The different properties of SCC are helping the engineers and designers to overcome many types of problems like segregation, etc.

The basic difference between the conventional concrete and SCC is the flowing properties; moreover, the ratio of aggregates in the SCC is also lower as compared to the conventional concrete. Besides having less proportion of aggregates, the aggregates play a very important role in the mechanical and durability properties of SCC. The economic and environmental benefits in using the demolished waste such as old

K. Kapoor (✉) · S. P. Singh
Dr. B.R. Ambedkar National Institute of Technology, Jalandhar, Punjab, India
e-mail: kapoork@nitj.ac.in

A. K. Agnihotri et al. (eds.), *Sustainable Engineering*, Lecture Notes in Civil Engineering 30, https://doi.org/10.1007/978-981-13-6717-5_38

used concrete as an alternative of Natural Aggregates (NA) for the manufacturing of concrete open the new path for the concrete industry. Manufacturing of concrete using CRA has many gains such as the load on natural aggregates will be decline; the problem of disposal of waste can be solved, etc. Subsequently, the CO_2 releases and road traffic congestions that are associated with the movement of the aggregates from the natural quarries or crushing plants to construction sites also minimized. The potential use of the CRA (Revathi et al. 2013) in SCC not only a sustainable alternative of construction but also minimizes the liability on natural resources.

The existence of weaker and permeable adhered mortar is the major culprit in the old concrete aggregate for being poorer as compared to the new aggregates, thus this weaker property needs to be addressed before using the CRA in the SCC. The volume of adhered mortar in CRA particle varies from 20 to 60% depending upon the size of the particle (Hansen and Narud 1983; Sánchez et al. 2009). The permeable nature of this adhered mortar is responsible for the high water immersion of CRA particle which is two–three times higher than the water immersion of NA (Butler et al. 2011; Katz 2003). The water immersion capacity of CRA particle reveals the water immersion capacity of Recycled Aggregate Concrete (RAC) as water penetrability of RAC is two–three times higher than that of the Natural Aggregate Concrete (NAC) (Ujike, n.d.; Zaharieva et al. 2003). The carbonation depth of the RAC has also been found to elevate with the growing quantity of CRA (Kou and Poon 2012). As reported, the carbonation depth of RAC was up to three times greater than that of the NAC (Abbas et al. 2009; Rao et al. 2007). The previous investigations also reported that the CRA has a negative effect on the resistance to chloride penetration (Kou et al. 2011), as an increase of approximately 73% in the chloride penetration was found at the maturing age of 28 days when complete percentage of NA was swapped by CRA (Olorunsogo and Padayachee 2002). Furthermore, the properties like compressive strength, split tensile strength, and drying shrinkage are also influenced by the physical, mechanical properties, and amount of CRA in the concrete mixtures. The compressive strength of the RAC is immensely dependent on the swapping ratio of NA with CRA. Up to 40% decline in the compressive strength was broadly reported upon the complete swapping of NA with CRA (Butler et al. 2011; Chen et al. 2003; Katz 2003; Padmini et al. 2009; Topçu and Şengel 2004). Furthermore, up to 24% decline in the split tensile strength was found with the different swapping levels of NA with CRA (Ajdukiewicz and Kliszczewicz 2002; Yang et al. 2008; Rao et al. 2011). The drying shrinkage of the RAC is more than that of the NAC due to the adhered mortar of the CRA particle (Kou and Poon 2012; Limbachiya et al. 2012). Up to 70% elevation in the drying shrinkage of RAC was reported as related to the NAC (Sagoe-Crentsil et al. 2001; Tavakoli and Soroushian 1996).

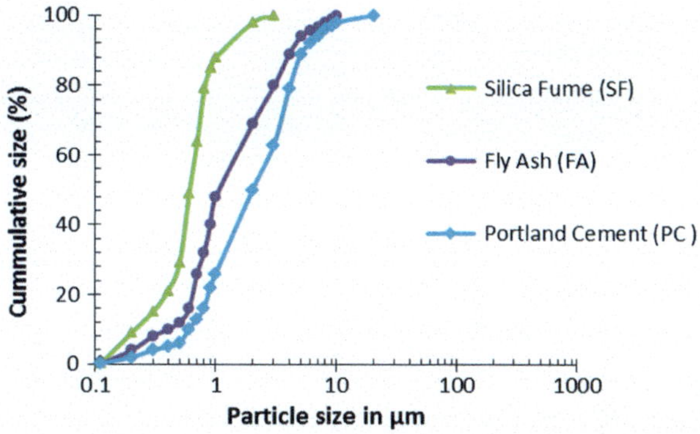

Fig. 1 Particle size distribution of the powders used

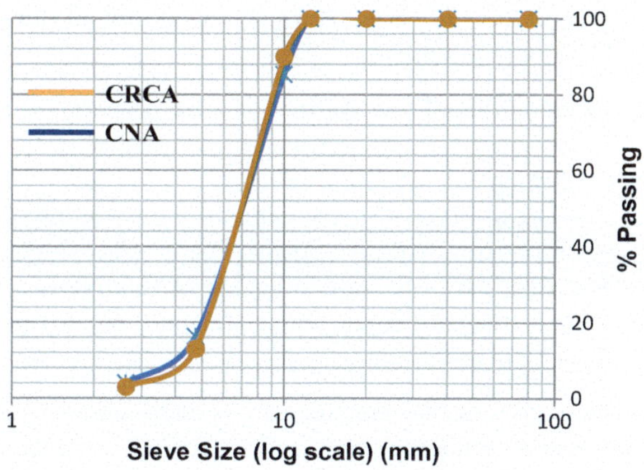

Fig. 2 Comparison of grading curves of CRA and CNA

2 Experimental Programme

2.1 Materials

Portland cement confirming IS 8112: 1989 was used throughout the experimental work in this investigation. Locally available CNA was used, whereas CRA were obtained by splitting the waste specimens available. Fly ash and SF were also combined in different quantities as cement supplements to manufacture various SCC mixtures. The particle size distribution of powders, i.e., PC, FA, and SF are plotted in Fig. 1. The CRA was sized and sieved in such a manner that the gradation of both the materials will be the same and in the limits conforming IS 383:1970 (Indian Standard 1970). The relative demonstration of the grading curves of CRA with CNA is shown in Fig. 2. A Polycarboxylic Ether (PCE) based super plasticizer was used in the present investigation to get the desired workability and viscosity modifying agent named master matrix-II that was used to get the desired viscosity of the concrete mixtures.

Table 1 Velocity criterion for concrete quality

Measured pulse velocity (m/sec)	Concrete quality
More than 4500	Excellent concrete
Between 3500 and 4500	Good concrete
Between 3000 and 3500	Medium concrete
Less than 3000	Doubtful concrete

2.2 *Testing Methods*

2.2.1 Fresh Concrete Properties Test

For a good SCC mixture, the freshly prepared mixture should satisfy the following workability requirements

1. **Filling ability/Flow ability**—It is the property of fresh SCC mixture that defines flows under its own self-weight and ability to fill the difficult spaces in dense reinforcement, avoiding stability.
2. **Passing ability**—It is the capability of SCC mixture to clear through crowded reinforcement without applying the external force.
3. **Stability**—It is the capability of SCC mixture to remaining homogenous by counterattacking segregation, bleeding, and air popping during transportation, placement at the site.

In the present investigation, the workability and stability of SCC mixtures were examined according to specifications for Self-compacting Concrete of the European recommendations of American Concrete Institute ACI 237 R (2007) committee for SCC. For each mixture, tests such as slump flow test, J-ring test, L-box test, and V-funnel test were performed to measure the passing ability, flowability, and filling ability of fresh SCC mixtures. The stability of the mixtures was observed through Visual Stability Index test suggested by ACI 237 R (2007).

2.2.2 Ultrasonic Pulse Velocity Test

The test was performed in the present study is as per the Indian Standard (1992).

The velocity criterion for concrete quality is given in Table 1.

3 Results and Discussion

3.1 *Fresh Properties*

It was observed from the result of the fresh properties that with the increase in the CRA content in the SCC mixtures the slump flow found to be declined. Furthermore, the V-funnel time is found to be increased. This clearly shows that the increase in the CRA in the SCC mixtures increase the resistance in the flow. The different fresh properties of SCC mixtures are tabulated in Table 2.

Table 2 Fresh properties of SCC mixtures

Mixture ID	Mixture description	Slump flow test		V-funnel test (s)	L-box test (h_2/h_1)	Pass/fail
		T50 (s)	D (mm)			
C-R0	70% Cement + 30% Fly Ash + 0% CRA	2.5	720	6	0.95	Pass
C-R50	70% Cement + 30% Fly Ash + 50% CRA	2.8	710	7.1	0.92	Pass
C-R100	70% Cement + 30% Fly Ash + 100% CRA	3.2	700	7.5	0.82	Pass
C-SFR0	70% Cement + 20% Fly Ash + 10%SF + 0% CRA	2.6	730	6.3	0.93	Pass
C-SFR50	70% Cement + 20% Fly Ash + 10%SF + 50% CRA	2.9	710	6.8	0.91	Pass
C-SFR100	70% Cement + 20% Fly Ash + 10%SF + 100% CRA	3.2	680	7.1	0.85	Pass

3.2 Ultrasonic Pulse Velocity Test

The observed values of various UPV tests conducted on the SCC mixtures with half and complete replacement of CNA with CRA at the maturing ages of 28, 56, and 120 days are illustrated in Fig. 3. In general, it was perceived that the UPV values were declining with the growing substitution percentage of CNA with CRA. For example, relative to the control SCC mixture a marginal decline of the order of 3%, 3.2%, and 1.5% was observed in the UPV values for partial replacement (50%) of CNA with CRA at the maturing ages of 28, 56, and 120 days, respectively. Subsequently, a further decline in the UPV values was observed for the complete replacement of CNA with CRA. In this case, the decline was 8.2%, 8.6%, and 8.7% relative to the control SCC at the maturing ages of 28, 56 and 120 days, respectively.

As per UPV-based concrete quality classification given in IS 13311(Part-1): 1992, all the SCC mixtures containing CRA fall in the quality classification category—excellent. This is indicative of the fact that the integrity of the hardened SCC mixtures was not compromised by the inclusion of the CRA. SF (series-3) was added to the SCCs. The influence of this addition on the SCC mixtures with partial and full replacement of CNA with CRA is shown in Fig. 4. It may be noted in this figure that, the introduction of SF does not prove to be effective in arresting the decline in UPV values due to the partial or complete replacement of CNA with CRA. For example, a decline of 1.8%, 1.5%, and 3% was observed in the UPV values of mixture C50F0S relative to the control SCC mixture C0F0, at the maturing ages of 28, 56, and 120 days, respectively. Further, a significant decline in the UPV values was observed when CNA was completely replaced with CRA. For example, UPV values of the SCC mixture C100F0S was observed to decline by 7.7%, 8.4%, and 8% at the maturing ages of 28, 56, and 120 days, respectively, relative to the SCC control mixture C0F0.

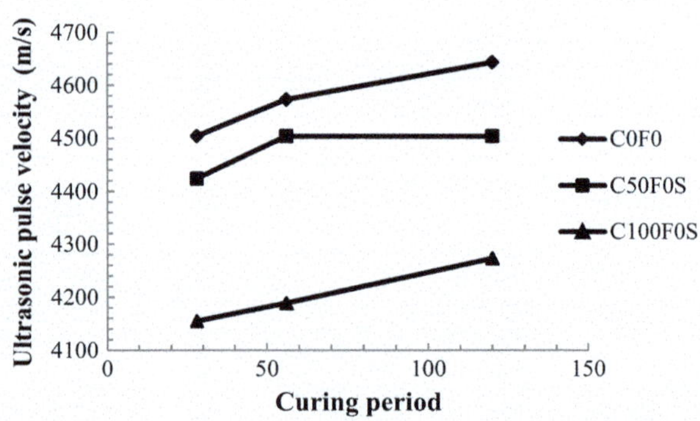

Fig. 3 UPV values of series-3 (blended with SF) SCC mixtures with partial (50%) and full (100%) replacement of CNA with CRA

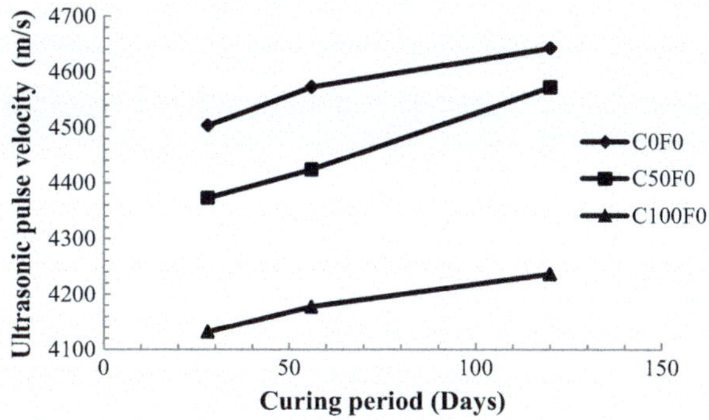

Fig. 4 UPV values of series-1 SCC mixtures with partial (50%) and full (100%) replacement of CNA with CRA

4 Conclusions

1. All the SCC mixtures successfully pass the required criterion for being SCC in all the parameters tested.
2. It is concluded that the SCC mixtures containing CRA as replacement of CNA behaves adequately, as the observed UPV values lie in the excellent concrete and good concrete groups at all substitution levels.
   These results recommend the use of CRA in SCC.

References

Abbas A, Fathifazl G, Isgor OB, Razaqpur AG, Fournier B, Foo S (2009) Durability of recycled aggregate concrete designed with equivalent mortar volume method. Cement Concr Compos 31(8):555–563. https://doi.org/10.1016/j.cemconcomp.2009.02.012

ACI Committee 237 (2007) Self consolidating concrete. ACI Emerging Technology Series, 237R–07, 1–30

Ajdukiewicz A, Kliszczewicz A (2002) Influence of recycled aggregates on mechanical properties of HS/HPC. Cement Concr Compos 24:269–279

Butler L, West JS, Tighe SL (2011) The effect of recycled concrete aggregate properties on the bond strength between CRA concrete and steel reinforcement. Cem Concr Res 41(10):1037–1049. https://doi.org/10.1016/j.cemconres.2011.06.004

Chen HJ, Yen T, Chen KH (2003) Use of building rubbles as recycled aggregates. Cem Concr Res 33(1):125–132. https://doi.org/10.1016/S0008-8846(02)00938-9

Hansen TC, Narud H (1983) Strength of recycled concrete made from crushed concrete coarse aggregate. Concr Int 5(1):79–83

Indian Standard (1992) Non-destructive testing of concrete - methods of test. IS 13311 (Part 1). Bureau of Indian Standards, New Delhi

Indian Standard (1970) Specification for coarse and fine aggregates from natural sources of concrete. IS:383

Katz A (2003) Properties of concrete made with recycled aggregate from partially hydrated old concrete. Cem Concr Res 33(5):703–711. https://doi.org/10.1016/S0008-8846(02)01033-5

Kou SC, Poon CS (2012) Enhancing the durability properties of concrete prepared with coarse recycled aggregate. Constr Build Mater 35:69–76. https://doi.org/10.1016/j.conbuildmat.2012.02.032

Kou SC, Poon CS, Agrela F (2011) Comparisons of natural and recycled aggregate concretes prepared with the addition of different mineral admixtures. Cement Concr Compos 33(8):788–795. https://doi.org/10.1016/j.cemconcomp.2011.05.009

Limbachiya M, Meddah MS, Ouchagour Y (2012) Use of recycled concrete aggregate in fly-ash concrete. Constr Build Mater 27(1):439–449. https://doi.org/10.1016/j.conbuildmat.2011.07.023

Olorunsogo FT, Padayachee N (2002) Performance of recycled aggregate concrete monitored by durability indexes. Cem Concr Res 32:179–185

Padmini AK, Ratmamurthy K, Mathews MS (2009) Influence of parent concrete on the properties of recycled aggregate concrete. Constr Build Mater 23(2):829–836

Rao A, Jha KN, Misra S (2007) Use of aggregates from recycled construction and demolition waste in concrete. Resour Conserv Recycl 50(1):71–81

Rao MC, Bhattacharyya SK, Barai SV (2011) Influence of field recycled coarse aggregate on properties of concrete. Mater Struct 44(1):205–220

Revathi P, Selvi RS, Velin SS (2013) Investigations on fresh and hardened properties of recycled aggregate self compacting concrete. J Inst Eng (India): Ser A 94(3):179–185. https://doi.org/10.1007/s40030-014-0051-5

Sagoe-Crentsil KK, Brown T, Taylor AH (2001) Performance of concrete made with commercially produced coarse recycled concrete aggregate. Cem Concr Res 31(5):707–712. https://doi.org/10.1016/S0008-8846(00)00476-2

Sánchez M, Juan D, Alaejos P (2009) Study on the influence of attached mortar content on the properties of recycled concrete aggregate. Constr Build Mater 23(2):872–877. https://doi.org/10.1016/j.conbuildmat.2008.04.012

Tavakoli O, Soroushian P (1996) Drying shrinkage behavior of recycled aggregate concrete. Concr Int 18(11):58–61

Topçu İB, Şengel S (2004) Properties of concretes produced with waste concrete aggregate. Cem Concr Res 34(8):1307–1312. https://doi.org/10.1016/j.cemconres.2003.12.019

Ujike I (n.d.) A study on air and water permeability of concrete using aggregate recycled from crushed concrete. In: RILEM proceedings, pp 194–198. http://cat.inist.fr/?aModele=afficheN&cpsidt=14172024

Yang KH, Chung HS, Ashour AF (2008) Influence of type and replacement level of recycled aggregates on concrete properties. ACI Mater J 105(3):289–296

Zaharieva R, Buyle-Bodin F, Skoczylas F, Wirquin E (2003) Assessment of the surface permeation properties of recycled aggregate concrete. Cement Concr Compos 25(2):223–232. https://doi.org/10.1016/S0958-9465(02)00010-0

Correction to: TiO_2-Assisted Photocatalytic Degradation of Herbicide 4-Chlorophenoxyacetic Acid: Slurry and Fixed-Bed Approach

Anoop Verma, Amrit Pal Toor, Palak Bansal, Vikas Sangal and Amit Sobti

Correction to: Chapter "TiO_2-Assisted Photocatalytic Degradation of Herbicide 4-Chlorophenoxyacetic Acid: Slurry and Fixed-Bed Approach" in: A. K. Agnihotri et al. (eds.), *Sustainable Engineering*, Lecture Notesin Civil Engineering 30, https://doi.org/10.1007/978-981-13-6717-5_14

The original version of the chapter was inadvertently published with incorrect affiliation in Chapter 14. The affiliation "School of Energy and Environment, Thapar Institute of Engineering and Technology, Patiala, India" of authors Amit Sobti and Amrit Pal Toor has been changed to "Dr. SSB University Institute of Chemical Engineering and Technology, Panjab University, Chandigarh-160014, India".

The erratum chapter has been updated with the changes.

The updated version of this chapter can be found at https://doi.org/10.1007/978-981-13-6717-5_14

A. K. Agnihotri et al. (eds.), *Sustainable Engineering*, Lecture Notes in Civil Engineering 30, https://doi.org/10.1007/978-981-13-6717-5_39

The manufacturer's authorised representative in the EU is Springer Nature Customer Service Centre GmbH, Europaplatz 3, 69115 Heidelberg, Germany. If you have any concerns regarding our products, please contact ProductSafety@springernature.com

Printed and bound by CPI Group (UK) Ltd, Croydon, CR0 4YY
30/09/2026
02225304-0014